Eric Lifshin (Editor)

X-ray Characterization of Materials

Related titles from WILEY-VCH

S. Amelinckx, D. van Dyck, J. van Landuyt, G. van Tendeloo (Eds.)
Handbook of Microscopy
3 Vols., ISBN 3-527-29444-9

S. N. Magonov, M.-U. Whangbo
Surface Analysis with STM and AFM
ISBN 3-527-29313-2

D. Brune, R. Hellborg, H. J. Whitlow, O. Hunderi
Surface Characterization
ISBN 3-527-28843-0

Eric Lifshin (Ed.)

X-ray Characterization of Materials

WILEY-VCH

Weinheim · New York · Chichester · Brisbane · Singapore · Toronto

Editor:
Dr. Eric Lifshin
Characterization and Environmental
Technology Laboratory
GE Corporate Research and Technology
1 Research Circle
Niskayuna, NY 12309
USA

This book was carefully produced. Nevertheless, authors, editor and publisher do not warrant the information contained therein to be free of errors. Readers are advised to keep in mind that statements, data, illustrations, procedural details or other items may inadvertently be inaccurate.

Library of Congress Card No. applied for.

A catalogue record for this book is available from the British Library.

Deutsche Bibliothek Cataloguing-in-Publication Data:
X-ray characterization of materials / Eric Lifshin (ed.). Robert L. Snyder ... – Weinheim ; New York ; Chichester ; Brisbane ; Singapore ; Toronto : Wiley-VCH, 1999
  ISBN 3-527-29657-3

© WILEY-VCH Verlag GmbH, D-69469 Weinheim (Federal Republic of Germany), 1999

Printed on acid-free and chlorine-free paper.

All rights reserved (including those of translation in other languages). No part of this book may be reproduced in any form – by photoprinting, microfilm, or any other means – nor transmitted or translated into machine language without written permission from the publishers. Registered names, trademarks, etc. used in this book, even when not specifically marked as such, are not to be considered unprotected by law.
Composition, Printing and Bookbinding: Konrad Triltsch, Druck- und Verlagsanstalt GmbH, D-97070 Würzburg
Printed in the Federal Republic of Germany

Preface

It is now just over 100 years since W. C. Roentgen (1898) first discovered x-rays. His work followed by that of H. G. Mosely (1912), W. L. and W. H. Bragg (1913), and other pioneers led the way to the development of many techniques essential to the characterization of metals, ceramics, semiconductors, glasses, minerals and biological materials. X-ray diffraction, fluorescence and absorption methods provide both qualitative and quantitative information about structure and composition essential to understanding material behavior. These methods are not only used in the course of basic research, but are also critical to the development of new materials required by society as well as understanding why materials fail in service. X-ray equipment is now found in laboratories all over including facilities that support steel mills, art museums, semiconductor fabrication facilities to cite just a few examples. Although it is not the main focus of this volume, many major advances in medicine can be linked to the findings of x-ray crystallography and various forms of radiography. Today, three-dimensional reconstruction of the human body is possible in minutes utilizing the latest in computerized tomographic clinical instrumentation.

The ability to do such remarkable diagnostic work is the result of the continuing evolution of x-ray science and technology that has drawn heavily on advances in electronics, materials science, mechanical engineering and computers. As a result, x-ray generators are more stable, tubes capable of much higher intensities, spectrometers more versatile and accurate, and detectors and associated electronics are more sensitive and capable of higher count rates. Most modern instruments also incorporate some degree of automation making control of instruments and unattended collection of data possible. A wide range of software is also readily available for phase and elemental identification, determination of strain, texture measurement, particle size distribution, single crystal structure and thin film characterization. Both commercial and "home-made" x-ray instrumentation can be found in every major industrial, academic and government laboratory.

Progress does stop, however, and over the past few decades there has been even greater interest in x-ray methods arising from the use of multi-user synchrotron facilities that provide very intense sources of radiation. Synchrotron laboratories have opened the door to the practical application of a wide variety of additional characterization techniques including x-ray absorption fine structure (EXAFS), x-ray topography and both micro-scale x-ray fluorescence and diffraction. EXAFS, for example, provides information about local atomic environments and is particularly useful in the study of catalysts even those present in concentrations below hundreds of parts per million.

This volume also covers small angle x-ray scattering (SAX), a method that can be performed with either conventional or synchrotron sources. Data obtained at low angles is indicative of grain size and shape, i.e. structure with slightly larger dimensions than atomic separation distances, which are difficult to determine in other ways. An excellent example is the determination of the radius of gyration as a function of molecular weight for polymers. Other examples include studies of phase separation in alloy systems.

The authors of the various articles present are all experts in their fields. They have done an excellent job of acquainting readers with the history, underlying principals, instrumentation, capabilities and limitations of x-ray methods as well as numerous examples of their use, and have also suggested related reading. I think all readers will find this volume a unique source of information.

Eric Lifshin
Voorheeseville, NY
5/10/99

List of Contributors

Dr. Andrea R. Gerson
University of London
Department of Chemistry
King's College
The Strand
London WC2R 2LS
UK

Dr. André Guinier
Université de Paris-Sud
Laboratoire de Physique des Solides
F-91405 Orsay Cedex
France

Dr. Peter J. Halfpenny
University of Strathclyde
Department of Pure and Applied Chemistry
295 Cathedral Street
Glasgow G1 1XL
Scotland

Dr. Ronald Jenkins
JCPDS
International Centre for Diffraction Data
1601 Park Lane
Swarthmore, PA 19801-2389
USA

Dr. Roland P. May
Institut Laue-Langevin
Avenue des Martyrs 156X
F-38042 Grenoble Cedex 9
France

Dr. Stefania Pizzini
Centre Universitaire Paris-Sud
LURE
Bâtiment 209 D
F-91405 Orsay Cedex
France

Dr. Radoljub Ristic
University of Strathclyde
Department of Pure and Applied Chemistry
295 Cathedral Street
Glasgow G1 1XL
Scotland

Dr. Kevin J. Roberts
SERC Daresbury Laboratory
Warrington WA4 4AD
UK

Dr. David B. Sheen
University of Strathclyde
Department of Pure and Applied Chemistry
295 Cathedral Street
Glasgow G1 1XL
Scotland

Professor John N. Sherwood
University of Strathclyde
Department of Pure and Applied Chemistry
295 Cathedral Street
Glasgow G1 1XL
Scotland

Prof. Robert L. Snyder
Department of Materials Science and
Engineering
Ohio State University
477 Watta Hall, 2041 College Rd.
Columbus, OH 43210
USA

Dr. Claudine E. Williams
Laboratoire de Physique
CNRS URA 792
Collège de France
11, Place Marcelin Berthélot
F-75231 Paris Cedex 05
France

Contents

List of Symbols and Abbreviations XI

1   X-Ray Diffraction 1
           Robert L. Snyder

2   Application of Synchrotron X-Radiation to Problems in Materials Science . 105
           Andrea R. Gerson, Peter J. Halfpenny, Stefania Pizzini, Radoljub Ristic,
           Kevin J. Roberts, David B. Sheen, and John N. Sherwood

3   X-Ray Fluorescence Analysis 171
           Ronald Jenkins

4   Small-Angle Scattering of X-Rays and Neutrons 211
           Claudine E. Williams, Roland P. May, and André Guinier

Index   255

List of Symbols and Abbreviations

a, b, c	crystal unit cell parameters
$\boldsymbol{a}, \boldsymbol{b}, \boldsymbol{c}$	unit cell edge translation vectors
$\boldsymbol{a}^*, \boldsymbol{b}^*, \boldsymbol{c}^*$	reciprocal cell translation vectors
A	sample area
$A(\boldsymbol{q})$	scattering amplitude
b_i	coherent scattering length of atom i
B	bending magnet strength in Tesla (Chapter 2)
B, b	background (Chapter 3)
B, B_{ij}	Debye-Waller temperature factor and tensor components (Chapter 1)
B_i	spin-dependent scattering length of atom i (Chapter 4)
c	solute concentration (Chapter 4)
c	speed of light (Chapter 1)
c	velocity of light (Chapter 2)
C	concentration
d	interplanar spacing (Chapter 3)
d	lattice plane spacing (Chapter 2)
d	sample thickness (Chapter 4)
d_0	Bragg spacing
$\boldsymbol{d}_{hkl}$	interplanar spacing vector
$\boldsymbol{d}^*_{hkl}$	reciprocal cell interplanar spacings
D	particle dimension
$\mathcal{D}$	fractal dimension
e	charge on the electron
$\boldsymbol{e}, \boldsymbol{e}_0$	unit vectors along the diffracted and incident beams
e_i	energy to produce one ion pair
E	energy
E	energy of the beam (Chapter 2)
E_p	energy of the particle
f, f_0	atomic scattering factor
$\Delta f', \Delta f''$	anomalous dispersion scattering components
F	Fano factor
F_{hkl}	structure factor (Chapter 1)
F_i	amplitude of the backscattering factor
F_N	Smith, Synder figure of merit evaluated at line N
F_{nkl}	modulus of the structure factor (Chapter 2)
$g(r)$	radial distribution function
G	Gaussian function
h	Planck's constant
hkl	Miller indices

List of Symbols and Abbreviations

I	integrated diffracted intensity (Chapter 2)
I	intensity (Chapter 1)
$\mathbf{I}$	nuclear spin
I_0	incident intensity (Chapter 1)
I_0	incoming flux (Chapter 2)
$I_{i\alpha}$	intensity of reflection i from phase α
$I(q)$	detector counts
$I(q)$	scattered intensity
I^{rel}	relative intensity, usually on a scale of 100
I_t	transmitted flux
$I(\lambda)$	photon intensity
J	total angular momentum
k	magnitude of the photoelectron wave vector (Chapter 2)
k	wave vector (magnitude: $2\pi/\lambda$) (Chapter 1)
$\mathbf{k}, \mathbf{k}_0$	scattering vectors along the diffracted and incident beams (Chapter 4)
K_0	bulk modulus
$K_{\alpha 1}, K_{\alpha 2}, K_\beta$	characteristic X-ray emission lines
l	angular quantum number
L	Avogadro's number (Chapter 1)
L	Lorentzian function (Chapter 1)
L	orbital angular momentum (Chapter 3)
L	sample to source distance (Chapter 2)
LLD	lower limit of detection
Lp	Lorentz and polarization corrections
m	magnetic quantum number (Chapter 3)
m	mass
m	sensitivity of X-ray fluorescence method (Chapter 3)
m_0	rest mass of the electron
m_e	mass of the electron
M	molecular mass (Chapter 4)
M	multiplicity of a plane (Chapter 1)
M_{20}	de Wolff figure of merit
n	principal quantum number
n_b, n_p	number of counts on peak (p) and background (b)
N	number of electrons (Chapter 3)
N	number of measurements (Chapter 3)
N	number of particles in the sample (Chapter 4)
N_A	Avogadro's number
N_i	co-ordination number for atoms of type i
$p(r)$	pair-distance distribution function
P	profile due to instrumental effects, the convolution of $W * G$ (Chapter 1)
P, p	peak (Chapter 3)
$P(r)$	Patterson function
$P(\lambda)$	photon flux
q	wave vector (magnitude)

List of Symbols and Abbreviations

q	momentum transfer, $\|q\| = (4\pi/\lambda)\sin\theta$
Q	Porod's invariant
r	real-space distance
r	shell distance
r_i	radial distance from absorbing atom
R	counting rate (Chapter 3)
R	radius of a sphere (Chapter 4)
R	ratio (Chapter 3)
R	refinement factor (Chapter 2)
R	resolution (Chapter 3)
R, r	distance (Chapter 1)
$R(E)$	reflectivity coefficient
R_b, R_p	background and peak counting rates
R_g	geometrical resolution factor in X-ray topography
R_G	radius of gyration
$RIR_{\alpha,\beta}$	reference intensity ratio of phase α with respect to β
R_s	radius of the synchrotron storage ring in meters
R_t	theoretical resolution
s	spin quantum number
s	neutron spin
S	profile from diffraction by the sample (Chapter 1)
S	source size (Chapter 2)
S_0	damping term for multibody effects in EXAFS analysis
S_α	Rietveld scale factor for phase α
t	sample thickness (Chapter 2)
t	time
t_b	background counting time
t_p	peak counting time
T	transmission coefficient
u	root mean square amplitude of vibration
v	partial specific volume
V	accelerating voltage (Chapter 1)
V	irradiated sample volume (Chapter 4)
V	unit cell volume (Chapter 1)
V	voltage (Chapter 3)
V_c	critical excitation potential
V_p	particle volume
W	atomic weight (Chapter 1)
W	weight fraction (Chapter 3)
$W * G$	wavelength and instrumental profiles
x	sample to film distance (Chapter 2)
x	thickness
x, y, z	atomic fractional coordinates
X	weight fraction
z	charge on the nucleus (Chapter 1)

List of Symbols and Abbreviations

z	number of molecules in the unit cell (Chapter 2)
Z	atomic number (Chapter 3)
Z	number of asymmetric units per unit cell (Chapter 1)
α	total absorption
α, β, γ	cell parameters (Chapter 2)
α, β, γ	interaxial angles (Chapter 1)
$\alpha^*, \beta^*, \gamma^*$	reciprocal cell interaxial angles
β	full width at half maximum of a diffraction peak
$\beta_\varepsilon, \beta_\tau$	peak broadening due to strain and size
$\gamma(r)$	correlation function
Γ	shear gradient
δ	deviation parameter for an incommensurate phase
ε	detector efficiency (Chapter 4)
ε	residual lattice stress (Chapter 1)
θ	Bragg diffraction angle
2θ	scattering angle
θ_m	diffraction angle of monochromator
Θ	vertical divergence of the beam
Θ_B	Bragg angle
λ	wavelength
λ_c	critical wavelength
λ_d	damping factor used in EXAFS analysis to allow for inelastic scattering effects
λ_{SWL}	short wavelength limit from an X-ray tube
μ	linear absorption coefficient
μ_0	absorption of an atom in the absence of neighbors (Chapter 1)
μ_0	background absorption (Chapter 2)
μ/ϱ	mass absorption coefficient
ν	frequency
$\bar{\nu}$	wave number
ϱ	density
$\varrho(r), \varrho(xyz)$	electron density at location r or xyz
σ	counting error
σ	shielding constant
σ	standard deviation
$d\sigma(q)/d\Omega$	scattering cross section per particle and unit solid angle
$d\Sigma(q)/d\Omega$	macroscopic differential cross section
σ_i	Debye-Waller type factor used in EXAFS analysis (Chapter 2)
σ_i	displacement between absorbing atoms (Chapter 1)
σ_{net}	net counting error
$\sigma_{(N)}$	random error
τ	crystallite size
ϕ	fixed incident glancing angle (Chapter 2)
ϕ	phase angle (Chapter 1)
ϕ	volume fraction occupied by matter (Chapter 4)

List of Symbols and Abbreviations

ϕ_c	critical angle for total external reflection
Φ_i	phase shift function used in EXAFS analysis
ψ	binding energy (Chapter 3)
ψ	wave function (Chapter 1)
ω	fluorescent yield
$\Delta\Omega$	solid angle subtended by a detection element
χ	EXAFS interference function
$\chi(k)$	EXAFS function
ADP	ammonium dihydrogen phosphate
ASAXS	anomalous small-angle X-ray scattering
b.c.c.	body-centred cubic
BNL/NSLS	Brookhaven National Laboratory National Synchrotron Light Source
CD-ROM	compact disk read only memory
CVD	chemical vapor deposition
DCD	double-crystal diffractometer
EDD	electron diffraction database
EDS	energy dispersive spectroscopy
EDXRD	energy dispersive X-ray diffraction
EISI	elemental and interplanar spacings index
EXAFS	extended X-ray absorption fine structure
f.c.c.	face-centred cubic
FET	field effect transistor
FOM	figure of merit
FWHM	full width at half maximum
ICDD	international centre for diffraction data
IFT	indirect Fourier transformation
ITO	indium/tin oxide
IUPAC	international union of pure and applied chemistry
KZC	K_2ZnCl_4
LSM	layered synthetic micro-structure
MBA-NB	(−)-2-(α-methylbenzylamino)-5-nitropyridine
MBE	molecular beam epitaxy
MCA	multichannel analyzer
ML	monolayers
NF	nickel formate dihydrate
PC	desktop computer
PDF	powder diffraction file
PHA	pulse height analyzer
PIXE	proton excited X-ray fluorescence
PSD	position sensitive detector
PTS	2,4-hexadiynediol-bis-(p-toluene sulfonate)
QEXAFS	quick-scanning EXAFS
RDF	radial distribution function
ReflEXAFS	reflectivity EXAFS

SANS	small-angle neutron scattering
SAS	small-angle scattering
SAXS	small-angle X-ray scattering
SR	synchrotron radiation
SSXRF	synchrotron source X-ray fluorescence
TAP	thallium acid phtalate
TEM	transmission electron microscopy
TOF	time of flight
TRXRF	total reflection X-ray fluorescence
WDS	wavelength dispersive spectroscopy
XAS	X-ray absorption spectroscopy
XANES	X-ray absorption near-edge structure
XRD	X-ray diffraction
XRF	X-ray fluorescence
XSW	X-ray standing waves
ZBH	zero background holder

1 X-Ray Diffraction

Robert L. Snyder

Institute for Ceramic Superconductivity, New York State College of Ceramics, Alfred University, Alfred, NY, U.S.A.

1.1	**Introduction**	4
1.2	**The Nature of X-Rays**	4
1.3	**The Production of X-Rays**	4
1.3.1	Synchrotron Radiation	5
1.3.2	The Modern X-Ray Tube	5
1.3.3	High Intensity Laboratory X-Ray Devices	7
1.4	**Interaction of X-Rays with Matter**	8
1.4.1	No Interaction	8
1.4.2	Conversion To Heat	8
1.4.3	Photoelectric Effect	9
1.4.3.1	Fluorescence	9
1.4.3.2	Auger Electron Production	10
1.4.3.3	Fluorescent Yield	11
1.4.4	Compton Scattering	11
1.4.5	Coherent Scattering	11
1.4.6	Absorption	12
1.5	**The Detection of X-Rays**	13
1.5.1	Non-Electronic Detectors	14
1.5.1.1	Photographic Film	14
1.5.1.2	Fluorescent Screens	14
1.5.1.3	Human Skin	14
1.5.2	Gas-Ionization Detectors	14
1.5.3	Solid State Detectors	16
1.5.4	The Electronic Processing of X-Ray Signals	17
1.6	**Crystallography**	18
1.6.1	Point Groups	20
1.6.2	Bravais Lattices	21
1.6.3	Space Groups	22
1.6.4	Space Group Notation	23
1.6.5	Reduced Cells	23
1.6.6	Miller Indices	24
1.7	**Diffraction**	24
1.7.1	Bragg's Law	25
1.7.2	The Reciprocal Lattice	26
1.7.2.1	Relationship between d_{hkl}, hkl and Translation Vectors	29

1.7.2.2	The Ewald Sphere of Reflection	30
1.7.3	Single Crystal Diffraction Techniques	32
1.7.3.1	X-Ray Topography	34
1.7.3.2	Laue and Kossel Techniques for Orientation Determination	35
1.7.4	Preferred Orientation	37
1.7.5	Crystallite Size	37
1.7.6	Residual Stress	38
1.7.7	The Intensities of Diffracted X-Ray Peaks	39
1.7.7.1	Scattering of X-Rays by a Bound Electron	39
1.7.7.2	Scattering of X-Rays by an Atom	39
1.7.7.3	Anomalous Dispersion	40
1.7.7.4	Thermal Motion	41
1.7.7.5	Scattering of X-Rays by a Crystal	42
1.7.8	Calculated X-Ray Intensities	43
1.7.8.1	Systematic Extinctions	44
1.7.8.2	Primary and Secondary Extinction and Microabsorption	45
1.7.9	Vision, Diffraction and the Scattering Process	46
1.7.10	X-Ray Amorphography	47
1.8	**X-Ray Absorption Spectroscopy (XAS)**	48
1.8.1	Extended X-Ray Absorption Fine Structure (EXAFS)	48
1.8.2	Reflectometry	50
1.8.3	X-Ray Tomography	51
1.9	**X-Ray Powder Diffraction**	51
1.9.1	Recording Powder Diffraction Patterns	52
1.9.1.1	Diffractometer Optics and Monochromators	55
1.9.1.2	The Use of Fast Position Sensitive Detectors (PSD)	56
1.9.1.3	Very High Resolution Diffractometers	57
1.9.2	The Automated Diffractometer	60
1.9.2.1	Background Determination	60
1.9.2.2	Data Smoothing	62
1.9.2.3	Spectral Stripping	63
1.9.2.4	Peak Location	63
1.9.2.5	External and Internal Standard Methods for Accuracy	63
1.9.3	Software Developments	65
1.9.3.1	The Accuracy of Diffraction Intensities	66
1.9.3.2	The Limit of Phase Detectability	66
1.10	**Phase Identification by X-Ray Diffraction**	67
1.10.1	The Powder Diffraction Data Base	68
1.10.2	Phase Identification Strategies	68
1.10.3	The Crystal Data Database	72
1.10.4	The Elemental and Interplanar Spacings Index (EISI)	72
1.11	**Quantitative Phase Analysis**	73
1.11.1	The Internal-Standard Method of Quantitative Analysis	74
1.11.1.1	$I/I_{corundum}$	74
1.11.1.2	The Generalized Reference Intensity Ratio	75

1.11.1.3	Quantitative Analysis with RIR^s	75
1.11.1.4	Standardless Quantitative Analysis	76
1.11.1.5	Constrained X-Ray Diffraction (XRD) Phase Analysis – Generalized Internal-Standard Method	76
1.11.1.6	Full-Pattern Fitting	77
1.11.1.7	Quantitative Phase Analysis Using Crystal Structure Constraints	78
1.11.2	The Absolute Reference Intensity Ratio: RIR_A	80
1.11.3	Absorption-Diffraction Method	81
1.11.4	Method of Standard Additions or Spiking Method	82
1.11.4.1	Amorphous Phase Analysis	82
1.12	**Indexing and Lattice Parameter Determination**	83
1.12.1	Accuracy and Indexing	83
1.12.2	Figures of Merit	83
1.12.3	Indexing Patterns with Unknown Lattice Parameters	84
1.12.4	The Refinement of Lattice Parameters	86
1.13	**Analytical Profile Fitting of X-Ray Powder Diffraction Patterns**	88
1.13.1	The Origin of the Profile Shape	89
1.13.1.1	Intrinsic Profile: S	90
1.13.1.2	Spectral Distribution: W	90
1.13.1.3	Instrumental Contributions: G	90
1.13.1.4	Observed Profile: P	91
1.13.2	Modeling of Profiles	92
1.13.3	Description of Background	92
1.13.4	Unconstrained Profile Fitting	93
1.13.5	Establishing Profile Constraints: The P Calibration Curve	94
1.13.6	Modeling the Specimen Broadening Contribution to an X-Ray Profile	94
1.13.7	Fourier Methods for Size and Strain Analysis	96
1.13.8	Rietveld Analysis	96
1.14	**Crystal Structure Analysis**	96
1.14.1	Structure of $YBa_2Cu_3O_{7-\delta}$	97
1.14.2	The Structures of the γ and η Transition Aluminas	98
1.14.2.1	Profile Analysis	98
1.14.2.2	Rietveld Analysis	99
1.15	**References**	100

1.1 Introduction

X-ray diffraction has acted as the cornerstone of twentieth century science. Its development has catalyzed the developments of all of the rest of solid state science and much of our understanding of chemical bonding. This article presents all of the necessary background to understand the applications of X-ray analysis to materials science. The applications of X-rays to materials characterization will be emphasized, with particular attention to the modern, computer assisted, approach to these methods.

1.2 The Nature of X-Rays

X-rays are relatively short wavelength, high energy electromagnetic radiation. When viewed as a wave we think of it as a sinusoidal oscillating electric field with, at right angles, a similar varying magnetic field changing with time. The other description is that of a particle of energy called a photon. All electromagnetic radiation is characterized by either its energy E, wavelength λ (i.e., the distance between peaks) or its frequency v (the number of peaks which pass a point per second). The following are useful relationships for interconverting the most common measures of radiation energy.

$$\lambda = \frac{c}{v} \tag{1-1}$$

$$E = h v \tag{1-2}$$

where c is the speed of light and h is Planck's constant. Spectroscopists commonly use wavenumbers particularly in the low energy regions of the electromagnetic spectrum, like the microwave and infrared. A wave number ($\bar{v}$) is frequency divided by the speed of light

$$\bar{v} = \frac{v}{c} = \frac{c/\lambda}{c} = \frac{1}{\lambda} \tag{1-3}$$

The angstrom (Å) unit, defined as 1×10^{-10} m, is the most common unit of measure for X-rays but the last IUPAC vention made the nanometer (1×10^{-9} m) a standard. However, here we will use the traditional angstrom unit. Energy in electron volts (eV) is related to angstroms through the formula,

$$E\,(\text{eV}) = \frac{hc}{\lambda_{\text{cm}}} = \frac{12\,396}{\lambda_{\text{Å}}} \tag{1-4}$$

Electron volts are also not IUPAC approved in that the standard energy unit is the Joule which may be converted by

$$1\,\text{eV} = 1.602 \times 10^{-19}\,\text{J} \tag{1-5}$$

It should be noted that despite the IUPAC convention, Joules are never used by crystallographers or spectroscopists, while a few workers have adopted the nanometer in place of the angstrom. Table 1-1 lists the various measures across the electromagnetic spectrum.

1.3 The Production of X-Rays

There are four basic mechanisms in nature which generate X-rays. These are related to the four fundamental forces that exist in our universe. Any force when applied to an object is a potential source of energy. If the object moves kinetic energy is generated. The weak and strong nuclear forces combine to produce not very useful X-rays, along with many other wavelengths and subatomic particles, in high energy nuclear collisions. The force of gravitation also produces X-rays which are not useful in the materials characterization

Table 1-1. Values of common energy units across the electromagnetic spectrum.

Quantity	Units	IR	UV	Vacuum UV	Soft X-ray	X-ray	Hard X-ray	γ
Wavelength	Å	10 000	1000	100	10	1	0.1	0.01
Wavelength	nm	1 000	100	10	1	0.1	0.01	0.001
Wavenumber	cm^{-1}	10^4	10^5	10^6	10^7	10^8	10^9	10^{10}
Energy	eV	1.24	12.4	124	1239.6	12.4 keV	124 keV	1.24 MeV
Energy	J	2×10^{-19}	2×10^{-18}	2×10^{-17}	2×10^{-16}	2×10^{-15}	2×10^{-14}	2×10^{-13}

laboratory, by giving rise to neutron stars and black holes which, in the process of accreting matter, produce X-rays visible at astronomical distances. However, it is the Coulombic force which produces the X-rays we harness in the laboratory.

1.3.1 Synchrotron Radiation

Particle accelerators operate on the principle that as a charged particle passes through a magnetic field it will experience a force perpendicular to the direction of motion, in the direction of the field. This causes a particle to curve through a "bending magnetic" and accelerate. As long as energy is supplied to the magnets, a beam of particles can be continuously accelerated around a closed loop. Accelerating (and decelerating) charged particles will give off electromagnetic radiation. When the particles are accelerated into the GeV range, X-radiation will be produced. A synchrotron is a particle acceleration device which, through the use of bending magnetics, causes a charged particle beam to travel in a circular (actually polyhedral) path.

Today there are a number of synchrotron facilities around the world which are dedicated to the production of extremely intense sources of continuous (white) X-radiation ranging from hundredths to hundreds of angstroms in wavelength. In recent years there has been a burst of activity in the use of these sources. The wavelength tunability and very high brightness of these sources has opened a wide range of new characterization procedures to researchers. The addition of magnetic devices to make the particle beam wiggle up and down on its path between bending magnetics, called wigglers and undulators, have raised the intensity of X-rays available for experiments by as much as a factor of 10^{12}. In addition, since the X-rays are only produced as the charged particles fly by the experimenter's window every few nanoseconds, time-resolved studies in the nanosecond range have become accessible. See the chapter on synchrotron radiation by Sherwood et al. in this volume for more information on synchrotron techniques.

1.3.2 The Modern X-Ray Tube

The conventional method of producing X-rays in a laboratory is to use an evacuated tube invented by Coolidge (1913). Figure 1-1 shows a modern version of this tube whose function is illustrated in Fig. 1-2. This tube contains a tungsten cathode filament which is heated by an AC voltage ranging from 5 to 15 V. The anode is a water-cooled target made from a wide range of pure elements. Electrons are accelerated in vacuum under potentials of 5000 to 80 000 volts and produce a spectrum of the type shown in Fig. 1-3. As the accelerated electrons reach the target they are repelled by the electrons of the target atoms, causing a slowing down or breaking. To slow down an electron and conserve energy, the electron must lose its energy in the only manner available to it – radiation. The German word for breaking is brems and for radiation is strahlung. Most of the early discoveries concerning X-rays oc-

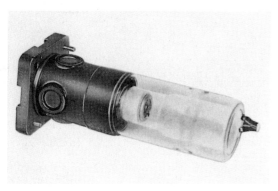

Figure 1-1. The modern sealed X-ray tube (Courtesy of Siemens AG).

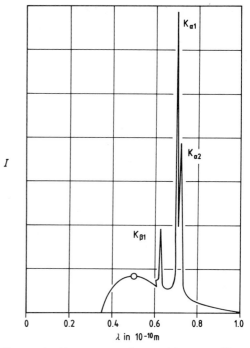

Figure 1-3. The spectrum from a Mo target X-ray tube.

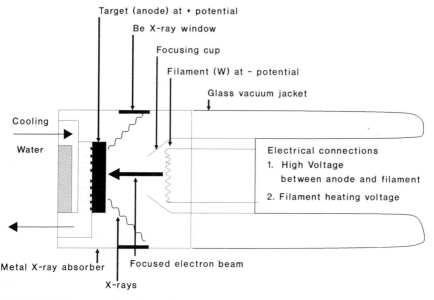

Figure 1-2. Schematic of an X-ray tube.
$(\lambda) = $ m, $(v) = (c/\lambda) = $ s^{-1}, $(hv) = (E) = $ eV,
$(c) = $ m/s, $(E) = $ J or eV, $\left(h\dfrac{c}{\lambda}\right) = $ eV, $(hc) = $ eV · m,
$(hc)/\text{eV} = $ m $= (\lambda)$.

curred in Germany; for example, Röntgen, the discoverer of X-rays, worked at the University of Munich (although the discovery was actually made in Würzburg), where others like von Laue and Ewald were to make dramatic advances. Since Germans seem to have a running competition to form the world's longest words, they called this continuous spectrum *bremsstrahlung*; which we have adopted as a rather odd sounding English word.

The maximum energy of a photon from such an X-ray tube would arise from a single dead stop collision of an accelerated electron with a target electron. The kinetic energy of the electron is the product of e and V, where e is the charge on the electron and V is the accelerating voltage. If this energy is completely converted to a photon of energy hv then the *short wavelength limit* (λ_{SWL}) of the photons in the continuous spectrum will be

$$\lambda_{SWL}(\text{Å}) = hc/(eV) = 12\,398/V \quad (1\text{-}6)$$

Superimposed on the white radiation from an X-ray tube are some very narrow spikes. The wavelengths of these lines were first shown by Moseley to be a function of the atomic number of the target material. They arise from billiard ball like collisions which eject inner shell electrons from the target atoms. This process is described more fully below. It should be noted that it makes no difference whether an inner shell "photoelectron" is ejected by an electron, as in an X-ray tube or, by a photon as in an X-ray spectrometer, the resulting emission lines will be the same. It is these nearly monochromatic emission lines which we employ for most of our X-ray experiments.

1.3.3 High Intensity Laboratory X-Ray Devices

The conventional modern X-ray tube uses a cup around the tungsten filament held at a potential of a few hundred volts more negative than the cathode so that the electrons are repelled and focused onto the target. The focal spot is actually a line about two centimeters in length, reflecting the length of the filament. Intensity is defined as the photon flux passing a unit area in unit time. Thus, focusing the electrons onto a smaller area increases the intensity. Various modifications of design parameters have produced "fine focus" and "long fine focus" X-ray tubes which take advantage of this fact to produce higher intensity. However, approximately 98% of the energy from the impacting electrons goes into producing heat. The limitation on the intensity which may be produced is the efficiency of the cooling system which prevents the target from melting.

Since the X-rays may be viewed from any of the four sides of the tube, two sides will produce X-rays from the line projection of the filament. The other two sides view the projection of the line from the end giving a focal spot (actually a rhombus) of about 1 mm^2 in size, when viewed from the usual take-off angle of from 3° to 6°. The take-off angle is the angle at which an experiment views the X-ray tube target. The higher the angle the more divergent X-rays will be present and the lower will be the resolution of any experiment. On the other hand, decreasing the angle decreases intensity but, by limiting the amount of angular divergence in the beam, increases the experimental resolution.

Microfocus tubes use the focusing cup to squeeze the electron beam down to a spot focus with a diameter as small as 10 μm. These units are used for experiments requiring extremely intense beams and can accept the small area of illumination. Such tubes usually have replaceable targets. Another, more popular, method to increase the intensity of an X-ray tube is to increase the power on the target and avoid melting it by rotating it. These rotating anode tubes continuously bring cool metal into the path of the focused electron beam. Such units can typically be run as high as 18 kW compared to about 1.8 kW for a sealed tube. They produce very intense X-ray beams. However, owing to the mechanical difficulties of a high speed motor drive which must feed through into the vacuum, there are difficulties in routine continuous operation. In recent years these units have become more common and more reliable.

The last laboratory method for generating X-rays is to charge a very large bank of capacitors and to dump the charge, in a very short time, to a target. These *flash X-ray* devices can reach peak currents of 5000 A in the hundreds of kV range. The extremely intense X-ray flash lasts for only a few nanoseconds but this has not stopped workers from performing some very clever experiments within this incredibly small time window.

1.4 Interaction of X-Rays with Matter

Consider the simple experimental arrangement shown in Fig. 1-4. Any mechanism which causes a photon, in the collimated incident X-ray beam, to miss the detector is called absorption. Most of the mechanisms of absorption involve the conversion of the photon's energy to another form; while some simply change the photon's direction. For the purposes of this discussion it is best to consider I_0 a monochromatic beam and that the detector is set only to detect X-rays of that energy. We may place the possible fates of an X-ray photon, as is passes through matter, in the following categories.

1.4.1 No Interaction

The fundamental reason for all X-ray–atom interactions is the acceleration experienced by an atom-bound electron from the oscillating electric field of the X-ray's electromagnetic wave. The probability of any interaction decreases as the energy of the wave increases. The probability of interaction is approximately proportional to the wavelength cubed. Thus, short wavelength photons are very penetrating while long wavelengths are readily absorbed. There is always a finite probability that an X-ray will pass through matter without interaction.

The simple cubic relationship of interaction probability is disturbed by the phenomenon of *resonance absorption*. When the energy of the incident radiation becomes exactly equal to the energy of a quantum allowed electron transition between two atomic states, a large increase is observed in the probability of a photon's being absorbed. The dramatic increase in absorption as photons reach the ionization potential of each of the electrons in an atom results in a series of absorption edges shown in Fig. 1-8.

1.4.2 Conversion To Heat

Heat is a measure of atomic motion. Heat may be stored in the quantum allowed translational, rotational and vibrational energy states of the atoms or molecules in a material. It also can be stored in the various excited electronic

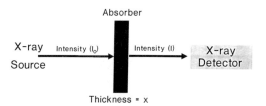

Figure 1-4. A simple absorption experiment.

states allowed to an atom and in the motion of the relatively free electrons in metals. The principal mechanism for converting photons to heat in insulators is the stimulation of any of the modes of vibration of the lattice.

There are two classes of vibrational modes allowed to any lattice. One is the acoustic modes of vibration which may be stimulated by a mechanical force such as a blow or an incident sound wave. The other class is the optic modes of vibration. Optic modes are characterized by a change in dipole moment as the atoms vibrate. This change in electrical field in the lattice allows these modes to interact with the electric field of a photon. Thus, an X-ray photon may stimulate an optic lattice vibrational mode which we observe as heat. The efficiency of the coupling between the lattice vibrational modes, called phonons, and photons, depends both on the lattice itself and on the energy of the incident photon. Thus, we observe sample heating in an X-ray beam to be higher in some samples than in others.

In fact X-rays can also gain energy by absorbing a phonon. The energy of the lattice vibrational modes is on the order of 0.025 eV, while a Cu K_α photon has an energy of 8 keV. Thus, the modification of the incident X-ray beam is rather small, and of course can be studied to understand the phonon structures of solids. However, Raman spectroscopy and thermal neutron scattering are better for these types of studies. Photons whose energy has been modified by a phonon interaction contribute to experimental background as thermal diffuse sactterring.

1.4.3 Photoelectric Effect

In a photon–electron interaction, if the photon's energy is equal to, or greater than, the energy binding the electron to the nucleus, the electron may absorb all of the energy of the photon and become ionized as shown in Fig. 1-5. The free electron will leave the atom with a kinetic energy equal to the difference between the energy of the incident photon and the ionization potential of the electron. This high energy electron can, of course, go on to initiate a number of photon creating events. However, any secondary photon must have a lower energy. The experiment illustrated in Fig. 1-4 assumes that the detector is set only to count pulses of the same energy as in the incident monochromatic beam. Thus, these secondary or fluoresced photons of lower energy do not get included in the measurement of intensity.

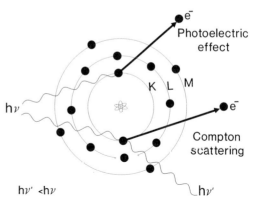

Figure 1-5. Photoelectric and Compton effects.

1.4.3.1 Fluorescence

An atom, ionized by having lost one of its innermost K or L shell electrons, is left in an extremely unstable energy state. If the vacancy has occurred in any orbital beneath the valence shell, then an immediate rearrangement of the electrons in all of the orbitals above the vacancy will occur. Electrons from higher orbitals will cascade down to fill in the hole. This process, illustrated in Fig. 1-6, causes the emission of

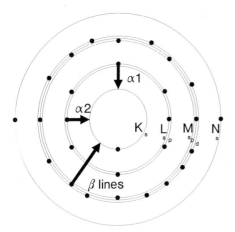

Figure 1-6. Fluorescence from an ionized atom.

secondary fluorescent photons. The energy gaps between the various electron orbitals are fixed by the laws of quantum mechanics. Thus, the photon emitted by an electron falling to lower energy (getting closer to the nucleus) will have a fixed energy, depending only on the number of protons in the nucleus. The photons fluoresced by any element will thus have X-ray wavelengths characteristic of that element.

If the ionized electron comes from the K shell, then there is a certain probability that an L_p, L_s or an M electron will fall in to replace it. The names of the resulting emitted photons are the $K_{\alpha 1}$, $K_{\alpha 2}$ and K_β, respectively. For a Cu atom the transition probabilities are roughly 5:2.5:1, respectively. The energies of any of these lines must, of course, be less than that of the original incident X-ray which caused the ionization. The study of the fluoresced photons is called X-ray fluorescence spectroscopy (XRF). This technique allows the rapid qualitative analysis of the elements present in a material and with more work, the quantitative analysis of the elemental composition. See Chap. 3 for a complete description of this method.

1.4.3.2 Auger Electron Production

There is a special tertiary effect of photoelectron production called the emission of Auger (pronounced oh-jay) electrons. Sometimes the removal of an inner-shell electron produces a photon which in turn gets absorbed by an outer-shell, valence electron. Thus, the incident X-ray gets absorbed by, for example, a K shell electron which leaves the atom. An L shell electron can fall to the K shell to fill in the hole and thereby causes the emission of a K_α X-ray photon. However, before this photon can leave the atom it gets absorbed by a valence electron which ionizes and flies off leaving a doubly charged ion behind. This process is illustrated in Fig. 1-7.

The kinetic energy of the Auger electron is not dependent on the energy of the initial X-ray photon which ionized the K electron. Any X-ray with sufficient energy to create the initial K hole can be responsible for the subsequent production of an Auger electron of fixed kinetic energy. This very specific kinetic energy is equal to the difference in energy between the fixed-energy K_α or K_β photon which ionized the Auger

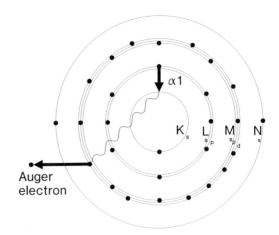

Figure 1-7. Auger electron emission from an ionized atom.

electron and the fixed binding energy of the valence electron to the nucleus. The study of these Auger electrons, called Auger Spectroscopy, allows us to measure the precise energy of the chemical bonds which involve the valence electrons. It also allows us to see subtle energy differences between chemical bonds. A full description of this characterization method may be found in Chapter 13 by Lou in Volume 2 B.

1.4.3.3 Fluorescent Yield

The efficiency of the production of characteristic X-rays is a function of the atomic numbers, z, of the elements in the absorber. We measure this efficiency with a quantity called the fluorescent yield (ω) which for a given spectral series is the ratio of the number of photons produced to the number of vacancies created

$$\omega = \frac{z^4}{A + z^4} \qquad (1\text{-}7)$$

where A is a constant on the order of 10^6 for the K series and 10^8 for L series X-rays. The fluorescent yield becomes poorer approaching zero, as the atomic number decreases. Thus, traditional X-ray fluorescence was seldom used for elements lighter than sodium. The reason for this poor efficiency is that as the atomic number of the absorber decreases it is more and more likely that the characteristic X-ray photon will be reabsorbed and cause the emission of an Auger electron. Therefore the efficiency of the production of Auger electrons is inversely proportional to the efficiency of producing (fluorescent) characteristic X-rays as a function of atomic number. It should be noted however, that modern high vacuum wavelength dispersive spectrometers (WDS) are able to observe, and even quantitatively analyze, elements as low in atomic number as beryllium.

1.4.4 Compton Scattering

This phenomenon, illustrated in Fig. 1-5, amounts to an inelastic collision between a photon and an electron. Part of the energy of the incident photon is absorbed by an electron and the electron is thus excited. However, instead of the remaining energy of the original photon converting to kinetic energy of the excited photoelectron, some of it is re-emitted as an X-ray photon of lower energy. Not only has the energy of this Compton photon been lowered but it loses any phase relationship to the incident photon. For this reason the process is called *incoherent scattering*. The increase in wavelength of the scattered photon is dependent on the angle between the scattered photon and Compton electron,

$$\Delta \lambda = \frac{h}{m_e c}(1 - \cos 2\theta) \qquad (1\text{-}8)$$

The intensity of incoherent radiation, I_{Compton} is given by,

$$I_{\text{Compton}} = Z - \sum_{j=1}^{\text{number of atoms in unit cell}} f_j^2 \qquad (1\text{-}9)$$

where f is a measure of the X-ray scattering from an atom and related to the number of electrons on an atom as described in Sec. 1.7.7.2 below. f is equal to Z at zero degrees θ and falls off as θ increases. As Z increases, f^2 will increase more rapidly, thus Compton scattering decreases in intensity as the atomic number of the scatterer increases.

1.4.5 Coherent Scattering

The last important mechanism of X-ray absorption in matter is the one which leads to the phenomenon of diffraction. Coherent scattering is analogous to a perfectly elastic collision between a photon and an electron. The photon changes direction after colliding with the electron but transfers none of its energy to the electron. The result is that the scattered photon leaves in a new direction but with the same phase and energy as that of the incident photon. It is coherent scattering that leads to the phenomenon of diffraction. However, since the incident photon changes direction, and therefore will miss the detector in Fig. 1-4, it is considered an absorption mechanism.

A minor point could disturb the reader here. The rules of quantum mechanics appear to restrict the energy of an atom-bound electron to fixed values. How, therefore, can such an electron accelerate from the electric field of a photon of random energy and temporarily absorb its energy? The answer lies in the uncertainty principle which, in this context, states that an electron can have a large energy within a time interval approaching zero. For coherent scattering, since there is no phase shift, the electron must absorb and re-emit the photon's energy in exactly zero time. In zero time the electron may take on any value of energy.

1.4.6 Absorption

A great number of experiments of the type illustrated in Fig. 1-4 have shown that when electromagnetic waves, of any fixed wavelength, are absorbed by any form of matter, the following general equation holds

$$I = I_0 e^{-\mu x} \qquad (1\text{-}10)$$

or

$$\ln \frac{I}{I_0} = -\mu x \qquad (1\text{-}11)$$

where:

I_0 is the intensity of incident X-ray beam,
I is the intensity of beam at the detector,
μ is the linear absorption coefficient (in cm^{-1}) and
x is the thickness of material (in cm).

The difference between I and I_0 for a fixed wavelength is therefore dependent on the thickness of the absorber and on the linear absorption coefficient μ. μ is a constant related to the absorbing material. Since all of the absorption processes described above ultimately depend on the presence of electrons, then it seems clear that the ability of a material to absorb electromagnetic radiation will relate to the density of electrons. In turn, the electron density of a material will be determined by the types of atoms composing the material and the closeness of their packing. The linear absorption coefficient of a material will therefore depend on the types of atoms present and the density of the material. However, if we eliminate the functional dependence on density, which is determined by the type and strength of the chemical bonds in a material, we will obtain a true constant, (μ/ϱ), for each element at any specified wavelength.

The quantity (μ/ϱ) is called the *mass absorption coefficient*. The (μ/ϱ) of Zr is shown as a function of wavelength at the top of Fig. 4-8. The sharp discontinuity in the curve is called *absorption edge*. They correspond to the entrance of a new absorption mechanism. The highest energy edge corresponds to the ionization potential of a K electron and is called the K edge. Other edges correspond to the L and M electron ionizations. The exact location of each of the edges corresponds to the energy required for the removal of that electron. Thus, absorption edges yield fundamental information about the structure of the atom, its environment and the amount of energy required to excite that series of characteristic fluoresced radiation. The

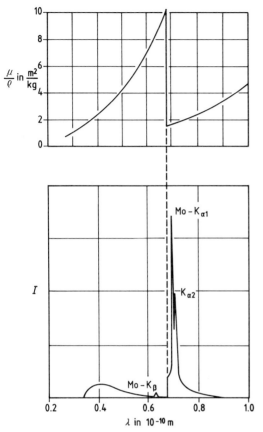

Figure 1-8. The absorption spectrum of Zr and the effect of a Zr filter on a Mo X-ray spectrum.

fine structure of these edges gives information on the exact energies of the electrons within the atom. These energies are slightly affected by the bonding environment around the atom. To explain the fine structure would require a full molecular orbital description of the atoms in the absorber and this fact dampened the spirits of many early researchers.

When it is necessary to find the mass absorption coefficient of a compound, solution, or mechanical mixture, containing more than one element, it may be computed by taking the weighted average of the mass absorption coefficients of the constituent elements

1.5 The Detection of X-Rays

$$\left(\frac{\mu}{\varrho}\right)_{\text{mixture}} = X_1 \left(\frac{\mu}{\varrho}\right)_1 + X_2 \left(\frac{\mu}{\varrho}\right)_2 + \ldots \quad (1\text{-}12)$$

or

$$\left(\frac{\mu}{\varrho}\right)_m = \sum_{j=1}^{\text{number of elements}} X_j \left(\frac{\mu}{\varrho}\right)_j \quad (1\text{-}13)$$

where ϱ is the density, $(\mu/\varrho)_j$ are the mass absorption coefficients of the constituent elements, and X_j are the weight fractions of the elements present.

An interesting application of the absorption edge is when the edge of one element (A) is located between the K_α and K_β lines for another element (B). When this occurs the K_β from the B atoms will be very strongly absorbed while the longer wavelength K_α will be only slightly absorbed. This means that a suitable thickness of element A can act as a *beta-filter* for characteristic radiation from element B. The bottom of Fig. 1-8 shows the effect of putting a Zr foil in a beam of radiation generated from a Mo X-ray tube. This method of emphasizing the K_α to make a pseudo-monochromatic beam for diffraction experiments has been used for many years. On most modern diffractometers a graphite crystal is used as a monochromator for the diffracted beam, eliminating the use of beta-filters.

1.5 The Detection of X-Rays

The nature of X-ray detectors and the electronics used with them is an important part of understanding the approaches to, and limitations of, X-ray applications. Beginning with the introduction of the Geiger-Müller counter (1928) and its use for the measurement of diffracted X-rays by LeGalley (1935), the modern X-ray diffractometer has developed primarily with the use of scintillation detectors, while fluorescence spectrometers more commonly use proportional counters. Most recently position sensitive proportional counters and high energy resolution solid state detectors have gained popularity. The principles of each of the methods used to detect X-rays will be described here, illustrating both the types of electronics and the physics employed.

1.5.1 Non-Electronic Detectors

1.5.1.1 Photographic Film

This is one of the oldest methods of detecting X-rays. When exposing the silver halide grains in a film so that they will become developable (i.e., reduced to pure silver on the plastic substrate) the grain size is important. Each particle in the emulsion which absorbs an X-ray photon becomes sensitized. Once sensitized the entire particle will reduce to silver in the development process. Thus, if a region of the film is exposed to an intense X-ray flux the same particles may be struck more than once. This causes a loss of information analogous to the dead time in electronic counters. The film darkening is only proportional to the intensity of the exposing X-rays over what is called the linear range of the film. To allow for this effect, when measuring intense sources, one must use a film of smaller grain size to extend the linear range or reduce the X-ray intensity with filters in front of the film or reduce the incident beam flux.

1.5.1.2 Fluorescent Screens

ZnS doped with Ni will emit a greenish visible light when struck by X-rays. Screens made of this material are useful in aligning X-ray cameras and intensifying X-rays to be recorded with a film sensitive to visible light.

1.5.1.3 Human Skin

Just to be complete we should consider this material which will redden and ulcerate under exposure to X-rays. Owing to its very slow response time and high dead time it is not a very useful detector. It is recommended that workers keep their skin out of the X-ray beam. In this context the phrase "dead time" has a more macabre meaning. The conventional meaning is the amount of time an electronic detector is inactive while processing a photon, before it can respond to another photon. The dead time of over-exposed human skin is infinite.

1.5.2 Gas-Ionization Detectors

The creation and collection of gas ions and electrons in an electrostatic field has led to the development of a number of different X-ray detectors:

(1) Ionization chamber.
(2) Sealed gas proportional counters.
(3) Flowing gas proportional counters.
(4) Geiger-Müller detectors.
(5) Position sensitive proportional counters.

All of these detectors are based on the same basic principle illustrated in Figs. 1-9 and 1-10. Figure 1-9 is a simple schematic

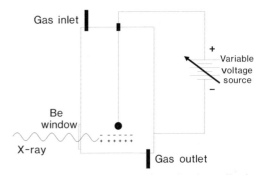

Figure 1-9. A schematic diagram of an ion collection device.

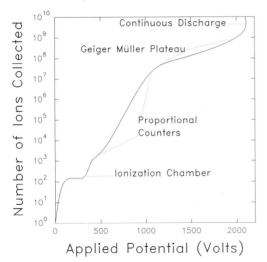

Figure 1-10. Gas amplification in ion collection devices.

of an ion collection device. The container has a beryllium window to allow the entrance of X-rays. The gas in the container is varied depending on the exact application the user has in mind however, P10 counting gas (10% methane and 90% Ar) is the most common gas used. The entering X-ray, on striking an argon atom, will, by the photoelectric effect, create an ion pair (an Ar^+ ion and an electron). The valence electron ejected from the Ar atom leaves with a kinetic energy equal to the energy of the ionizing X-ray minus its binding potential. The photon continues to lose its energy by colliding with other Ar atoms and creating a number of secondary ion pairs. The number of ion pairs created is proportional to the energy of the incident photon.

The fate of the ions and electrons will depend on the potential applied between the central wire and the outer cathode container. If, for example, no potential is applied, all of the electrons will recombine with ions and we have accomplished nothing very interesting! The five types of ion collection detectors mentioned above result from differing amounts of applied potential to the circuit in Fig. 1-9 as illustrated in Fig. 1-10.

1. *Ionization chamber.* If the voltage applied to the ion collection chamber is on the order of 100 to 200 V, then the electrons and ions will drift to their respective electrodes and a current pulse will flow through the circuit. A CuK_α photon (8040 eV) and an effective ionization energy of Ar of about 25 eV means that about 300 ion pairs can be produced and collected. This extremely small current must be amplified before a useful detector can be built. The ionization chamber is no longer in common use.

2. *Sealed gas proportional counters.* If the voltage applied to the ion collector is in the 700 to 800 V range then, the perhaps 300 electrons accelerate so rapidly to the anode that they collide with other Ar atoms, causing secondary ionizations. This effect is called gas amplification. At a constant applied potential the amount of current in the pulse that results will be proportional to the energy of the incident X-ray, hence the name proportional counter. The *energy resolution* ($\Delta E/E$ of about 20%) is quite good: about 3–4 times better than a scintillation detector. The dead time in these detectors is primarily the result of the time it takes for the relatively slow moving Ar^+ ions to move to the cathode and discharge. This process is slowed even more by the fact that the electrons move about 1000 times faster than the Ar^+ ions, leaving the gas with a net positive space charge. Due to mutual ion repulsions within this space charge, it takes a few micro seconds for discharge to complete, determining the dead time.

3. *Flowing gas proportional counters.* The principles are exactly as described for the sealed counter. The advantages are a nearly unlimited lifetime via occasional replacement of the anode wire, and much

wider range of sensitivity to X-ray energies made possible by the use of thinner X-ray detector windows.

4. *Geiger-Müller detectors.* When about 1000 V are applied to the sealed chamber a full Townsend avalanche occurs, where a single ionization causes every Ar atom in the chamber to ionize. This detector needs little external amplification due to the large current pulse which results from any X-ray entering the chamber. The principle disadvantage is a complete loss of information on the energy of the incident X-ray. In addition, since the Townsend avalanche is complete in this type of detector the space charge effect is extreme, producing a dead time of about 200 μs.

5. *Position sensitive proportional counters.* In recent years a number of designs have been developed which use the principles of the flow proportional counter in a number of clever ways which tell where on the anode wire the ionizing X-ray photon hit. The position sensitive detector (PSD) shown in Fig. 1-11 uses a graphitized wire to slow the current pulses which travel in both directions, to a sensitive time discriminator which looks at the time difference between the pulses arriving from each end of the wire and determines the location where the photon struck. When X-rays scatter from a sample to a detector, the angle between the incident and scattered beams is called (2θ). The wire of the PSD can be placed to sense the X-rays scattering from more than ten degrees of 2θ simultaneously. The photon location information can be used to address the appropriate channels in a multichannel analyzer (MCA) where the full scattering (or diffraction) pattern can be stored, while the PSD scans across 2θ (Göbel, 1982a). A number of PSD designs allow one to view a few to over a hundred degrees around an X-ray experiment.

1.5.3 Solid State Detectors

1. *Lithium-doped Si or Ge crystals.* This type of detector must be immersed in liquid nitrogen in order to reduce the diffusion of the Li out of the semiconductor and to reduce the electrical background noise so that a field effect transistor (FET) can amplify the delicate signal. The principle of operation is almost exactly like that of the ion collection devices described above. An

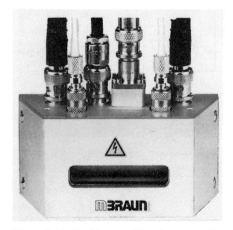

Figure 1-11. Position sensitive detector with its associated electronics (Courtesy of Braun).

X-ray photon enters the Si and is absorbed creating a photo electron whose energy is equal to that of the X-ray minus the binding energy of the electron. The energetic photoelectron then loses its energy in 3.8 eV increments by creating a string of electron hole pairs. Thus, the number of electrons created are proportional to the energy of the incoming photon. An electrical bias across the detector causes the electrons to be collected for amplification at the FET. The energy resolution (5%) is the best of any of the current detectors: about 15 times better than a scintillation detector. The dead time is determined primarily by the electronic amplification process and is on the order of 50 µs.

2. *Scintillation Detectors* are composed of a crystal of NaI with some Tl doped into it. When an X-ray photon hits the atoms in this crystal it get absorbed by producing a photo electron and a few Auger electrons. These electrons activate the production of fluoresced photons (scintilla) of blue light with a wavelength of 4100 Å, at the thallium sites. The total number of blue light photons produced are absorbed by a photoelectric material, typically a cesium-antimony intermetallic alloy. Approximately one electron is produced for each ten photons which strike the photocathode. The electrical signal in turn, is proportional to the energy of the original X-ray which entered the crystal. Thus, we have an energy discriminating detector. Unfortunately, owing to a rather large number of losses, the overall resolution is on the order of 75% of the energy of the incident photon. Thus, the resolution is not such as to be able to resolve the difference between Cu K_α (8.047 keV) and Cu K_β (8.9094 keV). This is made up to some extent by the high quantum efficiency (most entering photons are counted) and a low dead time of about 1 µs.

1.5.4 The Electronic Processing of X-Ray Signals

To operate a scintillation detector we need a high voltage source to put an electrical bias across the photocathode material and the very sensitive series of amplifiers (dynodes), which build the signal from the photon to a level that it can be amplified by a conventional amplifier and transported through a wire for processing. In a proportional counter we also have to apply a potential which will determine the gas amplification factor.

The preamplified signal brought back from most X-ray detectors is proportional to the energy of the X-ray photon which caused it. The only electronic detector still in occasional use for which this is not true is the Geiger-Müller detectors. The output from these detectors can be sent directly to a rate meter or scalar/timer for display. The electronics associated with the detection of X-rays on most modern X-ray equipment are illustrated in Fig. 1-12.

The initial preamplified voltage pulse arriving at the signal processing equipment is first sent to a linear amplifier to adjust its level to one appropriate for pulse height discrimination. An electrical circuit known as a pulse height analyzer (PHA), which is like a band pass filter, is the device we use to throw away extraneous pulses caused by any event other than a photon in the energy range of interest. Anyone not familiar with an electronic band pass filter should think of the very common graphic equalizer, or spectrum analyzer, attached to most modern stereo systems. This device breaks the music signal into a series of frequency bands which may be individually amplified. The PHA acts in an analogous manner, rejecting all pulses above and below voltages selected by the user.

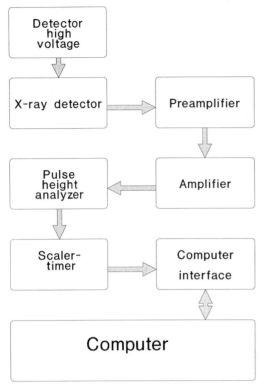

Figure 1-12. The electronics of signal processing.

The advantage of this device is clear. For example, cosmic rays send showers of very high energy particles into our detectors, also contamination of our tube target with another metal (like tungsten vaporized from the filament) will produce other characteristic radiation. In addition, there is a dark current, which is the phrase we use for electrons which spontaneously boil off of the photocathode in a scintillation detector, or spontaneous ionization in a proportional detector. These are a few of the many sources of background counts which will produce pulses of significantly higher or lower voltage than those caused by the radiation of interest. A pulse height analyzer can be set to reject all but the pulses from the radiation desired, within its and the detector's energy resolution.

After a pulse has been accepted by the PHA, it is passed on to two independent circuits. A scalar/timer, which allows us to count the number of pulses arriving in any time interval, and a rate meter. The rate meter is a circuit which takes in the random arrival of pulses and puts out an average signal, which can be displayed on a calibrated volt meter or a strip chart recorder.

In energy dispersive spectrometry (EDS) the high energy resolution of the Si(Li) crystal allows a different electronic approach. In place of the PHA and the scalar timer there is a multichannel analyzer (MCA). This device acts like an audio graphic equalizer with, from 4000 to 64 000 separate filters. The typically 4000 channels of an MCA each act as a separate scalar with its own PHA. An MCA allows us to store and then view photons of many different energies which strike the detector. Each energy will be counted into one of the scalars in the MCA. This is accomplished by converting the voltage associated with each ionization event to a digital number, using an analog to digital converter, and then using the resulting number to address a channel in the MCA. For wavelength dispersive spectrometry (WDS) and normal diffraction work we are only interested in photons of a single energy and therefore use the system shown in Fig. 1-12.

1.6 Crystallography

Symmetry can be thought of as an invisible motion of an object. If an object is hidden from view and moved in such a manner that when it reappears you cannot tell that it has been moved, then the object is said to possess *symmetry*. For example, if a perfect snowflake is covered and then rotated about its center by 60°, an observer will not be able to tell that any movement has occurred. The rotation of an object around an imaginary axis is the simplest symmetry element to visualize and is referred to as rotational symmetry. To get the complete set of all possible symmetry elements that an object may possess we must also allow for the possibility that, on removing the object from view, no movement at all was made. This of course, produces an effect identical to rotating the object about any of its symmetry axis and must also be a symmetry element. This element is called the identity operator by mathematicians but is more simply referred to as a 360° rotation by crystallographers who refer to it with the symbol 1.

Rotational symmetry is said to occur around a *proper axis* when the "handedness" of an object does not change. For example think of four right hand arranged in a circle each separated by 90°. The 4-fold axis relating them is proper, in that the hands remain "right" during each rotation. A rotation which changes the "handedness" after each operation is called *improper*. An improper rotation is a rotation followed by the operation of inversion through the origin of a coordinate system. Inversion causes an object located at x, y, z to change coordinates to $\bar{x}, \bar{y}, \bar{z}$. This, of course, turns a right hand into a left. Table 1-2 shows the complete set of crystallographic proper and improper axes of rotation.

The symbols used (n and $\bar{n}$) are called Hermann-Mauguin (Hermann, 1931; Mauguin, 1931) symbols and are uniformly used by crystallographers. The value of n refers to a rotation of $360/n$ degrees. Spectroscopists prefer a different set of symbols called Schönflies (1891) notation. Schönflies symbols only apply well to the shapes of objects or point groups and have serious ambiguity when applied to repeat-

Table 1-2. Crystallographic symmetry elements.

Degrees of rotation	Proper axis	Improper axis
$360°/n$	n	$\bar{n}$
360	1	$\bar{1}$
180	2	$m (= \bar{2})$
120	3	$\bar{3}$
90	4	$\bar{4}$
60	6	$\bar{6}$

ing arrays of objects or space groups which are described below.

The reader will note that the symmetry elements in Table 1-2 do not include a 5-fold rotation or any rotation greater than 6. In crystallography and mathematical group theory we always seek the *irreducible representation* of an object's symmetry (i.e., the simplest set of symmetry elements which describe the full symmetry of an object). The only classical irreducible rotations which can fill space are 1-, 2-, 3-, 4- and 6-fold rotations.

The absence of a 5-fold axis from Table 1-2 is due to the very old principle that objects with 5-fold symmetry cannot fill space, and therefore may not be observed as a crystallographic symmetry. The reader may easily verify this by trying to arrange cutouts of pentagons into a space filling array in two dimensions. As with most age-old principles, given sufficient time we must begin to add qualifications. Beginning in 1984, Shechtman et al. (1984) opened a whole new field of crystallography by discovering five fold symmetry in an electron diffraction pattern of a rapidly quenched Al_6Mn alloy. In their pattern there is not one repeating pentagon (which clearly cannot fill space) but two different sized pentagons in a pattern which, while filling space, does not produce any simple repetition. The ratio of the sides of these two pentagons is 1.6 ± 0.02, which is a number referred to as the "golden mean" of Hellenic architecture $(\sqrt{5} - 1)/2$. A two dimensional arrangement of two different 4-sided rhombs having the same, golden mean, ratio of their perimeters can also fill space when arranged in a pattern known as a *Penrose tile*. Penrose tilings have become the most widely studied of two dimensional space filling patterns which have no repeating unit cell. These patterns show local regions of five-fold symmetry. Figure 1-13 shows, previously thought to be impossible, ten fold symmetry. In fact examples of eight and twelve fold axis have now also been reported. None of these materials qualify as a crystal in the conventional sense in that, they are only quasi-periodic. Hence they have been called quasi-crystals. However, this name should not lead one to conclude that these materials are not real crystals, in that the best of these give diffraction patterns as sharp as any conventional crystal.

The full nature of quasicrystals, is still a subject of active investigation, with no clear structural model able to explain the growing number of examples. For further information see the recent comprehensive review by Steurer (1990) and the book by Hargittai (1990).

1.6.1 Point Groups

Now that we have examined the possible types of simple symmetry that an object, which can fill space may possess, we may next ask how many unique combinations of symmetry these space-filling objects may have. The simplest symmetry group is 1, or no symmetry at all. The addition of any other symmetry element to group 1 will, of course, raise the total symmetry and have to be classified as a new group. The addition of a center of symmetry puts us into group $\bar{1}$. Increasing the symmetry further, we must look at objects containing only a 2-fold axis (group 2) or a mirror (group *m*). At this point we see the first opportunity of combining two symmetries in an object to form a unique new group, $2/m$, which refers to objects which have a 2-fold axis with a mirror perpendicular. The two axes in the plane of the mirror cannot meet at 90° or they would be forced to become 2-fold axes forcing the total symmetry into the group 222, *mm*2 or *mmm*.

Similar arguments lead to the conclusion that there are only 32 unique ways of combining symmetry elements in objects which can repeat in three-dimensions to fill space. These are called point groups. We should comment here on the type of coordinate systems required to specify these various symmetries. In the groups 1 and $\bar{1}$ there need be no relationship between the three coordinate axes or the angles between them. We refer to this coordinate system as *triclinic* or *anorthic*. For the groups 2, *m* and $2/m$, there need be no relation between the lengths of the three axes

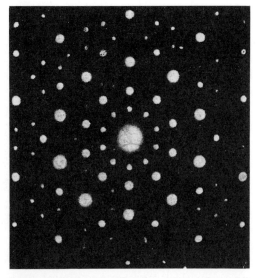

Figure 1-13. Electron diffraction pattern of a rapidly quenched Ni–Ti–V alloy (from Zhang et al., 1985).

but this symmetry requires that one axis meet the plane of the other two at right angles. This system with one unique axis is called *monoclinic*. The groups 222, *mm*2 and *mmm* require that all three axes be orthogonal. This class is called *orthorhombic*. To increase the symmetry further will require that one or more of the axes must be equal in length. In the *hexagonal* and *tetragonal* systems the unique axis is called *c* by convention. In *cubic* all three orthogonal axes are equal. In the last crystal class *rhombohedral*, the three axes are equal in length but each are separated by a common non-90° angle α. Rhombohedral differs a bit from the crystal systems in that we can represent it on a hexagonal axis; thus, we really only need six coordinate systems to represent all symmetries. Table 1-3 shows the relationships among the axes and interaxial angles demanded by the symmetry of the various point groups. The seven different coordinate systems are called crystal systems.

1.6.2 Bravais Lattices

It is convenient for us to imagine that the atoms in a crystal are arranged in groups which repeat three-dimensionally to make up a crystal structure. If we look at each symmetrically unique group of atoms as if they were a single point, then we can imagine the entire crystal structure as a three-dimensional lattice of these points. This concept is called a space lattice and the points which define it are lattice points. It is important to understand that the lattice points which we use to abstractly think about a crystal structure are not atoms, but represent a collection of atoms (which are sometimes, but definitely not always, a molecule or a formula unit). The lattice point represents the *asymmetric unit* or *basis* of the crystal structure which is the smallest group of atoms upon which the crystallographic symmetry operates to produce the complete crystal structure.

We can recognize symmetry in the various possible ways of arranging points into three-dimensional arrays. These periodic arrays can be sorted into seven unique arrangements. These seven basic patterns are called crystal systems and define the coordinate systems described in Table 1-3. In a periodic array of points we can think of another type of symmetry: translational symmetry. Imagine that there are translation vectors extending between the lattice points along the three principal axes of the coordinate system. Connecting all of these vectors will produce a collection of boxes each called a unit cell. The three vectors which define the unit cell have lengths of *a*, *b* and *c*, making angles of α, β and γ with each other. The number of asymmetric units in each unit cell is called *Z*. In two dimensions we can recognize patterns cor-

Table 1-3. The seven crystal systems and fourteen Bravais lattices.

Crystal class	Axis system	Lattice symmetry	Bravais lattice
Cubic	$a = b = c, \alpha = \beta = \gamma = 90°$	$m3m$	P, I, F
Tetragonal	$a = b \neq c, \alpha = \beta = \gamma = 90°$	$4/mmm$	P, I
Hexagonal	$a = b \neq c, \alpha = \beta = 90°, \gamma = 120°$	$6/mmm$	P
Rhombohedral	$a = b = c, \alpha = \beta = \gamma \neq 90°$	$\bar{3}m$	R
Orthorhombic	$a \neq b \neq c, \alpha = \beta = \gamma = 90°$	mmm	P, C, I, F
Monoclinic	$a \neq b \neq c, \alpha = \gamma = 90°, \beta \neq 90°$	$2/m$	P, C
Triclinic	$a \neq b \neq c, \alpha \neq \beta \neq \gamma \neq 90°$	$\bar{1}$	P

responding to a general rhombus, 60° rhombus, square, rectangle and a rectangle with an additional lattice point in the center of its face or face centered. If we chose only the simple cell in the last case it would have no symmetry. The choice of the face centered cell allows us to specify *mm* symmetry.

When the unit cell, which repeats in space, is defined by one point at each corner, the unit cell is called *primitive* and given the symbol **P**. If there is an additional lattice point in the middle of each cell it is called *body centered* and given the symbol **I**. When the cell has points in each corner and in the center of each face, it is called *face centered* and the symbol **F** is used. The rather strange rhombohedral cell, in which the three interaxial angles are equal and acute, is given the symbol **R**. The last type of Bravais lattice is when only one face is centered. The symbol used to describe this type of cell depends on the name of the face that is centered. The plane perpendicular to the ***a*** and ***b*** directions of the lattice is called **C**. Likewise, the face perpendicular to the ***a*** and ***c*** directions is called **B**. On considering the seven basic lattice symmetries and the various types of unit cells allowed to each symmetry, there are fourteen Bravais lattices which describe the only ways of filling space in a periodic manner. These are also indicated in Table 1-3.

1.6.3 Space Groups

We have seen that there are 14 space lattices and 32 point groups. If we arrange the 32 point groups in the various patterns allowed by the 14 Bravais lattices we will be able to distinguish 230 unique three-dimensional patterns which we call space groups. Every crystal structure can be classified into one of these 230 symmetry groups. It has long been recognized that the notation we would need to describe these 230 space groups would need to be extremely complex. However, if we recognize that space groups can be thought to result from the combination of simple primitive symmetry elements (proper and improper rotations and reflection) with the translational symmetries of the Bravais lattices, we can recognize two new combinational symmetry elements: glide planes and screw axes. These non-primitive symmetry elements allow for the succinct space group notation used in the solid state sciences. This notation is most comprehensively described in the *International Tables for Crystallography*, Volume A (1983).

The combination of a translational movement with a rotation changes the rotational axis into a *screw axis*. The symbol for a screw axis is N_j, where N stands for the type of rotational axis (i.e., $360/N =$ the number of degrees in the rotation), and j tells the fraction of the cell translated (i.e., j/N). So 2_1 means a rotation of 180° and a translation of 1/2 of the unit cell along the direction of the rotation axis. A 3_2 means a rotation of 120° followed by a translation of 2/3 of the cell.

The combination of a translation with a mirror reflection is called a *glide plane*. The location of the symbol for the glide plane in the space group symbol tells which mirror the reflection occurs across, while the actual symbol (*a*, *b*, *c*, *n* or *d*) tells the direction of the translation. After reflection across a mirror, an "*a*" glide translates in the ***a*** direction. Similarly for "*b*" and "*c*" glides. An "*n*" glide translates along the face diagonal of the unit cell while the "*d*" glide moves along the body diagonal after reflection.

1.6.4 Space Group Notation

The direction of the mirror plane before the translation of a glide operation, introduces the subject of space group notation. Space group symbols are made of two parts: the first is a capital letter designating the Bravais lattice type [**P**, **C**, (**B** or **A**), **I**, **F** or **R**]. The second part of the symbol tells the type of symmetry the unit cell possesses. The symbols used conventionally only specify the minimum symmetry to uniquely identify the space group. All implied symmetry is not mentioned in the symbol. This fact along with a bit of arbitrariness in our choice of origin and definition of directions in a unit cell leads us to one of the great generalizations of modern crystallography: "Look up your space group in Volume A of the International Tables for Crystallography" (1983). This invaluable reference work lists each possible view of a unit cell and all of the implied symmetry along with a wealth of other information.

Since the only possible symmetry that can occur in triclinic is 1 and $\bar{1}$, there are only two space groups: **P**1 and **P**$\bar{1}$. In order to increase the symmetry from triclinic we must add a 2-fold axis or a mirror. To do this requires that we bend one of the axis so that it makes a 90° angle with the plane defined by the other two. This places us in the monoclinic crystal class. Since there can be only one symmetry plane or 2-fold axis, its name is purely arbitrary. We will adopt the most common ***b*** axis unique notation here.

The various monoclinic space groups are composed of all possible combinations of the point group symmetries 2, *m*, 2/*m* and the corresponding translational elements 2_1, *c*, $2_1/m$, 2/*c* and $2_1/c$ with the Bravais lattices **P** and **C**. These are shown in Table 1-4. Note that in Table 1-4 there are three blank entries for the groups **C**2_1,

Table 1-4. The 13 monoclinic space groups.

Point group	P	C
2(C_2)	P2	C2
	P2_1	
m(C_s)	P*m*	C*m*
	P*c*	C*c*
2/*m*(C_{2h})	P2/*m*	C2/*m*
	P2_1/*m*	
	P2/*c*	C2/*c*
	P2_1/*c*	

C2_1/*m* and **C**2_1/*c*. This is because these combinations of symmetry are contained by implication in the groups **C**2, **C**2/*m* and **C**2/*c* respectively. This means that there are only 13 unique monoclinic space groups, if any more symmetry is added to a cell the cell is forced into a higher symmetry crystal class and the rest of the 230 allowed space groups result.

1.6.5 Reduced Cells

Any lattice can be described by a large number of choices of unit cells. A cubic lattice may easily be described by three translation vectors which will give monoclinic or even triclinic symmetry. Since any three non-collinear reciprocal lattice vectors will define a unit cell which will repeat in space to give the space lattice, there in fact are an extremely large number of unit cells which may be chosen to describe any lattice (most of them non-primitive). Which is the correct cell? To answer this question we must examine what we mean by "correct". Since the recognition of symmetry saves a great deal of work in describing crystals we follow the usual crystallographic convention of choosing the unit cell with the highest symmetry. However, since we have a huge number of cells to choose from for any lattice, recognizing this cell may, at times, be difficult.

Niggli (see Mighell, 1976) has described a mathematical procedure which will always produce the same reduced cell from any of the large number of descriptions for a lattice. The reduced cell is defined as that cell whose axes are the three shortest noncoplanar translations in the lattice; consequently there is only one such cell in any lattice. It is, by convention, the standard choice for the triclinic cell. This Niggli cell is a good place to start, once three potential translation vectors have been found which describe the lattice. Sublattices and superlattices may sometimes be obtained by halving or centering the "correct" cell. The Niggli procedure will show that these cells also reduce to the same reduced cell. In fact, Niggli showed in 1928, that there are only 28 unique reduced cells. Mighell (1976) has taken advantage of this fact in designing the Crystal Data, database which classifies all materials with known unit cells according to their reduced cells. Program TRACER II of Lawton and Jacobson (1965), carries out the Niggli procedure. Program LATTICE, by Mighell and Himes (1986), will not only find the reduced cell but also will attempt to identify the cell in the Crystal Data computer database.

1.6.6 Miller Indices

The unit cell is a very useful concept and we use it not only to characterize the symmetry of a crystal, but also to specify crystallographic directions and even interatomic distances. To describe directions and distances, we imagine planes, with various orientations, intercepting the translation vectors at various points. We imagine that these planes are members of sets which cut all cells identically. Each plane may be characterized by the integers which result when the reciprocal of the axial intercepts are taken. For example, the plane which passes through each unit cell intercepting the a axis at 1/2 and parallel to the b and c axes has Miller indices of $1/2, 1/\infty, 1/\infty$ or (200). To find the Miller indices of any set of planes, only one trick is needed: we must use the mathematical relation that parallel planes will meet at infinity. Thus, a plane parallel to an axis has an intercept of infinity and the reciprocal of infinity gives a Miller index of zero. The following general rule will always, unambiguously, allow the determination of the Miller indices of any plane: Locate the first plane of the family away from the origin of the unit cell and take the reciprocal of its axial intercepts. Figure 1-14 illustrates a series of planes parallel to the c axis with their Miller indices.

Since the imaginary planes, characterized by Miller indices, are defined in terms of the lengths of the unit cell edges, then the perpendicular distances between the planes in any family are characteristic of the unit cell size and hence, the crystal structure. The distances (d_{hkl}) between these planes are on the same order as the distances between atoms. We should note that these **d**-values are vectors, having direction and magnitude. It should also be noted that Miller indices defining a plane are usually written in parenthesis, as ($h\,k\,l$), while crystallographic directions which are the normal to any plane, are written in square brackets as [$h\,k\,l$].

1.7 Diffraction

Diffraction occurs when waves scattering from an object constructively and destructively interfere with each other. Waves are characterized by their wavelength (λ) which is defined as the distance between

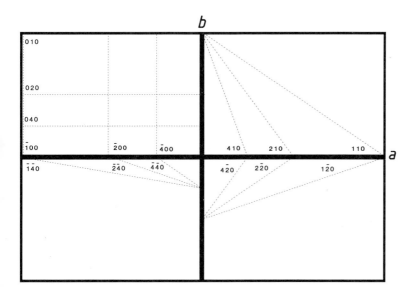

Figure 1-14. Planes, indicated by dashed lines, parallel to the **c** axis. The coordinate system origin is in the center with four unit cells shown about it.

peaks. If a wave scatters from an object it will do so in all directions. If a second wave scatters from another object, displaced from the first by a distance on the order of the wavelength, then there will be some angle at which we can view the two scattered waves such that they will be in phase. This phenomenon is illustrated in Fig. 1-15.

Visible light is an electromagnetic wave with a wavelength of about 50 000 Å. If a surface is ruled with periodic grooves that are 50 000 Å apart and visible light falls on this surface, the light will scatter from each of the rulings and be diffracted. The reason for the diffraction is that the path difference between rays scattering from each adjacent groove is on the order of λ which causes all of the scattered waves to be in phase at some angle. For a fixed scatterer spacing, the angle will depend on the wavelength, with short wavelengths diffracting at higher angles than longer ones. If the incident light is white (i.e., containing all wavelengths), an observer sees this effect as a rainbow; the different wavelengths of the visible light diffract from this ruled surface or grating at different angles, depending on their wavelength.

1.7.1 Bragg's Law

W. L. Bragg (1913) was the first to show that the scattering process which leads to diffraction can be equally visualized as if the X-rays were "reflecting" from the imaginary planes defined by Miller indices. This view pushes the reflection analogy rather hard; however, this rather strained analogy allows a simple derivation of the overall controlling law which describes the phenomena of diffraction. Of course, the same law can be derived without need of the reflection analogy but requires a bit more effort.

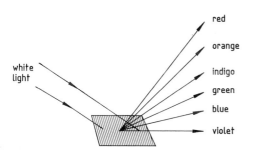

Figure 1-15. Diffraction from an optical grating.

In Fig. 1-16 X-rays impinge on a set of atomic planes, with indices (h k l), from the left, making an angle θ with them. The distance between the planes is d_{hkl}. It helps if you consider atoms to be located on the planes acting as sources of scattering and thus, for these planes the d_{hkl} would correspond to an interatomic distance. However, all of the atoms in a unit cell will be bathed in the X-ray beam and, in fact, all will be scattering in all directions. It is not necessary that there be any atoms on the planes for diffraction to be thought of as occurring from them but, it makes our picture a little easier to understand. If we look only at beam 1 and beam 2 we see that beam 2 must travel the distance ABC farther than beam 1. If beam 1 and 2 start out in phase, meaning lined up peak to peak and valley to valley, then the extra distance ABC will cause beam 2 to be misaligned with 1 after reflection.

The degree of phase shift or misalignment is equal to the distance ABC. Constructive interference or diffraction will occur when the two waves (and all the waves scattering from deeper planes in the crystal) come out in phase. This of course will happen when the distance ABC = 1λ or 2λ or 3λ or, in general, when

$$n\lambda = ABC \tag{1-14}$$

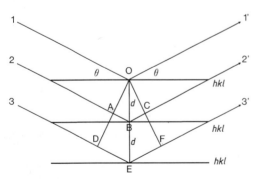

Figure 1-16. Diffraction of X-rays from the planes in a crystal.

where n is an integer. All that remains is to get the distance ABC in terms of the measurable angle θ. To do this we look at the right triangle ABO and notice that $d \sin \theta = AB$, or $2 d \sin \theta = ABC$. Then the condition for diffraction to occur is

$$n\lambda = 2 d \sin \theta \tag{1-15}$$

This is the famous equation first derived by W. L. Bragg and called Bragg's law. This equation allows us to relate the distance between a set of planes in a crystal and the angle at which these planes will diffract X-rays of a particular wavelength. It is usually more convenient to divide both sides of the Bragg equation by n and to define d/n as d_{hkl}. This makes Bragg's law look like

$$\lambda = 2 d_{hkl} \sin \theta \tag{1-16}$$

In the original notation Eq. (1-15) we consider first order ($n = 1$) or second order ($n = 2$) diffraction from any plane like (111). With the modified Eq. (1-16), we consider all diffraction to be first order from each of the families of planes in the crystal [i.e., from (111) and (222) etc.]. An examination of the Bragg equation shows that when λ is known and θ is measured, we can calculate d_{hkl} and discover the dimensions of the unit cell and even of atoms and their bonds.

The artificial reflection analogy used in obtaining the Bragg equation does not, in any way, compromise its validity. It can be derived in the same form by a rigorous analysis of the Huygens scattering from each of the atoms in the crystal.

1.7.2 The Reciprocal Lattice

P. P. Ewald (1913) devloped what is by far the most useful method for describing and explaining diffraction phenomena. Its wide acceptance is an indication of its basic simplicity. The concept introduced by Ewald has the name reciprocal lattice or reciprocal space.

When a particular arrangement of an X-ray source, sample and detector is proposed, we would like to be able to predict the various motions which will have to be applied to the three components in order to see particular diffraction effects. If, for example, we wish to see diffraction from the 100 planes of a LiF single crystal which has observable and identifiable cleavage, we simply use the Bragg reflection analogy and orient the crystal so that its 100 surface makes an equal angle with the beam and detector. The angle can be calculated from Bragg's law using the lattice parameter for LiF, which we must look up (i.e., in the Crystal Data database), and the wavelength of the X-rays to be used. However, if we wish to look at diffraction from the (246) planes, the visualization of the required orientation of the crystal becomes formidable and even the computation of the magnitude of d_{246} is non-trivial. Ultimately, the problem can be thought of as the difficulty of visualizing two dimensional planes intersecting a three dimensional unit cell.

In general, problems become easier to understand if we visualize them with fewer dimensions than are required in our three dimensional world. When one or more dimensions can be removed from an analysis the mathematics and conceptualization usually become much easier. For this reason many derivations begin in one dimension and only at the end expand to three.

Our problem, in diffraction theory, is that the Bragg planes in a space lattice are inherently three dimensional and restricting our analysis to a two dimensional plane within the lattice really doesn't simplify their three dimensional relationships with the other planes in the lattice. So we use another method to try and remove a dimension from the problem; we will represent each two dimensional plane as a vector: d_{hkl} is defined as the vector pointing from the origin of the unit cell to a point perpendicular to the first plane in the family (hkl) as illustrated for the 110 plane in Fig. 1-17. If we plot all of the d_{hkl} vectors as points in the appropriate crystallographic coordinate system, we observe that these points begin on each axis at the fractional coordinate 1 and increase in density as we approach the origin. (Fractional coordinates are simply the distance expressed as a fraction of the unit cell edge.) Figure 1-18 shows the vectors representing the planes shown in Fig. 1-14. If the outermost points $(100, 010, 001, \bar{1}00, 0\bar{1}0, 00\bar{1})$ are connected, we will observe the three dimensional shape of the unit cell. Thus, this construction has the advantage of representing two dimensional planes as points while maintaining the symmetry relationships between planes. However, Fig. 1-18 shows a very serious hindrance: as we approach the origin of this vector diagram the density of points, representing planes, increases to infinity. Rather than simplifying our real space picture this vector space construction has definitely made it more complex.

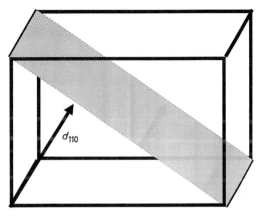

Figure 1-17. d_{110} as a vector in a unit cell.

The full three dimensional vector representation of the planes in a unit cell will be a sheaf of vectors projecting out from the origin in all directions, getting thicker and thicker as we approach the center. On examining Fig. 1-18 we see that the vectors are approaching the origin according to the reciprocals of the d_{hkl} values. Ewald, on noting this relationship, proposed that we plot not the d_{hkl} vectors, but instead the reciprocal of these vectors. We will define this reciprocal vector as

$$d^*_{hkl} = \frac{1}{d_{hkl}} \qquad (1\text{-}17)$$

Let us now redraw the rather complicated Fig. 1-18 but plotting d^*_{hkl} vectors instead of d'^s_{hkl}. Figure 1-19 shows this construction. The units are in reciprocal Ångströms and the space is therefore a reciprocal space. Notice that the points in this space repeat at perfectly periodic intervals defining another space lattice only this one we will call a reciprocal lattice. The repeating translation vectors in this lattice are called a^*, b^* and c^*. The interaxial angles are α^*, β^* and γ^*, where the reciprocal of an angle is defined as the complement or 180° minus the angle.

The concept of the reciprocal lattice makes the visualization of the Bragg planes extremely easy. To establish the index of any point in the reciprocal lattice one simply has to count out the number of repeat units in the a^*, b^* and c^* directions. Figure 1-19 only shows the $hk0$ plane through the reciprocal lattice, however, the concept is of a full three dimensional lattice, which extends in all directions in reciprocal space. If the innermost points in this lattice are connected we will see the three dimensional shape of the reciprocal unit cell which is directly related to the shape of the real space unit cell (in fact, the surfaces of this figure define the well known Brillouin zone). Thus, the symmetry of the real space lattice propagates into the reciprocal lattice.

The reciprocal lattice has all of the properties of a real space lattice. Any vector in

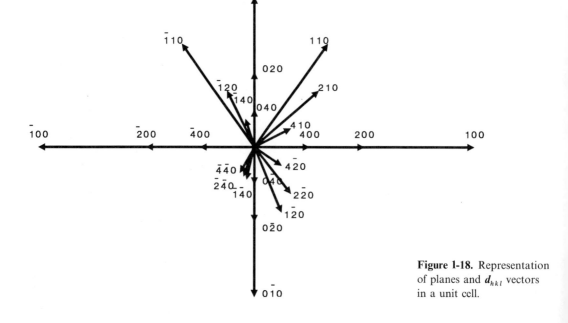

Figure 1-18. Representation of planes and d_{hkl} vectors in a unit cell.

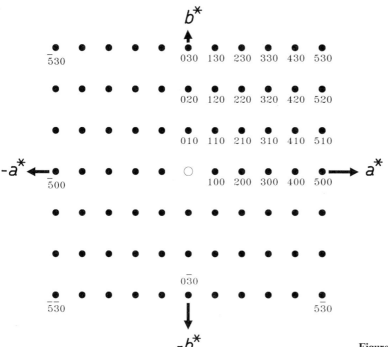

Figure 1-19. The reciprocal lattice.

this lattice, representing a set of Bragg planes, can be resolved into its components

$$d^*_{hkl} = h\mathbf{a}^* + k\mathbf{b}^* + l\mathbf{c}^* \quad (1\text{-}18)$$

An important point here is that the integers in Eq. (1-18) are in fact, equal to the Miller indices. We shall see in the next section that the process of X-ray diffraction can be thought of as recording the reciprocal lattice. It is this fundamental relation between the size, shape and symmetry of the real and reciprocal lattices which allows X-ray diffraction to determine the crystallographic information about a material and hence, gives rise to the name of this discipline as X-ray crystallography.

1.7.2.1 Relationship between d_{hkl}, hkl and Translation Vectors

There are an infinite number of sets of planes which can be thought of as intersecting a unit cell. We characterize these planes with their hkl values and the d_{hkl} interplanar spacings. The d_{hkl} values are a geometric function of the size and shape of the unit cell. The relationship between d_{hkl} and real unit cell is cumbersome and usually stated in a different form for each crystal system. However, the functional relation between the square of the reciprocal lattice vectors and the size and shape of the reciprocal unit cell is easily derived, has a simple form and applies easily to all crystal systems. The d^{*2}_{hkl} equation coupled with the relations given in Table 1-5 will allow simple computations relating d_{hkl} to the lattice parameters in any crystal system.

Table 1-5 shows the equations relating the real and reciprocal lattice translation vectors for the various crystal systems.

To derive d^{*2}_{hkl} we simply square Eq. (1-18). To get the square of a vector we must multiply it by itself, and since it is a

Table 1-5. Direct and reciprocal space relationships [a].

System	a^*	b^*	c^*
Orthogonal systems	$a^* = 1/a$	$b^* = 1/b$	$c^* = 1/c$
Hexagonal	$a^* = 1/a \sin \gamma$	$b^* = 1/b \sin \gamma$	$c^* = 1/c$
Monoclinic	$a^* = 1/a \sin \beta$	$b^* = 1/b$	$c^* = 1/c \sin \beta$
Triclinic	$a^* = \dfrac{bc \sin \alpha}{V}$	$b^* = \dfrac{ac \sin \beta}{V}$	$c^* = \dfrac{ab \sin \gamma}{V}$

[a] Note that the following relationships hold and are especially useful for triclinic:
$\cos \alpha^* = \cos \beta \cos \gamma - \cos \alpha/(\sin \beta \sin \gamma)$,
$\cos \beta^* = \cos \alpha \cos \gamma - \cos \beta/(\sin \alpha \sin \gamma)$,
$\cos \gamma^* = \cos \alpha \cos \beta - \cos \gamma/(\sin \alpha \sin \beta)$,
$V^* = a^* b^* c^* \sqrt{1 - \cos^2 \alpha^* - \cos^2 \beta^* - \cos^2 \gamma^* + 2 \cos \alpha^* \cos \beta^* \cos \gamma^*}$.

vector the multiplication takes the form of a dot product

$$d^{*2}_{hkl} = d^*_{hkl} \cdot d^*_{hkl} \tag{1-19}$$

$$d^{*2}_{hkl} = h^2 a^{*2} + k^2 b^{*2} + l^2 c^{*2} +$$
$$+ 2hk\, a^* b^* \cos \gamma^* + 2hl\, a^* c^* \cos \beta^* +$$
$$+ 2kl\, b^* c^* \cos \alpha^* \tag{1-20}$$

This expression relates the square of the inverse of d_{hkl} to the size and shape of the reciprocal unit cell for any plane in any crystal system. It is useful to point out that in orthogonal crystal systems the final three terms of the equation include a cos 90° term and therefore go to zero. For many applications then, Eq. (1-20) reduces to a particularly simple form. For example, Eq. (1-20) becomes for

cubic $\quad d^{*2}_{hkl} = (h^2 + k^2 + l^2) a^{*2}$,

tetragonal $d^{*2}_{hkl} = (h^2 + k^2) a^{*2} + l^2 c^{*2}$,

hexagonal $d^{*2}_{hkl} = (h^2 + hk + k^2) a^{*2} + l^2 c^{*2}$,

orthorhombic $d^{*2}_{hkl} = h^2 a^{*2} + k^2 b^{*2} + l^2 c^{*2}$.

1.7.2.2 The Ewald Sphere of Reflexion

In Fig. 1-20 we define a sphere with a radius of $1/\lambda$ and place it such that the origin of the reciprocal lattice is tangent to it. If we now imagine a real crystal at the center of this sphere and an X-ray beam entering from the right, we have all the components needed to visualize the diffraction process geometrically instead of mathematically, using the Bragg equation. It is clear that a rotation of the crystal (and its associated real space lattice), will also rotate the reciprocal lattice because the reciprocal lattice is defined in terms of the real lattice. We now examine this arrangement, at a specific time in the rotation of the crystal, when the 230 point in the reciprocal lattice is brought in contact with the sphere.

This orientation is illustrated in Fig. 1-21 where, $\overline{CO} = 1/\lambda$. From our definitions, $\overline{OA} = d^*_{230}/2$, hence

$$\sin \theta = \frac{\overline{OA}}{\overline{CO}} = \frac{d^*_{230} \lambda}{2}$$

or

$$\lambda = \frac{2 \sin \theta}{d^*_{230}}$$

but by definition,

$$d_{230} = \frac{1}{d^*_{230}}$$

1.7 Diffraction

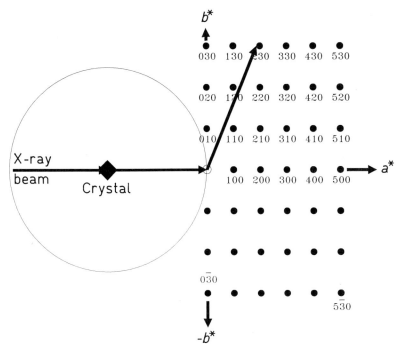

Figure 1-20. The Ewald sphere of reflection with a crystal in the center and its associated reciprocal lattice tangent to the sphere, at the point where the direct X-ray beam emerges.

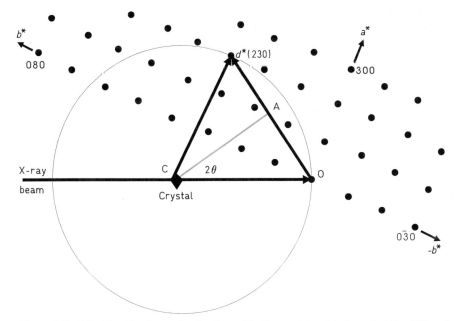

Figure 1-21. The Ewald sphere of reflection with the crystal rotated so that the 230 reciprocal lattice point touches it, permitting it to diffract.

therefore,

$$\lambda = 2 d_{hkl} \sin \theta$$

Thus, we see that the reciprocal lattice and sphere of reflection concepts incorporate Bragg's law. As each lattice point, representing a d^*_{hkl} value, touches the sphere of reflection, the condition for diffraction is met and the corresponding real space lattice plane "reflects" in the direction $C-d^*_{hkl}$. If we look at the non-imaginary components of our construction, we see the sample and the incident and diffracted X-ray beams. Thus, the Ewald construction allows us to easily analyze an otherwise complex diffraction geometry, showing what motion must be applied to a crystal in order to produce a diffracted beam in a particular direction.

The Ewald sphere construction is very useful in explaining diffraction phenomena in any type of geometry. The principal advantage is that it avoids the need to do calculations to explain any phenomena and instead allows us to visualize an effect using a pictorial, mental model. In addition, it permits the simple analysis of the otherwise complex relationships among the crystallographic axis and planes.

1.7.3 Single Crystal Diffraction Techniques

There have been a number of techniques developed to record the diffraction pattern of single crystals. Each may be viewed as a method of visualizing the reciprocal lattice with various distortions introduced by a particular experimental geometry. For example, if we set up the experiment shown in Fig. 1-22 with a crystal rotating around its a axis, in an X-ray beam, and place a cylindrical film around the crystal, we will record each of the $0\,k\,l$ points of the reciprocal lattice, projected onto a line on the film. This technique, called an oscillation or rotation photograph permits rapid determination of the length of the axis being rotated about. Each of the resulting parallel lines of points contains all of the diffraction points in each $n\,k\,l$ plane of the reciprocal lattice, when n is constant in each line. The intensities of the reciprocal lattice

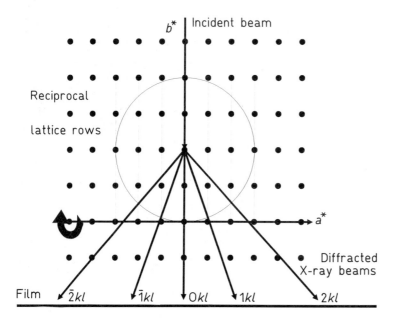

Figure 1-22. The experimental arrangement for a rotation photograph.

points on the film are related to the types and locations of the atoms in the unit cell and will be discussed later. This recorded pattern of spots is referred to as the intensity weighted reciprocal lattice.

Since all of the reflections in each reciprocal lattice plane are projected onto a line, the identification of individual reflections is quite difficult. One of the earliest procedures allowing the easy indexing of individual spots and subsequent measurement of their intensity was realized in the Weissenberg camera. This device allows the isolation of a single plane of the reciprocal lattice by placing an absorbing screen over all of the rows of points in the rotation photograph, other than the row to be measured. The film is then oscillated left and right as the crystal is rotated in the X-ray beam. This causes each diffraction spot in the row to be recorded in a different location on the film. Figure 1-23 shows a Weissenberg camera and Fig. 1-24 the $0\,k\,l$ zone of a crystal spread over the film. This distortion of the reciprocal lattice is readily analyzed permitting indexing.

Martin Buerger (1964) designed a single crystal X-ray diffraction camera, called the precession camera, which will take pictures of the reciprocal lattice exactly the way we imagine it. This camera records the diffraction pattern while the crystal is precessed around the X-ray beam. If the film is kept parallel to the reciprocal lattice plane being photographed, during this motion, an undistorted picture of the reciprocal lattice plane can be made. Figure 1-25 shows such a camera with the layer-line screen in place, allowing only the reciprocal lattice layer desired to pass through the annulus. The photograph of the $0\,k\,l$ layer of ammonium oxylate shown in Fig. 1-26 was taken in this manner. It is interesting to think of this procedure as taking a photograph of a mental construction.

Figure 1-23. The Weissenberg camera with the crystal hidden from view inside the film cassette on its movable mount (courtesy of Siemens AG).

Today most single crystal diffraction is done on automated single crystal diffractometers. These devices have four individual degrees of freedom allowing the positioning of each reciprocal lattice point on the sphere of reflection and then scanning an electronic detector past it to record its

Figure 1-24. A Weissenberg photograph of the $0\,k\,l$ reciprocal lattice net.

Figure 1-25. The Buerger precession camera with layer line screen behind the crystal and in front of the film cassette (Courtesy of Siemens AG).

intensity. In this manner a few thousand data points can be measured per day. For very large unit celled materials, two dimensional array detectors have been developed to permit rapid measurement of hundreds of thousands of diffraction intensities. The rapid recording of large quantities of diffraction intensities is currently the subject

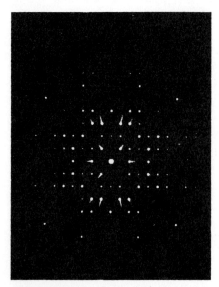

Figure 1-26. A precession photograph of the $0kl$ zone of ammonium oxalate.

of intensive development. An example of this thrust is the two dimensional array of charge coupled devices (CCD) which can record X-rays over a region of about 30 cm^2 with a resolution on the order of 100 μm.

Another thrust area of applied crystallography is the use of the high brightness of synchrotron sources to collect full single crystal intensity data-sets from tiny crystallites. Normally the crystals required for diffractometer measurement are on the order of 0.1–0.3 mm in dimension. Recent studies have obtained full diffraction data-sets from crystals with volumes on the order of 10–100 μm^3. In a recent study by Newsom et al. (1991) single crystal intensity data was collected from a 7 μm zeolite crystal (about 2500 unit cells on an edge).

1.7.3.1 X-Ray Topography

X-ray topography is the imaging of single crystals from the viewpoint of a particular crystallographic plane. In the Lange camera procedure a narrowly collimated X-ray beam is directed at a sample oriented at the Bragg angle for, say, the (111) reflection. The diffracted beam is then directed in a narrow line across a piece of film. Then, both the crystal and film are moved in unison so that each band of the crystal, diffracting the (111) X-rays, is recorded on a corresponding band of the film. The resulting photograph, recorded on high resolution film or a nuclear emulsion plate, shows all of the crystal imperfections, grain boundaries and domain structure. The resolution of such a topograph is on the order of 1 μm which is three orders of magnitude less than electron microscopy. However, the X-ray topograph examines large areas of the crystal, is non-destructive and can be run in non-ambient conditions. Higher intensity synchrotron sources have allowed the use of full parallel beam geometry, eliminating the need to scan the crystal and film. These procedures are widely used in the materials industries involved with single crystals, silicon in particular.

1.7.3.2 Laue and Kossel Techniques for Orientation Determination

Two other single crystal techniques are particularly important to materials science and technology. The Laue technique is so named because it is the procedure used by Friedrich, Knipping and von Laue (1912) to record the very first X-ray diffraction pattern. The procedure is to place a single crystal in front of a white X-ray beam and allow the various planes, which happen to be in diffracting position, to pick out the wavelength, which meets the Bragg condition, and diffract it. The reciprocal space way of thinking about this technique is to picture the white X-ray beam as a series of Ewald spheres within each other, each showing a common surface point which is the origin of the reciprocal lattice. Each sphere corresponds to the $1/\lambda$ of the longest and shortest wavelengths in the beam. This solid Ewald sphere has individual onion-skin like layers, some of which will intersect each of the points in the reciprocal lattice within the solid sphere, putting them all into diffracting position. This situation is illustrated in Fig. 1-27. This geometry implies that the reflections in each crystallographic zone will lie on the edge of a cone surrounding the zone axis. Depending on the angle at which this cone intersects the film, we will observe zonal reflections to lie along various conic sections, like hyperbolas and ellipses.

This technique can have the film placed in transmission, behind the crystal or in back-reflection, bringing the incident X-ray beam through a hole in the film. The

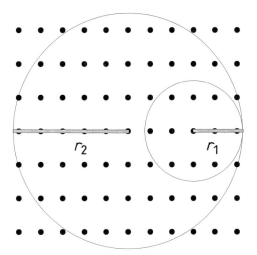

Figure 1-27. The reciprocal lattice model for the Laue method. All reciprocal lattice points between the two limiting Ewald spheres will diffract.

patterns obtained are rather difficult to interpret quantitatively in that the positions of the diffraction spots on a film will be completely dependent on, and extremely sensitive to, the orientation of the crystal. However, the pattern of spots will immediately show the symmetry of the reciprocal axis direction pointing at the film. Thus, this method has found wide popularity in both industry and research in aligning large single crystals along specific crystallographic directions. Since the Laue method is the only practical way of orienting large industrial crystals, the difficulty of interpreting the Laue pattern quantitatively has been solved by automating the procedure. A number of algorithms are available which interpret Laue patterns and enable this technique to be very widely used.

The other method of single crystal diffraction important to materials science is Kossel (1936) photography. Kossel diffraction is a single crystal technique but the crystal is usually part of a polycrystalline aggregate in a material. It results from the stimulation of characteristic X-ray emission by atoms within a tiny single crystal, which is part of a polycrystalline aggregate. This happens, for example, in a transmission electron microscope when the energy of the beam is higher than the K excitation potential of say, Ti atoms in the material being examined. The K_α radiation emitted by the Ti atoms in the few cubic micron volume of the electron beam, emerges as a spherical wave which will diffract from the various Bragg planes in the crystal below it. The spherical wave will diffract as a cone about each reciprocal lattice vector. A film will record circles, ellipses and the various conic sections depending on the angle each reciprocal lattice vector, and its associated cone of diffraction, makes with the film. The efficiency of this method is reflected in the short times (seconds) required to expose a Kossel photograph compared with those (hours) required for a Laue photograph. An example Kossel photo is shown in Fig. 1-28. The pseudo-Kossel method refers to a trick used to change the radiation in Kossel diffraction. For example, if Cu K_α is desired a thin layer of Cu can be deposited on the surface of the sample to produce the desired X-rays when the electron beam strikes it. For a good example of the use of this method see Bellier and Doherty (1977).

As with Laue diffraction, the Kossel technique is used both in transmission and reflection. However, this monochromatic diffraction method uses the very low intensity X-ray beam generated from the atoms in the small crystallite. Thus, samples for transmission Kossel diffraction usually need to be thinned to under 100 μm to permit the diffracted beam to pass. Like the Laue method, Kossel diffraction is most often applied to the study of orientation of crystallites, however here it is used in polycrystalline materials. In the Kossel technique

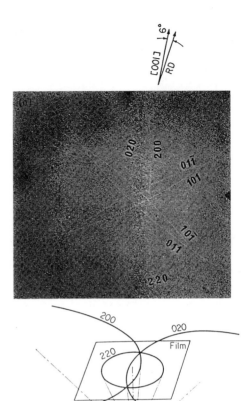

Figure 1-28. Kossel lines from a 10 μm thick single crystal of 3% Si-steel (100)[001] using the Fe K_α radiation stimulated from a 20 kV electron beam. Courtesy of Dr. Yukio Inokuti.

the monochromatic spherical wave is generated very close to the diffracting planes so the diffracted beam must travel a relatively long distance to the film. This very long lever arm, makes the method very sensitive to small orientation tilts or d_{hkl} value changes due to strain displacement and thus, can be used to evaluate residual stress in crystallites. A recent application by Inokuti et al. (1987) obtained approximately 1500 Kossel patterns from individual grains in a silicon containing transformer steel. The Goss texture was then displayed by coloring all grains with similar orientation.

1.7.4 Preferred Orientation

Preferred orientation of crystallites in bulk, polycrystalline materials, is a vital subject to many materials industries. A number of industrial materials are based on vector physical properties of the crystallites. For example the barium hexaferrite ceramic magnets, used commonly as seals on refrigerator doors, are polycrystalline materials in which only the $(00l)$ crystallite directions have a magnetic moment. Thus, the fabrication and quality control procedures must involve manipulating and measuring the degree of preferred orientation in the ceramic. Extruded wires show a characteristic preferred orientation as do most pressed powder materials. The most common way of evaluating the type and extent of preferred orientation is to measure the pole figure for a particular crystallographic direction. The pole figure is simply the intensity of a particular Bragg diffraction line plotted as a function of the three dimensional orientation of the specimen. It is determined on a pole figure diffractometer which is essentially the same as a single crystal diffractometer, able to rotate the specimen through all orientations while monitoring the diffraction intensity of a reflection. The results are displayed as a pole figure which is a two dimensional stereographic projection. Bunge (1982) has written a comprehensive treatment of preferred orientation evaluation techniques. A non-conventional method which avoids the requirement of an expensive automated pole figure device while often providing the required information should be mentioned. All of the preferred orientation on crystallites lying within a materials surface can be obtained by comparing the observed powder pattern diffraction intensities to the "ideal" intensities which can be obtained by calculating the powder pattern from known atomic positions. The trick is to find a way to compare these two patterns which are on different scales. Snyder and Carr (1974) devised a scale independent function which allows this comparison and permits direct computation of a vector properties enhancement or disenhancement in a particular crystallographic direction. For a modern summary of texture analysis, see Cahn (1991).

1.7.5 Crystallite Size

In Fig. 1-16 we saw that when diffraction occurs, the distance between adjacent planes, ABC, must exactly equal 1λ. Geometry requires that when this condition is met that the distance between three sets of planes, DEF, must be 2λ and all similar triangles from lower lying planes must also be an integral multiple of λ. If we increase the angle of incidence θ, so that the distance ABC becomes 1.1λ then DEF will be 2.2λ. The scattered waves from the sixth deepest plane will be 6.6λ and will therefore be exactly out of phase (0.5λ) with the waves scattered from the first plane. The scattering from the second and seventh planes will also be exactly out of phase. When all the planes from all of the unit cells are considered we see that no net scattering will occur; namely, there will be another unit cell at some depth within the crystal which will exactly cancel the scattering from any other cell except exactly at the Bragg angle where all scattering is in phase.

If θ is set closer to the Bragg angle so that the distance ABC is equal to 1.001λ then the scattering from the first plane will be cancelled by the scattering from the plane 500 layers deep in the crystal, with a phase shift of 500.5 λ. Similarly, if ABC is 1.00001 λ the scattering will be cancelled by a plane 50 000 layers deep in the crystal. Thus, it is clear that Bragg reflections should occur only exactly at the Bragg diffraction angle producing a sharp peak. However, if the crystal is only 1000 Å in size then the planes needed to cancel scattering from, for example the (100) plane, with an ABC distance of 1.0001 λ (i.e., the 5000th plane) are not present. Thus, the diffraction peak begins to show intensity at a lower θ and ends at an angle higher than the Bragg angle. This is the source of "particle size broadening" of diffraction lines. The observed broadening can be used to determine the crystallite size of materials less than one micrometer. Crystallites larger than one micron typically have a sufficient number of planes to allow the diffraction peak to display its inherent Darwin (1914) width (i.e., the width dictated by the uncertainty principle) additionally broadened by instrumental effects, with little contribution from size broadening (see Sec. 1.13).

The crystallite size broadening (β_τ) of a peak can usually be related to the crystallite size (τ) via the Scherrer equation (1918)

$$\tau = \lambda/(\beta_\tau \cos \theta) \tag{1-21}$$

The additional broadening in diffraction peaks beyond the inherent peak widths due to instrumental effects, can be used to measure crystallite sizes as low as 10 Å. However, a second cause of broadening, due to stress, can complicate the picture. The size of small particles dispersed in a matrix can be determined by small-angle scattering of X-rays (or neutrons).

1.7.6 Residual Stress

There are two types of effects strain in a material can produce on diffraction patterns. If the stress is uniformly compressive or tensile, it is called a macro-stress and, the distances within the unit cell will either become smaller or larger, respectively. This will be observed as a shift in the location of the diffraction peaks. These macro-stresses are measured by an analysis of the lattice parameters which will be described later in Sec. 1.12.4.

When the residual stress in a material produces a distribution of both tensile and compressive forces, they are called micro-stress and, the observed diffraction profiles will be observed to broaden about the original position. See Langford et al. (1988) and Delhez et al. (1988) for recent discussions of these effects. Both of the size and strain broadening effects generally produce a symmetric broadening, the observed asymmetry in diffraction profiles is usually due to instrumental effects (see Sec. 1.13 below). Micro-stress in crystallites can come from a number of sources: vacancies, defects, shear planes, thermal expansions and contractions, etc. (see Warren, 1969). Whatever the cause of the residual stress in a crystallite, the effect will cause a distribution of d-values about the normal, unstrained, d_{hkl} value. Figure 1-29 illustrates the expansion and contraction of d_{hkl} values caused by compression and extension of unit cells in micro-stress.

The broadening in a peak due to stress has been shown to be related to the residual stress ε, by

$$\beta_\varepsilon = 4\varepsilon \tan \theta \tag{1-22}$$

The fact, that stress induced diffraction peak broadening follows the tangent of θ while crystallite size broadening follows a $1/\cos \theta$ allows us to separate these effects.

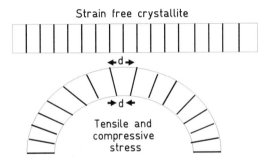

Figure 1-29. Stress expanding and contracting d's.

The most common procedure for accomplishing this was developed by Warren and Averbach (1950); this, along with a more recent procedure will be described later in Sec. 1.13 and 1.13.7. For a complete review of these and the preferred orientation methods mentioned above, and how preferred orientation affects determination of residual stresses, see Cahn (1991).

1.7.7 The Intensities of Diffracted X-Ray Peaks

All of the diffraction theory discussed until now has looked at the metric aspects of a diffraction pattern. That is, the positions of the diffraction maxima and how they are related to the size and shape of the unit cell. We have not considered why the intensities of the diffraction lines are what we observe. In order to understand what determines the intensity of a diffraction peak we need to examine a number of independent phenomena which determine the intensity. First we must consider how much intensity a single electron will coherently scatter (the more massive nucleus is a very inefficient scatterer and may be ignored). Next we need to consider the interference effects which will occur due to the electrons being distributed in space around atoms and the fact that atoms are not stationary in a lattice but vibrate in an anisotropic manner. Then the interference effects of atoms scattering from different regions of the unit cell must be allowed for, and finally optical and absorption effects must be considered.

1.7.7.1 Scattering of X-Rays by a Bound Electron

X-rays are electromagnetic radiation and from a fixed point will be seen as an oscillating electric field. The field will cause a bound electron to also oscillate (i.e., accelerate and decelerate) and, therefore, reradiate the incident radiation through a solid angle of 360°. This process occurs coherently, with the scattered waves having the same phase as the incident waves.

J. J. Thomson (1906) first showed that the intensity scattered from an electron is

$$I = \frac{I_0 e^4}{m_e^2 r^2 c^4} \frac{1 + \cos^2 2\theta}{2} \quad (1\text{-}23)$$

where I_0 is the intensity of the incident beam, e is the charge on the electron, m_e is the mass of the electron, c is the speed of light and r is the distance from the scattering electron to the detector. Note that the inverse square law is explicit in the Thomson equation. The final term involving the cosine function is called the *polarization factor* and results from the fact that the incident X-ray beam is unpolarized while the scattered beam will be decreasingly polarized as the angle of view is increased. The Thomson equation addresses only coherently scattered radiation; all other absorption mechanisms are ignored.

1.7.7.2 Scattering of X-Rays by an Atom

In Fig. 1-30 we see X-ray beams X and X' coherently scatter from electrons at A and B around an atom. When viewed at an angle of 0°, scattered waves Y and Y' are

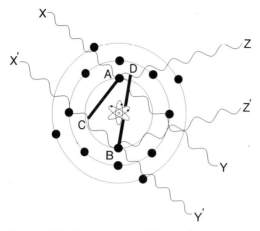

Figure 1-30. The scattering of X-rays from an atom.

exactly in phase, but when viewed from any non zero angle, we find that wave Z' has to travel CB-AD farther than wave Z. Because atomic dimensions are on the same order as the wavelength of X-rays, this path difference causes partial destructive interference and lowers the resultant amplitude. Therefore, the intensity of scattered radiation decreases with angle of view (θ). This phenomenon is described by the quantity f_0 which is called the *atomic scattering factor*. The function f_0 is normalized in units of the amount of scattering from a single electron. At zero degrees f_0 will be equal to the number of electrons surrounding any atom or ion. Figure 1-31 shows typical atomic scattering factor curves. Since the phase difference CB-AD depends on the wavelength and the angle of view, the function f is specified as a function of $\sin \theta / \lambda$.

The actual shape of the function f_0 must be calculated by integrating the scattering over the electron distribution around an atom as shown in Eq. (1-24)

$$f_0 = \int_0^\infty \varrho(r) \frac{\sin(kr)}{kr} dr \qquad (1-24)$$

where

$$k = 4\pi (\sin \theta)/\lambda \qquad (1-25)$$

The electron density function $\varrho(r)$ around the atom is related to the quantum mechanical wave function Ψ via

$$\varrho(r) = 4\pi r^2 |\Psi|^2 \qquad (1-26)$$

Every few years a quantum mechanic will produce a better set of wave functions describing the electron densities of atoms and the atomic scattering factors are recomputed. The wave functions for all atoms except hydrogen must be obtained by approximate methods and this introduces a small amount of error into the atomic scattering factors. Experience in using these f_0's has shown that the error in them is less than the typical errors from other sources and for most work can be considered negligible. The preferred atomic scattering factors today were computed by Cromer and Waber (1965) using Dirac-Slater orbitals.

1.7.7.3 Anomalous Dispersion

There are two other factors which influence the intensity of the scattering from an atom and it is convenient to consider these

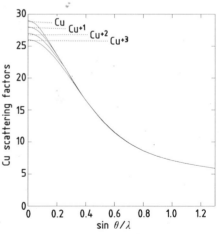

Figure 1-31. The atomic scattering factor f_0 for a copper atom.

as modifying the atomic scattering factor. The first is a phenomenon called *anomalous scattering*. In normal *Thomson scattering*, the electron acts as an oscillator under the stimulation of the oscillating electric field of the incident radiation. We can crudely picture this effect as a true movement of the electron back and forth away from the nucleus. When the frequency of the radiation gets high enough to cause the electron on its next oscillation away from the nucleus, to no longer feel the restoring attraction, ionization occurs. This is a wave description for picturing the photoelectric effect. Using his notation we can think of removal of an electron as a resonance phenomenon and the various absorption edges as natural frequencies of vibration.

Using this over-simplified model we can think of frequencies just short of the resonance frequency as lifting the electron from its inner orbital to the outer unfilled energy states of the atom. Since there is a finite quantum probability that the electron could exist in these states there will be a slight delay time as it oscillates back toward the nucleus. This delay will in turn cause a delay in the radiation being scattered by the oscillating electron. We will see the effects of this time delay as a phase shift in the scattered wave. Though a physicist might object to this simple argument it allows one to picture the origin of the observed phase lag. To properly describe anomalous scattering, we need to correct the normal scattering factor f_0 with a real ($\Delta f'$) and an imaginary ($\Delta f''$) term. The effective scattering from an atom will be

$$|f|^2 = (f_0 + \Delta f')^2 + (\Delta f'')^2 \quad (1\text{-}27)$$

1.7.7.4 Thermal Motion

Figure 1-30 shows that the size of the atom causes some destructive interference from the scattering of electrons that are separated by distances on the order of a fraction of the wavelength of the X-rays. If the atom in question is vibrating about its lattice site then its effective size is larger and the interference effects are, in turn, larger. We have seen, in Fig. 1-31, that the interference in a stationary atom causes the atomic scattering factor to fall off somewhat exponentially as a function of $(\sin\theta)/\lambda$. In order to describe the thermal motion induced enhancement of this fall off in f, Debye (1913) and later Waller (1928) defined the parameter B, which is related to the vibrational amplitude of the atom

$$B = 8\pi^2 u^2 \quad (1\text{-}28)$$

B is called the *Debye-Waller temperature factor*. It is directly related to u^2, the mean-square amplitude of vibration of an atom. The amount and direction of atomic vibration will depend on the temperature (i.e., the amount of kT, thermal energy available), the atomic mass and the direction and strength of the force constants holding the atom bonded in its location. Figure 1-32 shows the effect of increasing B on f

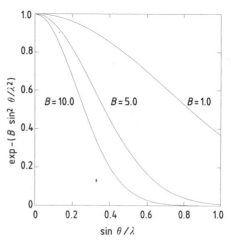

Figure 1-32. Effect of the temperature parameter B on f.

when B is used in the form

$$f = f_0 \exp\left(-\frac{B \sin^2 \theta}{\lambda^2}\right) \quad (1\text{-}29)$$

The temperature factor B is the same for all directions of vibration of an atom and is therefore called the isotropic temperature factor. In fact, most atoms in solids will have special directions in which they can vibrate with higher amplitudes. Thus, a more accurate description of the thermal vibration in a solid is to use a tensor to describe the anisotropic motion. To do this we recognize that the $\sin^2 \theta/\lambda^2$ term in Eq. (1-29) is simply $d^{*2}/4$. We have already derived the equation for d^{*2} in terms of its vector components as Eq. (1-20). Thus, we can substitute Eq. (1-20) into (1-29) and break B into six B_{ij} anisotropic terms,

$$\exp\left(-\frac{B\, d^{*2}}{4}\right) = \exp[-\tfrac{1}{4}(B_{11} h^2 a^{*2} +$$
$$+ B_{22} k^2 b^{*2} + B_{33} l^2 c^{*2} + 2 B_{12} h k\, a\, b +$$
$$+ 2 B_{13} h l\, a\, c + 2 B_{23} k l\, b\, c)] \quad (1\text{-}30)$$

In Eq. (1-30) the cosines of the reciprocal interaxial angles are included into the values of the B_{ij} cross terms.

1.7.7.5 Scattering of X-Rays by a Crystal

Since a crystal is made up of repeating identical unit cells, we need only consider diffraction from a single unit cell to see the effects from a whole crystal. Each atom in the unit cell will scatter an X-ray beam of intensity given by the atomic scattering factor f and with a phase depending on its location in the cell. If we consider an atom at the origin of the unit cell scattering with a phase angle of $0°$, then an atom half way along the a edge will scatter with a phase angle of $180°$ when we consider Bragg reflection from the 100 plane. However, for the 200 plane these two atoms will scatter in phase, each having a phase angle of $0°$. Our task is to find a mathematical notation that will allow us to add up the scattering contributions from all of the atoms in the unit cell allowing for the interference effects caused by their different locations.

The resultant scattering from all of the atoms in the unit cell for a particular diffraction line is called the structure factor F_{hkl}. In order to conveniently sum up the scattering amplitudes from all of the atoms in the cell, allowing for their individual phase angles, we represent this scattering as a complex vector. The Euler coordinate system which has an imaginary axis normal to the real axis is ideal for our current needs. When vectors in an Euler coordinate system are added, the resultant vector will have an amplitude and phase angle which properly reflects the interference between the waves that the vectors represent. In order to represent the scattered wave from each atom in a unit cell as a vector, we simply need to find a direction and magnitude to associate with each wave. The amplitude of the wave is equal to f the atomic scattering factor and this we will define as the length of the atomic scattering vector. We may use the phase angle to define the direction of the vector in polar coordinates.

Any vector in an Euler coordinate system can be resolved into its components as $f \cos \phi + f i \sin \phi$, where ϕ is the phase angle. The well known Euler relation allows us to express these vectors in exponential form

$$f \exp(i\phi) = f \cos \phi + f i \sin \phi \quad (1\text{-}31)$$

To generalize this notation we will look at a real three dimensional unit cell containing N atoms. Each atom will scatter X-rays with a different amplitude equal to its atomic scattering factor evaluated at the diffraction angle of the hkl planes in question, and for the wavelength being used. The phase angle of the scattered wave from each atom, j, will be

$$\phi = 2\pi(hx_j + ky_j + lz_j) \quad (1\text{-}32)$$

The problem of describing the scattering from a unit cell reduces to adding the waves scattered from each atom in the cell, represented by vectors in a complex coordinate system, each with its own amplitude and phase angle. The resultant diffracted wave for any set of Bragg planes is called the structure factor F_{hkl} and may now be written as

$$F_{hkl} = \sum_{j=1}^{\text{number of atoms}} f_j \cdot \exp[2\pi i(hx_j + ky_j + lz_j)] \quad (1\text{-}33)$$

Figure 1-33 illustrates the resultant scattering (F_{hkl}) form a unit cell containing two atoms, represented by their scattering vectors.

1.7.8 Calculated X-Ray Intensities

The intensity we observe for any Bragg reflection i, (i.e., hkl), is proportional to F_i^2.

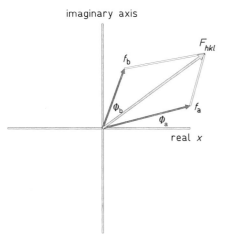

Figure 1-33. Resultant scattering (F_{hkl}) from a unit cell with two atoms whose scattering is represented by vectors f_a and f_b.

Therefore, if we know the position of the atoms in the unit cell (x, y, z) we can calculate the structure factor and relate this to the intensity. In addition to the Thomson Eq. (1-23), a few other terms are needed to complete this relationship. Two related to the sample itself and three to the type of diffraction experiment used to measure the intensities.

- The number of types of planes in the set i, called the plane *multiplicity factor* M_i, will directly affect the intensity. For example, the face diagonal reflection (110) and ($\bar{1}$10), (1$\bar{1}$0), ($\bar{1}\bar{1}$0) are equivalent in all unit cells where the a and b axis meet at 90°, making $M_{110} = 4$. When the a and b axis are not orthogonal the (110) and ($\bar{1}\bar{1}$0) lose their equivalence to the ($\bar{1}$10) and (1$\bar{1}$0), allowing two independent diffraction maxima each with $M_i = 2$.
- Dynamic scattering considerations require a scale factor of $1/(2V)^2$ where V is the volume of the unit cell.
- The *Lorentz factor* is a measure of the amount of time that a point in the reciprocal lattice remains on the sphere of reflection. For example, the high angle (longest) reciprocal lattice vectors cut the Ewald sphere in a manner approaching a tangent. This causes the high angle reflections to remain in diffracting position longer and increase their intensity relative to the low angle reflections. For a powder diffractometer this effect can be eliminated, and all reflections put onto the same intensity scale, by including the term $1/(\sin^2\theta\cos\theta)$ in our expression for calculating diffraction intensities. Usually, the Lorentz factor is combined with the polarization term from the Thomson equation and called the Lp correction.
- The additional polarization of a monochromator crystal will also effect the in-

tensity. For a diffracted beam monochromator the term is, $\cos^2 2\theta_m$. Where, θ_m is the Bragg angle of the monochromator crystal.

- The last term required to calculate a diffraction intensity allows for the differing path length that the X-ray beam takes in the sample, depending on diffraction angle and experimental geometry. The beam will be absorbed according to Eq. (1-10). For a powder diffractometer with a flat brickette sample, the volume of sample irradiated is independent of diffraction angle, so this absorption term reduces to a simple, $1/\mu$. For other geometries such as a cylindrical sample the absorption term will have a trigonometric component.

We are now ready to put all of the above considerations together and write the, rather formidable, equation for the diffraction intensity, for line i of phase α, for a powder diffractometer as

$$I_{i\alpha} = \frac{I_0 \lambda^3 e^4}{32 \pi r m_e^2 c^4} \frac{M_i}{2 V_\alpha^2 \mu} |F_{i\alpha}|^2 \cdot$$

$$\cdot \left(\frac{1 + \cos^2 2\theta_i \cos^2 2\theta_m}{\sin^2 \theta_i \cos \theta_i} \right) \quad (1\text{-}34)$$

where,

$$F_{hkl} = \sum_{j=1}^{\text{number of atoms in cell}} \left\{ (f_{jo} + \Delta f_j' + \Delta f_j'') \cdot \right.$$

$$\left. \cdot \exp\left[-B_j \sin^2\left(\frac{\theta}{\lambda}\right)^2 \right] \right\} \cdot$$

$$\cdot \exp[2\pi i (h x_j + k y_j + l z_j)] \quad (1\text{-}35)$$

It is a triumph of our understanding of physics that computations of diffraction intensities, using Eq. (1-34), produce the observed values. This triumph has been heavily exploited in determining the crystal structure of materials. This ability, first used by the Braggs in (1913), has established the basis of modern solid state science. To test that a particular model for a crystal structure is correct we use Bragg's law, i.e., Eq. (1-16), and the lattice parameters with Eq. (1-20), to calculate the positions of possible diffraction lines. Then Eq. (1-34) is used to compute the intensity of these lines. Even slight errors in the crystal structure model will be seen as discrepancies between the observed and calculated intensities. When each of the intensities can be independently measured, as in the case of single crystal diffractometry, they can be used in a least squares procedure to refine the atomic locations. When the reflections are difficult to separate as in the powder pattern of a low symmetry material, the least squares procedure is performed against the whole powder pattern. This *Rietveld* procedure will be further discussed in Sec. 1.13.8. The most common application of our ability to compute a powder pattern, in the materials laboratory, is to check any features of the pattern (like preferred orientation of the crystallites, solid solution effects, etc.) and any structural modifications which may have taken place. The most commonly used program for this purpose is POWD10 by D. K. Smith (1982).

1.7.8.1 Systematic Extinctions

A centered Bravais lattice permits us to divide the atoms in a unit cell into groups associated with each lattice point. Since a lattice point represents some grouping of atoms, each lattice point in a centered cell must represent exactly the same group. Thus, in a body centered cell, for each atom located at x, y, z there will be an identical atom located at $x + \frac{1}{2}, y + \frac{1}{2}, z + \frac{1}{2}$. Thus the structure factor can be factored into a sum over the atoms represented by each lattice point,

$$F_{hkl} = \left\{ \sum_{j=1}^{(\text{number of atoms in cell})/2} f_j \cdot \exp[2\pi i(hx_j + ky_j + lz_j)] \right\}$$

$$\cdot \left\{ \sum_{n=1}^{(\text{number of atoms in cell})/2} f_n \cdot \exp\left[2\pi i\left(hx_n + \frac{h}{2} + ky_n + \frac{k}{2} + lz_n + \frac{l}{2}\right)\right] \right\} \quad (1\text{-}36)$$

If the sum of $h + k + l$ is even, the second term in Eq. (4-36) will contain an exponent with an integer in it. An integral number of 2π's will have no effect on the value of this term and in this case the equation reduces to,

$$F_{hkl} = \sum_{j=1}^{(\text{number of atoms in cell})/2} 2f_j \cdot \exp[2\pi i(hx_j + ky_j + lz_j)] \quad (1\text{-}37)$$

However, if the sum of $h + k + l$ is odd then the second term in Eq. (1-36) will contain an exponent with a $2\pi(0.5)$ term. This causes each term in the second sum to be negative. Thus, for each atomic scattering vector in the first sum there is an equal and opposite scattering vector in the second sum. The result is that the structure factor (and hence the intensity for all reflections with the sum of $h + k + l =$ odd) is exactly equal to zero. This condition is called a *systematic extinction*. A similar analysis will show different extinction conditions for the other Bravais lattices. Screw axis and glide planes will also introduce systematic extinctions into various classes of reflections. Reflections which happen to have zero intensity due to the locations of the atoms in the unit cell and not due to a systematic symmetry condition are called accidently absent.

1.7.8.2 Primary and Secondary Extinction and Microabsorption

All of our theoretical development until now has referred to what is called the ideally imperfect crystal. This means a crystal with slight misalignments between the small mosaic blocks. These slight misalignments are usually due the relief of strain energy resulting from crystal imperfections. In some cases very perfect crystallites approach an ideally perfect crystal where the Bragg planes extend long distances with no variation in *d* value. In these very perfect crystals a dynamical effect reduces the diffracted intensity of a beam as it passes through the crystal. Each successively deeper Bragg plane in a crystal will reflect some portion of the intensity of the incident beam, through the Bragg angle back out of the crystal. If these planes are perfectly parallel, as they will be in a perfect crystal, then the beam heading out of the crystal will encounter more Bragg planes above it, which must be exactly in position to again reflect the diffracted beam back into the crystal. This doubly reflected wave will be exactly out of phase with the incident beam (because each reflection causes a phase shift of $\pi/2$) causing a net reduction in the observed intensity of a diffraction peak. This effect is called *primary extinction* and was first described by Darwin (1922). Zachariasen (1945) has developed a quantitative description of this effect.

There is another dynamical effect in perfect crystallites called *secondary extinction*. This results when the strongest reflections diffract most of the intensity in the incident beam, out of the crystal before the beam can penetrate to any significant depth. The lower lying planes thus never have a chance to diffract the amount of intensity they are capable of reflecting. This causes the most intense reflections to have a lower relative intensity than weaker reflections.

Both primary and secondary extinction effects cause the calculated intensities to differ from those observed in crystals with a high degree of perfection. The common advice to control this otherwise hard to quantitatively model phenomenon, is to introduce strains and imperfections into the crystallites by grinding. The traditional thinking is that crystallites in the tens of micron size range would not show appreciable extinction effects. However, recent studies by Cline and Snyder (1983, 1987) have shown that extinction will have significant effects on intensities even from crystallites as small as one micron.

There is one other effect which can disturb the relative intensities of diffraction peaks in a poly-phase mixture. Microabsorption occurs when crystallites of phase α lie above or below crystallites from phase β. Since the incident beam will spend some percent of its time inside an α crystallite, it will not be absorbed as if it passed through a medium having the average mass absorption coefficient of the mixture. Instead, if it spends more time in an α crystallite it will act as if it had been absorbed by a material with a mass absorption closer to that of pure α. This phenomenon was first described by Brindley (1945) and is of most concern in quantitative phase analysis applications. This, like extinction, will be reduced in importance as crystallite size decreases.

1.7.9 Vision, Diffraction and the Scattering Process

We have seen that one of the things that happens when electromagnetic radiation falls on matter is the process of coherent scattering. The incident radiation is scattered in all directions with no change in phase. Consider this process when visible light falls on a clock and a person's eye intercepts some of the scattered radiation. We of course, know that an image results from this process, projected on the retina. What is not so evident is that the image of the clock has been encrypted onto the scattered radiation and that the lens of the eye decrypts this image.

One must think about this process to be convinced that it has to be true. For example, placing a sheet of paper between the clock and the eye causes the image to disappear from the retina. Another person looking at a right angle between the clock and the first person (i.e., looking at the line of view to the clock rather than at the clock itself), sees nothing of the clock's image. Thus, the lens in the eye does not see the encoded information but instead transforms it to an image. We can describe this process mathematically: the encrypted image is the *Fourier transform* of the original image. The lens of the eye performs another Fourier transformation recreating the image to project on the retina. The process of holography captures this Fourier transform on film, so that when one looks through the film, the eye sees the encrypted information and transforms it to the image of the object whose Fourier transform had been captured. The result is the appearance of a three dimensional image behind the film.

When X-rays fall onto the atoms making up a crystal, the scattered waves carry the Fourier transform of the image of the crystal structure. If we had a lens that could bend x-rays we could use it to image the atoms in the crystal. In fact, if we had an X-ray laser we could use a visible light laser to create a hologram of the crystal structure which could be projected on a screen! However, we still await a method for producing practical coherent X-rays. While waiting, we continue to use the techniques which have evolved since 1913.

Since we do not have a material that can act as a lens to X-rays we must simulate this process with a mathematical lens on a computer. The Fourier transform of the electron density at the point in the unit cell, x, y, z (ϱ_{xyz}) is

$$\varrho_{(xyz)} = \frac{1}{V} \sum_h \sum_k \sum_l F_{hkl} \cdot \exp[-2\pi(hx + ky + lz)] \quad (1\text{-}38)$$

where V = Volume of the unit cell. Note: The term needed in this Fourier series is F_{hkl}, not F_{hkl}^2. Since the phase of each Bragg reflection is not measurable, the intensity only yields $|F_{hkl}^2|$.

Because F_{hkl} is a complex number, we can only measure $|F_{hkl}|$. This means that the phase angle of the structure factor $\{\exp[2\pi i(hx_j + ky_j + lz_j)]\}$ is lost during the measuring process and must be inferred from other techniques. The obtaining of this term is called the phase problem. Today, due to the early work of Patterson (1938) and the more recent work of Karle and Hauptmann (recognized in 1985 with the Nobel prize), the problem is solvable by direct computer calculations for about 90% of all crystal structures (when single crystal diffraction data are available). When F_{hkl} is known from solving the phase problem, the above Fouier series can be computed for all x, y, z positions in a unit cell to give a picture of the electron density (or nuclear spin density in the case of neutron diffraction.

1.7.10 X-Ray Amorphography

The observed intensities, without their respective phases, may be used as the coefficients in the Fourier synthesis shown in Eq. (1-38), in place of the F_{hkl} terms. This "$|F^2|$ synthesis" was first interpreted by Patterson (1935) to be a bond distance map of the unit cell rather than the atom map one obtains from the conventional Fourier. The *Patterson map* is directly computed from the observed intensities and contains peaks which correspond to all of the interatomic bond distance vectors between the atoms in the unit cell. Each of the bond distance vectors has its origin at the origin of the Patterson function. Thus, the Patterson map looks like a pin-cushion with different length vectors protruding in all directions from a common origin. The heights of the peaks in the Patterson function correspond to the product of the electron density of the two atoms comprising each vector. This three dimensional bond distance map has been widely used to establish the phases of the F_{hkl} terms, so that Eq. (1-38) can be used to produce an image of the atoms in the unit cell.

If we plot all of the vectors in the three dimensional bond distance map on a one dimensional axis as a function only of distance r, we will have

$$g(r) \, dr = 4 \pi^2 \varrho(r) \, dr \quad (1\text{-}39)$$

in which $g(r)$ is known as the *radial distribution function* (RDF). This function has maxima corresponding to each shell of neighboring atoms as one moves out in distance, r, from each atom in the material. Debye (1915) was the first to point out that the Fourier transformation of the scattering from an amorphous array of atoms would yield the RDF for a glass. This theory was brought to maturity by Warren and Gingrich (1934) and today represents one of the principal tools in the analysis of glass and liquid structure. For an excellent tutorial review of X-ray and neutron scattering from amorphous materials, see Wright (1992), for a complete description of amorphous scattering techniques, see Gaskell (1991).

1.8 X-Ray Absorption Spectroscopy (XAS)

1.8.1 Extended X-Ray Absorption Fine Structure (EXAFS)

In Sec. 1.4.6 the phenomenon of the absorption edge was described and shown in Fig. 1-8. The fine structure of this absorption spectrum, which can be seen with high resolution experiments directly at the absorption edge, is related to the slight energy shifts in the ground state of the K electron induced by the bonding molecular orbitals. For many years it had been noted that the absorption edge had an extended fine structure. This is illustrated in Fig. 1-34 which shows the absorption of copper atoms in a glass with the composition of the superconducting phase $Bi_2Sr_2CaCu_2O_8$ (Bayya et al., 1991). During the 1970's an explanation of this extended X-ray absorption fine structure (EXAFS) emerged. If we think of the ionized K electron emerging from an atom as a spherical wave, then we can picture how this wave will be reflected back onto itself by the atoms in the coordination sphere around the ionizing atom. Thus, the EXAFS will depend on the number and type of atoms in the coordination sphere.

The linear absorption coefficient μ, introduced in Sec. 1.4.6, must now be thought of as having two components. The first is μ_0, the absorption of an atom in the absence of any neighbors. The other term is χ, the oscillatory component of μ, arising from self-interference of the photo-electron reflected back onto itself by neighboring atoms. χ is related to μ and μ_0 by (Lengeler, 1990a),

$$\chi = \frac{\mu - \mu_0}{\mu} \tag{1-40}$$

It is clear that the amplitude of the back-reflected wave will be related to the repulsive field of (i.e., number of electrons on) the neighboring atoms and the number of neighboring atoms in each successive shell away from the central atom. The amount of interference will further depend on the phase of the back reflected wave which, in turn, will depend on the distance of the scattering neighbor atom from the central atom being ionized and the wavelength of the photo-electron. Scattering theory shows that the EXAFS signal χ is,

$$\chi(k) = \frac{1}{k} \sum_i \frac{N_i}{R_i^2} F_i(k) \cdot \tag{1-41}$$

$$\cdot \exp\{(-2\sigma_i^2 k^2)\sin[2kR_i + \phi_i(k)]\}$$

where N_i is the number of atoms in ith coordination shell at a distance R from the absorbing atom. F_i is the backscattering amplitude from neighboring atoms and ϕ is the phase shift experienced by the ejected photoelectron. σ_i is the mean relative displacement between the absorbing and scattering atoms (a Debye-Waller type factor).

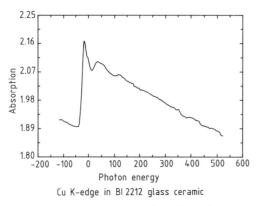

Figure 1-34. The copper K-edge absorption spectrum of a glass with composition $Bi_2Sr_2CaCu_2O_8$.

The radial distribution function Eq. (1-39) was described in Sec. 1.7.10 as the one dimensional projection of all of the bond distances in a structure. This function is obtained by taking the Fourier transformation of the X-ray scattering from either a crystal or an amorphous material. The Fourier transformation of the background corrected EXAFS signal, produces the atomic radial distribution function for the absorbing atom in a material. This function, which shows the probability of encountering other atoms as you move, in distance, away from the central atom, offers a powerful characterization tool for amorphous, disordered and quasi-crystalline materials. It also is considerably easier to interpret, in that it contains only the shells around the absorbing atom rather than all of the neighboring shells from all of the atoms, superimposed. Later we will discuss methods which use X-ray line broadening to determine the size of very small crystallites however, when these atom clusters get as small as a few hundred, the diffraction signal fades into background. For these cases the essentially amorphous clusters may be analyzed by EXAFS as has been done by Apai et al. (1979) for Cu and Ni clusters and Zhao et al. (1989) for Ag, Mn, Fe and Ge clusters.

As an example of amorphous analysis we will look at the processing of the Cu atom EXAFS signal from a glass with the composition of the superconducting compound $Bi_2Sr_2CaCu_2O_8$, shown in Fig. 1-34 (Bayya et al., 1991). This spectrum was measured at Brookhaven National Laboratory using the National Synchrotron Light Source Facility. Copper K-edge spectra were collected on the glass, the glass-ceramic which resulted from heating the glass, and CuO as a reference material.

EXAFS oscillations, measured by the interference function χ, were extracted from the absorption spectra shown in Fig. 1-34 by fitting a background function and normalizing using a spline fit (Teo, 1986). A Fourier transform of these oscillations was taken to obtain the atomic RDF which is shown in Fig. 1-35. The distances in this transformed data are displaced from the actual distances by a phase shift. The peak corresponding to the cation-oxygen coordination shell was back-transformed. Numerical single shell fitting was applied to the Fourier filtered data using a theoretically calculated back scattering amplitude factor for oxygen (Teo et al., 1977).

The first major peak in the spectra shown in Fig. 1-35 corresponds to the Cu–O distance. In order to determine the structural parameters of the glass and glass-ceramic, CuO was used as a model compound. Using the known values of the Cu–O bond distance and coordination number in CuO, the phase function for a Cu–O pair was determined. Theoretically calculated values of the back scattering amplitudes were used. The phase function determined from CuO and the theoretical amplitude function were then transferred to the Fourier spectra in the glass and glass-ceramic, for curve fitting, in order to

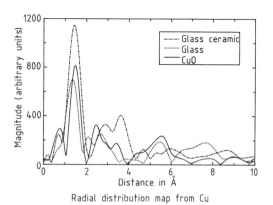

Radial distribution map from Cu

Figure 1-35. The Fourier-transform of the Cu K-edge EXAFS $k^3 \chi(k)$ in CuO, $Bi_2Sr_2CaCu_2O_8$ glass and glass-ceramic.

determine actual Cu–O distances and the coordination numbers. Figure 1-36 shows the curve fitting of $k^3 \chi(k)$ in these samples where $k = 2\pi/\lambda$ and is the magnitude of the wave vector.

The best fit least-squares refined bond-distances and coordination numbers are listed in Table 1-6. The Cu–O bond distance in the $Bi_2Sr_2CaCu_2O_8$ glass-ceramic was found to be 1.91 Å with 6 oxygens surrounding each copper atom. These values are in good agreement with those reported for the crystalline material via conventional X-ray structure analysis. However, a shorter Cu–O bond distance, 1.83 Å, was observed in the glass. This reduction of Cu–O bond distances in glass may be attributed to the possible changes in oxidation state of the Cu ion and to the absence of the crystal field energy which, if present, would pull the atoms apart from their minimum energy positions. For an excellent review of all of the techniques of X-ray Absorption Spectroscopy see Lengeler (1990b).

1.8.2 Reflectometry

In recent years the materials industries have turned more and more to the use of thin films. The required characterization of films, often less than 1 µm in thickness, involves chemical content, phase and crystal structure, thickness, surface roughness and density. All of these things may be learned by lowering the angle of the primary X-ray beam to a *grazing incidence* where the phenomenon of total reflection occurs. At angles, on the order of 0.5°, the incident beam sets up a standing wave in the surface of the specimen. It penetrates into the material to a depth ranging from 20 Å for highly absorbing material to 70 Å for water. A number of exciting applications have recently developed based on this phenomenon (Lengeler, 1990b).

Since the incident X-ray beam penetrates so shallowly into the material, all X-ray analytical techniques, like fluorescence, diffraction and absorption become inherently surface sensitive under the condition of total reflection. As the angle of incidence is raised from the critical angle to a few times the critical value, the incident beam increases its penetration into the specimen, up to about 1 µm. This fact has recently led a number of researchers to examine the fluorescence and diffraction signals from thin films, as a function of depth, almost Å by Å!

At grazing incidence *Fresnel reflection* will also occur from a specimen. This basic reflection phenomena is dependent on the

Table 1-6. Cu K-edge results for Cu–O bond distance and number of coordinating oxygens.

Compound	Cu–O length R (Å)	Coordination number, N
CuO (model compound)	1.96	4.0
$Bi_2Sr_2CaCu_2O_8$ glass	1.83	3.9
$Bi_2Sr_2CaCu_2O_8$ glass-ceramic	1.91	6.0

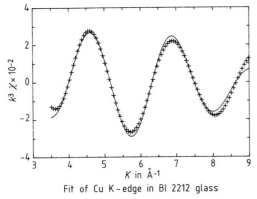

Figure 1-36. The Fourier-filtered $k^3 \chi(k)$ (solid lines) and simulated curve (dashed lines) of Cu K-edge EXAFS oscillations in $Bi_2Sr_2CaCu_2O_8$ glass.

density and surface roughness of the material and both of these parameters may easily be determined. If the specimen is a thin film deposited on a substrate of different composition, then the X-ray beam at grazing incidence will reflect off of both the film and substrate interfaces producing an interference pattern like that shown in Fig. 1-37. The interference pattern immediately yields the thickness of the film. Grazing incidence X-ray reflectometry is one of the most powerful new tools in the materials characterization repertoire. A recent summary of these techniques can be found in Huang et al. (1991).

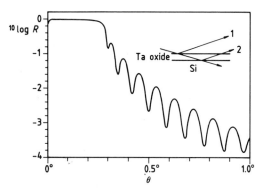

Figure 1-37. Reflectivity of 9800 eV photons from a Ta oxide layer on Si deposited at 410 °C by CVD (from Lengeler, 1990).

1.8.3 X-Ray Tomography

The use of tomographic imaging developed in the world of medicine from the need to do non-destructive testing on humans. The most common form of tomography passes radio frequencies through a brain and monitors their absorption due to hydrogen nucelar magnetic resonance. The absorption experiment is, in principal, exactly the same as that pictured in Fig. 1-4. The difference is that the brain (or the source and detector) is rotated through many orientations to collect three dimensional absorption data. The creative part of tomography developments was the writing of the computer algorithm which can reconstruct a complete three dimensional image of the brain or other organ of interest.

Recently X-ray crystallographers (Kinney et al., 1990) have taken advantage of the high intensity and wavelength tunability of synchrotron radiation to collect three dimensional absorption data on solid materials. The medical tomographic algorithms have been adapted to give a full three dimensional picture of the interior of solid objects, nondestructively. At the moment, the resolution of this technique approaches one micrometer, opening a wide variety of ceramic problems to scrutiny. There are only a few examples of full three dimensional quantitative microstructural examinations of ceramic bodies, due to the extreme difficulty of sectioning, polishing and characterizing an entire piece, so tomography offers a new window for complete microstructural analysis.

1.9 X-Ray Powder Diffraction

The preceding analysis has allowed us to predict the angle at which X-rays will diffract from a crystal. The intensity of a diffracted beam is another easily measured parameter and will be a function of the types and locations of atoms in the unit cell. Together the angle of diffraction and the intensity of the diffracted peak will be characteristic of a particular crystal structure. Since no two atoms have exactly the same size and X-ray scattering ability (i.e., number of electrons), the d_{hkl}'s between all of the possible sets of atomic planes and the intensities of diffracted beams will be unique for every material. This uniqueness allows us to identify any material, just as unique fingerprints allow the identification of any person.

We have seen that a single crystal with a particular set of atomic planes oriented toward the X-ray beam will diffract X-rays at an angle θ determined by the distance between the planes. However, most materials are not single crystals but are composed of billions of tiny crystallites. This type of material is referred to as a powder or a polycrystalline aggregate. Most materials in the world around us, ceramics, polymers and metals, are polycrystalline because they were fabricated from powders. In any polycrystalline material there will be a great number of crystallites in all possible orientations. Thus, when a powder is placed in an X-ray beam all possible interatomic planes will be seen by the beam but diffraction from each different type of plane will only occur at its characteristic diffraction angle θ. Thus, if we change the angle, 2θ, an X-ray detector makes with a specimen we will see all of the possible diffraction peaks which can be produced from the differently oriented crystallites in the powder. Figure 1-38 shows how a d^*_{hkl} from each crystallite becomes a cone of vectors which intercept the Ewald sphere forming a cone of diffraction. Thus, instead of a dot pattern, a powder pattern is a series of concentric rings. Since each point in the whole three dimensional reciprocal lattice produces a ring, a powder pattern is much more complex than its single crystal equivalent. In general, single crystal diffraction is more simple and preferred. However, since most of the high-interest phases of materials science are difficult to impossible, to grow as single crystals, we are forced use powder diffraction to study them.

1.9.1 Recording Powder Diffraction Patterns

The first powder diffraction patterns were recorded by directing an X-ray beam at a powdered specimen in a capillary tube or glued to the end of a thin glass rod, and recording the diffracted X-ray beams on a strip of film surrounding the sample. The cylindrical film wraps around the Ewald sphere so that it intercepts the cones, recording arcs. The camera which carries out this diffraction measurement is called a Debye-Scherrer camera. This was the standard way of recording powder diffraction patterns until the 1960's. The inherent poor resolution of this technique and the cumbersomeness of film techniques has caused it to fall into disuse. However, a form of this camera called the Gandolfi camera has found some amount of modern use. The Gandolfi camera is designed to obtain a powder pattern from a small single crystal. The crystal is mounted on the end of a glass rod and attached to a goniometer head. This goniometer is then turned through all of the motions needed

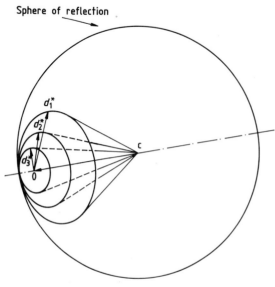

Figure 1-38. The Ewald construction for a powder.

to bring each reciprocal lattice vector onto the sphere of reflection, in all orientations. Thus, the crystallite is moved through all of the orientations that would be found in a powder and the Debye-Scherrer film arrangement records the diffraction pattern. The usefulness of this procedure is in obtaining full powder patterns from very small samples.

Beginning in 1945 and continuing through the 1950's William Parrish and his colleagues (Parrish and Gordon, 1945) developed the powder diffractometer for recording powder patterns electronically. The diffractometer is shown schematically in Fig. 1-39 and a modern realization of it in Fig. 1-40. The para-focusing geometry is called Bragg-Brentano but should more appropriately be called Parrish geometry. In order to maintain the para-focusing geometry, the detector has to move twice as fast as the sample in scanning through the diffraction angle. In the original, non-automated diffractometers, the detector is moved at a constant rate of speed, with a synchronous motor. The sample stage is coupled to the detector through a 2:1 reducing gear producing the required $\theta:2\theta$ motion. The signal from the detector is coupled to the pen of a chart recorder, whose paper is also moving at a constant speed, producing an X-ray diffraction recording like that shown in Fig. 1-41.

Today, most patterns are obtained from automated powder diffractometers with a few from high resolution focusing cameras. The para-focusing principle in the diffractometer comes from keeping the X-ray source (i.e., the X-ray tube target), the specimen and the receiving slit on a common circle, called the focusing circle. To meet the condition of Bragg's law we must move the sample through an angle of θ while the detector is scanned through an angle of 2θ. This θ–2θ motion implies that the radius of the focusing circle is continually changing throughout a diffraction pattern scan. The systematic errors for a powder diffractometer fall into two categories. The first is related to the instrument itself and include such things as the error in the setting of a true zero angle and the eccentricity in the gears which drive θ and 2θ. The second error category is related to the sample. By far the most serious of these is due to a displacement of the sample from the focusing circle. This, of course, can never be avoided in that even if the sample surface is perfectly tangent to the focusing circle, the X-rays penetrate into the sample to an av-

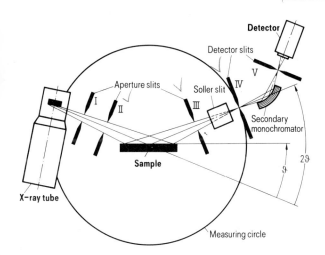

Figure 1-39. Schematic of an X-ray diffractometer of a Bragg-Brentano parafocusing diffractometer.

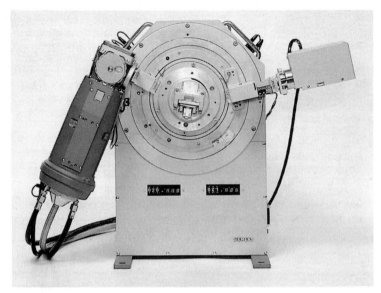

Figure 1-40. A modern automated diffractometer (Courtesy of Siemens AG).

erage depth dependent on the sample's absorption coefficient. Thus, there is always a "sample displacement" (or "sample transparency") error present. To correct the errors caused by this effect see Sec. 1.9.2.5.

Each peak, in the pattern shown in Fig. 1-41, corresponds to diffraction from a particular set of interatomic planes whose spacing d_{hkl}, may be calculated from the Bragg equation since we use X-rays of known wavelength (1.54060 Å for the $K_{\alpha 1}$ line from a Cu target X-ray tube). Although the pattern shown is characteristic of sodium chloride or table salt it would look different if we had used X-rays of a different wavelength. To remove this wave-

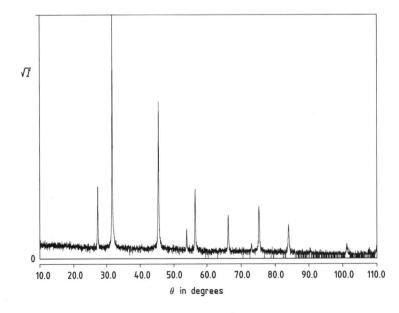

Figure 1-41. The diffraction pattern of NaCl.

length effect when the pattern is published we use only the wavelength independent d_{hkl} values, as shown in Fig. 1-58. Likewise, the intensities we observe are a function of the incident beam intensity which will vary from one laboratory and instrument to another, thus, the published values are normalized so that the highest peak is equal to 100.

1.9.1.1 Diffractometer Optics and Monochromators

The need to make the beam monochromatic follows directly from Bragg's law. If more than one wavelength is present in the beam, each set of planes will form a diffraction cone for each wavelength. This would hopelessly complicate the diffraction pattern. We have seen that the wavelength spectrum from a sealed X-ray tube contains the intense K_α and K_β lines and the continuous bremsstrahlung. The high energy continuous radiation and the beta peak can be reduced in intensity by the use of a beta filter, however, a crystal monochromator is much preferred. A crystal monochromator is a single crystal placed in the X-ray beam and oriented so that one set of its planes meet the Bragg condition. At the angle dictated by Bragg's law for the oriented d_{hkl}, of the monochromator crystal, only one wavelength will diffract. Thus, we use diffraction itself to get a monochromatic beam for our powder diffraction experiment. The most common arrangement for a diffractometer is to use a quasi-single crystal graphite monochromator which, due to its high *mosaicity*, allows the closely spaced $K_{\alpha 1}$ and $K_{\alpha 2}$ wavelengths through. Mosaicity is a measure of the misalignment of the small perfect crystalline domains which make up a single crystal. The elimination of all spectral wavelengths from the X-ray tube except the $K_{\alpha 1}$ and $K_{\alpha 2}$ greatly improves the quality of the powder pattern. This, most common type of monochromator, is placed in the diffracted beam, in front of the detector, as illustrated in Figs. 1-39 and 1-40. Placing the monochromator in the diffracted beam not only removes spectral impurities from the X-ray tube but also any fluoresced X-rays from the specimen.

Monochromators may also be placed on the incident X-ray beam. Figure 1-42a illustrates the conventional Parrish geometry diffractometer, which places the specimen in a reflecting position, with such a monochromator in place. A second diffractometer geometry, which places the sample in transmission, uses what is called Seemann-Bohlin optics and is illustrated in Fig. 1-42b. For this truly focusing geometry the powder sample is spread on a thin film with grease. Incident beam monochromators are usually made from single crystal of quartz, silicon or germanium, cut along a particular crystallographic direction. These very perfect crystals (i.e., with very low mosaicity) will separate the $K_{\alpha 1}$ and $K_{\alpha 2}$ components of the incident beam and allow complete rejection of the $K_{\alpha 2}$. These types of monochromators sacrifice intensity for wavelength resolution. To compensate for the loss in intensity, a position sensitive detector may be used in place of one that detects radiation at only one angle. The most common device employing such a monochromator is the Guinier camera, where the sample is mounted in transmission geometry and a position sensitive, piece of film is placed around the focusing circle. The true focusing condition produced by Seemann-Bohlin geometry gives a resolution slightly better than the parafocusing Parrish optics. However, the removal of the $K_{\alpha 2}$ spectral component is a much larger factor in improving the resolution over any inherent design feature.

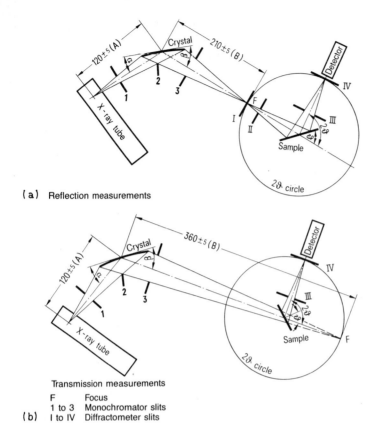

(a) Reflection measurements

Transmission measurements
F   Focus
1 to 3   Monochromator slits
(b)   I to IV   Diffractometer slits

Figure 1-42. Focusing incident beam monochromators used with (a) Bragg-Brentano optics and (b) Seemann-Bohlin transmission geometry.

1.9.1.2 The Use of Fast Position Sensitive Detectors (PSD)

The use of electronic PSD's in scanning mode has been developed by Göbel (1982a). In this mode the active window of perhaps 10° 2θ, is passed over the full 2θ range to be measured, the full pattern being summed into a multi channel analyzer (MCA). The attachment of such a scanning PSD to a Seemann-Bohlin transmission diffractometer, illustrated in Fig. 1-43, can produce a digital powder pattern, equivalent in resolution to a Guinier camera, in less than five minutes.

The development of fast PSD's has also led to the ability to record the diffraction peaks in a region of 2θ in as little as a few

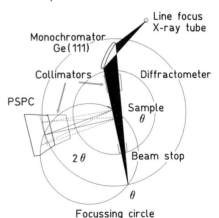

Focussing transmission diffractometer rapid data collection with PSPC

Figure 1-43. Diffractometer using Seemann-Bohlin transmission geometry and a position sensitive proportional counter (PSPC) (Göbel, 1982b).

1.9 X-Ray Powder Diffraction

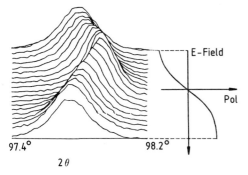

Figure 1-44. Dynamic XRD of the 400 peak of a lead zirconium titanate material as a function of electric field (from Zorn et al., 1985).

μs, when dealing with oscillatory phenomena. An elegant application of this speed is shown in Fig. 1-44 where a peak of a PZT piezoelectric was recorded dynamically, into 20 different banks of MCA channels, which were addressed according to the instantaneous value of the electric field (Zorn et al., 1985). The full electrostriction tensor was derived from measurements like these.

Another important application of PSD's is their use in scanning mode with a high-temperature stage (Zorn et al., 1987). This permits the following of a complex temperature profile, collecting diffraction patterns every few minutes. High-temperature reactions can be mapped over hundreds of different temperatures in an overnight run. Figure 1-45 shows such a series of patterns for the peritectic melting of the high temperature superconductor $YBa_2Cu_3O_{7-\delta}$. The dynamic observation of the peritectic melting of the $YBa_2Cu_3O_{7-\delta}$ superconducting phase shows the presence of $BaCuO_2$ beginning at 950 °C. Conventional quench studies missed this important observation (Snyder et al., 1992).

1.9.1.3 Very High Resolution Diffractometers

Angular resolution can only be achieved by sacrificing intensity. For this reason the highest resolution diffractometers have been built at high intensity synchrotron facilities. The simplest of these has been built at the National Synchrotron Light Source (NSLS) at Brookhaven National Laboratory by D. Cox and colleagues (1986). It is nothing more than two germanium crystals, one placed in the incident beam in

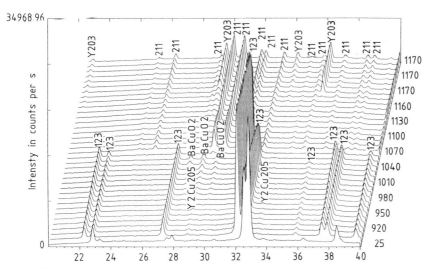

Figure 1-45. High temperature powder patterns of the decomposition of $YBa_2Cu_3O_{7-\delta}$.

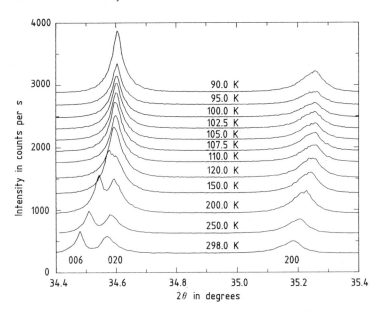

Figure 1-46. High-resolution synchrotron pattern of 123 showing complementary peak asymmetries in the 020 and 200 peaks due to martensitic strain distortion (Rodriguez et al., 1990).

front of the sample, the other in the diffracted beam after the sample. This "two monochromator" system, uses a parallel beam geometry, rather than focusing, and eliminates most of the instrumental effects which cause diffraction profiles to broaden, allowing measurement of the profile shape due only to the sample. Figure 1-46 (Rodriguez et al., 1990) shows the powder pattern of $YBa_2Cu_3O_{7-\delta}$ obtained at various temperatures on this extremely high-resolution diffractometer. The observed patterns clearly show the opposite shaped asymmetric broadening of the (020) and (200) reflections. This asymmetry is caused by the compression of **b**, and expansion of **a**, as the oxygens located at $0, 1/2, 0$, in a tetragonal cell, order on to the **a** axis of an orthorhombic cell. This observation permits the quantitative assessment of the strain energy which until now could only be qualitatively inferred from the martensitic twinned nature of $YBa_2Cu_3O_{7-\delta}$ crystals. A second design for a high-resolution diffractometer has been developed by Hart and Parrish (1989) and is shown in Fig. 1-47. This instrument, like the previous two crystal instrument, uses a parallel beam geometry rather than focusing or para-focusing. In addition, it uses a type of monochromator where two diffracting surfaces, cut exactly at the Bragg angle, have been carved into a single crystal of silicon. This *channel-cut* monochromator introduces no instrumental aberrations into the diffraction experiment.

Another use of a parallel beam diffractometer is illustrated in Fig. 1-48. Here the incident beam is directed at the sample at a very shallow glancing angle to meet the

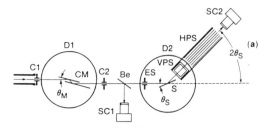

Figure 1-47. The Hart-Parrish (1989) parallel beam high-resolution diffractometer. CM is a channel-cut monochromator, S is the specimen, HPS a long soller slit assembly and SC2 the detector.

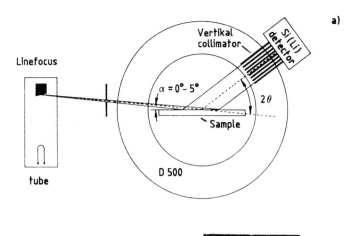

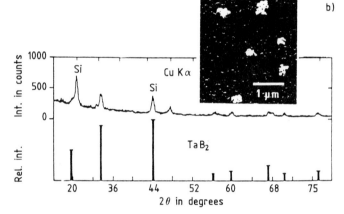

Figure 1-48. Diffraction geometry and pattern from 20 ng of TaB_2 on B-implanted, etched, $TaSi_2$ via glancing angle parallel beam experiment, from Göbel (1990).

condition of total reflection. The enhanced diffraction intensity, resulting from this geometry, allowed a full diffraction pattern to be obtained from 20 ng of sample (Göbel, 1990). A man-made multilayer monochromator has been used to monochromate the incident beam. These procedures are particularly suitable for thin film analysis. Grazing angle experiments have also been used to examine structural variations via diffraction from films as a function of depths of as little as 50 Å (Lim, 1987).

The development of high-resolution neutron diffractometers have also proved invaluable in materials characterization. These instruments have been particularly important for the case of high-temperature superconducting ceramic systems. The first significant insight into the cause and nature of the Cooper pair condensation temperature T_c, has come from ILL experiments at Grenoble (France) (Cava et al., 1990), using a high-resolution neutron diffractometer. Highly accurate bond distance measurements allowed the computation of the electron population on the ions, which showed that T_c directly relates to the charge transferred from the chain to the plane copper atoms of the $YBa_2Cu_3O_{7-\delta}$ superconductor.

1.9.2 The Automated Diffractometer

The automation of the X-ray powder diffractometer during the 1970's, led to a renaissance of this method. Many of the hardware advances already described are the direct result of the stimulation of this field brought about by automation. The other two broad areas of advances are in computer procedures and method development. We will survey each of these areas beginning with the automation of the diffractometer, which naturally leads to the evolution of software and finally to the development of new applications techniques.

The microelectronics revolution of the 1970's brought inexpensive computers and interfaces to the X-ray laboratory, although the principal thrust in the early 1970's was to develop the hardware interfaces needed to allow a computer to control a diffractometer. This work rapidly gave way to the much more serious problem of devising algorithms for the control of the instrument, and the processing of the digital data, which has occupied the subsequent decade and remains an active area of research.

There are four aspects to the automation of a diffractometer which are illustrated in Fig. 1-49:

(1) The replacement of the synchronous θ–2θ motor with a stepping motor and its associated electronics.
(2) The replacement of a conventional analog scalar/timer with one which can be remotely set and read.
(3) The conversion of the various alarms, limit switches and shutter controls to computer readable signals.
(4) The creation of a computer interface which will allow a computer to control items 1 through 3.

These four items, illustrated in Fig. 1-50, are easily obtained today by modules which often plug directly into the bus of a modern minicomputer.

The major impact of computer automation on improving the accuracy of the measurement of diffraction angles has come from the algorithms which bring much more "intelligence" to the process, than has been conventionally used in manual measurements. There are two areas here which need to be considered, the first is the algorithms which control the collection of data and the second, those which reduce the data to d values and intensities. The first generation of control algorithms were principally non-optimizing, move and count methods. Only limited progress has been made on the development of algorithms which bring full optimization to the various aspects of data collection. The techniques used to process the digitized step scan data produced by an automated diffractometer, or an automated film reading densitometer, represent a fundamentally new approach to materials characterization. The manual methods for most characterization techniques have very rapidly fallen into disuse due to the increased power brought by numerical analysis procedures. To appreciate this major development of the last twenty years, we will examine the procedures which have been developed for processing automated diffraction scans. A number of researchers and companies have taken somewhat different approaches to this problem. We will look at the first of the published algorithms (Mallory and Snyder, 1979) as representative of the problems and solutions to them.

1.9.2.1 Background Determination

The operation of differentiating peaks from background noise may be performed in two discrete steps (Snyder, 1983a). The first is to linearize the pattern in order to

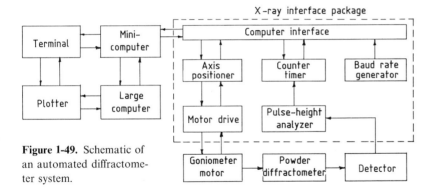

Figure 1-49. Schematic of an automated diffractometer system.

remove the low angle curvature due to the divergence of the X-ray beam and the broad maximum resulting from amorphous scattering. This is performed by connecting and then smoothing the minimum intensity points in each 0.25° 2θ segment of the pattern. A polynomial passed trough these minima is subtracted from the observed pattern, producing the linearized pattern shown in Fig. 1-51. The second step is to determine the threshold of statistically significant data (Mallory and Snyder, 1980). This is performed in a manner similar to the linearization step, but here the maximum intensity in each 0.25° pattern segment are collected and, after some values which clearly belong to peaks are rejected, they are fit to a polynomial which represents the threshold, above which points are significantly different from background. Both of these procedures are illustrated in Fig. 1-51. This pattern was obtained from a five milligram sample placed on a glass slide with a small depression etched in it to minimize sample displacement error. Due to the very small sample size the pattern was counted at 20 seconds per 0.04° step. The raw data shown in the

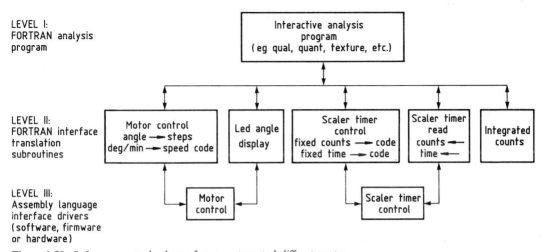

Figure 1-50. Software control scheme for an automated diffractometer.

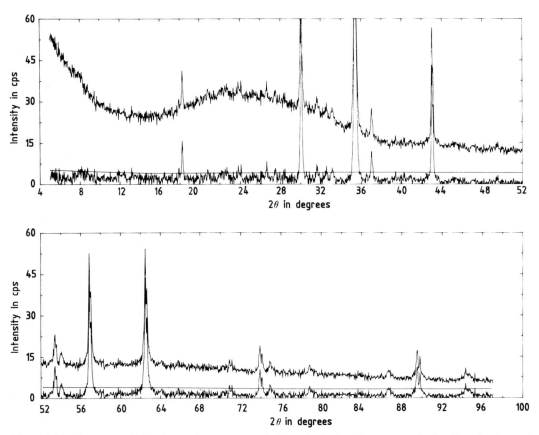

Figure 1-51. Upper trace is the observed pattern extended over both plots. Lower trace is after linearization and solid line is the threshold level.

upper curve exhibit both the effect of the amorphous sample holder and the increased low angle intensity due to the 1° divergence slit. The lower tracing shows the pattern after linearization. Note that all traces of the distortions have been removed. The smooth line on top of the background noise of the linearized pattern is the computer established threshold level. The peaks rising above this line are statistically significant.

1.9.2.2 Data Smoothing

Statistical fluctuations and the possible presence of noise spikes in the intensity measurements can lead to the detection of false peaks in the regions above the threshold. In order to avoid this, a quadratic or cubic polynomial is fit to an odd number of adjacent raw data points using a least squares regression. The point in the middle of the interval is replaced with the point computed from the interpolating polynomial. This process is illustrated in Fig. 1-52 for the linear case. Subsequent smoothed data points are produced in a similar manner by selecting the odd number of raw data points to start at each successive raw data point in the peak or peak group. As this "digital filter" slides over the data, statistical fluctuations are greatly reduced.

1.9.2.4 Peak Location

A second derivative method is used to locate peaks because the first derivative is insensitive to shoulders which indicate overlap. The Savitzky Golay (1964) polynomial smoothing procedure automatically produces a curve whose second derivative may be evaluated. The minima in the second derivative, indicate peaks in the raw data. The number of peaks found in a procedure will depend on the amount of data smoothing and the signal to background ratio, which is a simple function of the count time. The number of false peaks found can be minimized by correlating the smoothing parameters with the count time.

1.9.2.5 External and Internal Standard Methods for Accuracy

The accuracy of the peak locations will depend in some degree on each of the above steps (Snyder, 1984b). However, to achieve absolute accuracy it is essential that we know the X-ray wavelength exactly and that the data be corrected for the aberrations introduced by the instrumental measurement technique. The sample displacement error described above is by far

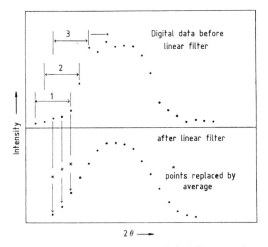

Figure 1-52. A five-point linear digital filter to reduce noise. The upper plot shows observed data, the lower shows the averaged data points.

However, there is also a corresponding loss in peak resolution. This loss increases with the 2θ step width and the number of points used in the filter. Savitzky and Golay (1964) have produced an extremely efficient computational method for applying this filter to digital data.

1.9.2.3 Spectral Stripping

For most work the $K_{\alpha 1}$ diffraction peaks can be readily recognized by a computer algorithm based on their location and height. However, when it is desired to completely remove the $K_{\alpha 2}$ peaks from the raw data, a modified Rachinger procedure is used (Ladel et al., 1970). The *Rachinger method* uses our knowledge of the exact wavelengths of the $K_{\alpha 1}$ and $K_{\alpha 2}$ lines and their intensity ratio of 2:1. As the computer scans through the raw data from low angle to high, half of the intensity at each 2θ is subtracted from the forward position calculated for the $K_{\alpha 2}$ peak. Figure 1-53 shows the results of this procedure for a portion of the quartz diffraction pattern.

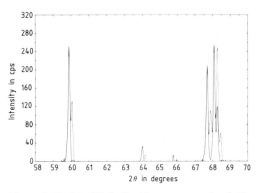

Figure 1-53. Modified Rachinger removal of $K_{\alpha 2}$ peaks. The outer enevelopes are the raw data. The inner peaks are the result after removal of the $K_{\alpha 2}$ peak.

the most serious. Displacements in the micron range can cause peak shifts of hundredths of a degree.

Wavelength Accuracy

In the early years of X-ray analysis the wavelength of the radiation was determined from Bragg's law by calculating the d_{hkl} values from the known NaCl lattice parameter and using the observed diffraction angles from the NaCl powder diffraction pattern. The lattice parameter is obtained from the density of NaCl (ϱ), the atomic weights of the Na and Cl atoms in the unit cell (ΣW) and Avagadro's number (L) via,

$$a = V^{1/3} = \left(\frac{\Sigma W}{L \varrho}\right)^{1/3} \quad (1\text{-}42)$$

The computed value of d_{200} was used in the Bragg equation with the observed θ for this peak, to calculate the wavelength. This method has evolved over the years through the use of ruled gratings (Bearden, 1967), to modern calibration in terms of the wavelength of the ^{3}He–^{20}Ne laser (Deslattes and Henins, 1973). The current value for the wavelength of the Cu K$_{\alpha 1}$ line is 1.54060 Å.

Internal Standard Method

The internal standard method uses a substance of very well known lattice parameters like the NBS Standard Reference Material Si 640 (Hubbard et al., 1987; Dragoo, 1986). Since the lattice parameter of the cubic Si standard has been very accurately determined (5.4060 Å), we can compute exact values for all of the d_{hkl} values and hence, the expected 2θ's of the diffraction peaks. To measure the pattern of a material accurately, some of this Si is mixed with the specimen and the combined pattern is measured. Since the average μ/ϱ of the mixture will determine the average depth from which diffraction will occur, then the sample displacement error will be the same for all lines. Thus, if we calculate the $\Delta 2\theta$ values between the observed and calculated positions for the Si lines, we can plot a correction curve of $\Delta 2\theta$ vs. 2θ. This curve is then used to correct the observed 2θ's of the material mixed with the internal standard.

External Standard Method

The use of computer automated procedures can affect accuracy in two ways: the first is to remove the computational drudgery from making internal standard corrections and the second is to allow the routine application of an external standard calibration curve (Mallory and Snyder, 1979). Approximately half of the error in peak location is due to instrumental error (Snyder et al., 1982). All instrumental error can be removed running a series of known standards and plotting $\Delta 2\theta$ (i.e., $2\theta_{obs} - 2\theta_{calc}$) vs. 2θ, as shown in Fig. 1-54. The least squares fitting of a polynomial to this curve allows the polynomial coefficients to be stored on a disk file. These may be used to correct all patterns for instrumentally induced error. However,

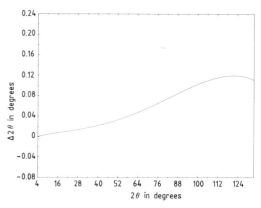

Figure 1-54. An instrumental calibration curve.

it should be emphasized that the error in peak location can be reduced by another factor of two, if an internal standard is used. This is illustrated in Table 1-7 using the F_N powder pattern quality figure of merit described below in Sec. 1.12.2.

Table 1-7. The effect of calibration on the figure of merit F_N. Notation: $F_N(\overline{|\Delta 2\theta|}, N_{poss})$.

Method	Arsenic trioxide F_{29}	Quartz F_{30}
No correction	9.9 (0.049, 59)	16.4 (0.052, 35)
External standard	15.4 (0.026, 59)	30.0 (0.028, 35)
Internal standard	42.0 (0.012, 59)	66.1 (0.013, 35)

Zero Background Holder Method

In recent years "zero background holders" (ZBH) have come into wide use. These are diffraction specimen holders made from a single crystal of quartz or silicon, which has been cut along a reciprocal lattice direction in which no diffraction will occur. The technique thus makes use of Bragg extinction to remove all X-ray scattering except that which is due to the specimen. If the specimen is carefully ground to less than 10 μm particle size, and spread on a thin (1 μm) layer of grease on the ZBH, the total specimen displacement can be held to less than 10 μm. Even the small 2θ error which would be introduced from such a displacement can be eliminated by preparing and using, an external standard calibration curve measured in exactly the same manner on a ZBH. It should also be noted that if only a monolayer of powder is exposed to the X-ray beam, all matrix absorption effects are eliminated.

1.9.3 Software Developments

During the 1960's as mainframe computers evolved (e.g., IBM 7094, CDC 6600), crystallographic computing was performed in a non-interactive environment, where cards containing coded data were input to a program operating in a batch stream. Due to the very high costs of computers, laboratory automation was nearly non-existent, so we may refer to these early developments as the zeroth generation of laboratory computational software. The 1970's saw the evolution of the laboratory computer (e.g., PDP 8 and PDP 11/10) and the first generation of process control software. Input, to control programs, was from cards and paper-tape in a non-interactive environment. The second generation of laboratory software evolved on the next generation of laboratory computers (e.g., PDP 11/34) and conducted an interactive dialogue with the user to setup and conduct an automated experiment and to analyze the data (Mallory and Snyder, 1979). This rapidly evolved into a third generation of software (ca. 1985) which employed forms and help screens on video terminals, rather than interactive dialog (Snyder et al., 1981 b). This generation saw the evolution of complex file structures containing all of the information required to process the data from an automated experiment.

A fourth generation of software is currently evolving where all interaction with the user is done through a "point and click" video interface with a mouse. The trend is to eliminate all keyboard activity and convert crystallographic methods, which used to rely on the user evaluating numbers, to visual evaluation of graphical data. For example Fig. 1-62 shows a screen from one of the first of these programs (Zorn, 1991). Most interaction is done with the mouse and all evaluation is done in the visual display window. The figure illustrates the overlaying of PDF reference patterns onto an experimental pattern. The program display is, of course, in color. The elimination of numerical d-I methods of phase identification have allowed a much larger volume of data to be processed per experiment. The problem here is that software development continues to be a scientific orphan done mostly on bootleg time. The trend to visual analysis, with mouse driven interfaces, requires substantial development efforts which both companies and universities have traditionally found difficult to muster.

1.9.3.1 The Accuracy of Diffraction Intensities

One of the traditional limitations of X-ray powder diffraction is the large uncertainty in observed intensity values (Snyder, 1983 b), due primarily to the preferred orientation of crystallites in powders. This effect arises from the tendency of crystallites to lie on the faces determined by their growth habits. This causes the powder to expose a non-random set of orientations to the X-ray beam thereby distorting the relative intensities of the diffraction lines. This problem has strongly limited the application of powder diffraction techniques to qualitative, quantitative and structural analysis. This problem can be seen in the very first published powder patterns and has remained in need of a general solution. The spray drying of powders has been shown to remove preferred orientation and produce intensity values which are accurate to a few percent (S. T. Smith et al., 1979; Cline and Snyder, 1985, 1987).

Spray drying is a technique in which a powder is suspended in a non-dissolving liquid. A small amount of an amorphous binder like polyvinyl alcohol is added to the suspension along with a defloculant to keep the particles in a fifty weight percent suspension. The slurry is then atomized through a spray nozzle into a heated chamber where the droplets dry before falling back onto a collector tray. The dried droplets, which are small spheres with a typical average agglomerate size of 50 μm, are shown in Fig. 1-55. Since a sphere is isotropic, the crystallites composing it show a random distribution of orientations to the X-ray beam. The powder pattern from a spray dried sample shows no intensity distortion from preferred orientation.

1.9.3.2 The Limit of Phase Detectability

The limit of phase detectability, by manual X-ray powder diffraction, is usually stated as being in the one to two percent concentration range. However, with automated systems, which can count for longer periods of time, the limit depends only on the number of counts collected and the signal to background ratio. Thus, to increase the sensitivity of the method we simply need to increase the number of counts collected. This may be done by increasing the count time an automated instrument

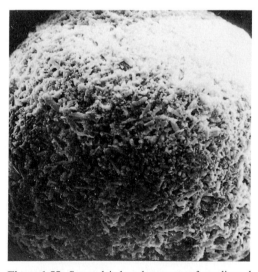

Figure 1-55. Spray dried agglomerates of acyclic wollastonite (200 ×).

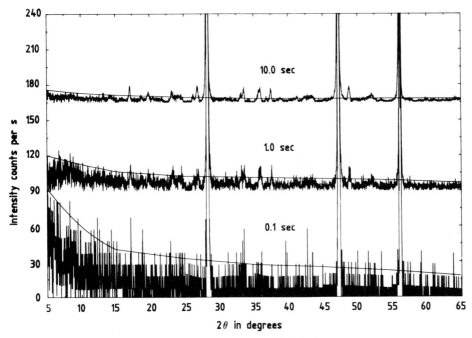

Figure 1-56. Scans of an impure Si sample at 0.1, 1.0 and 10.0 s/step.

spends at each angular step in a pattern. This is illustrated in Fig. 1-56 which shows the powder pattern of a hot pressed sample of silicon counted for 0.1, 1.0 and 10.0 seconds per point. These patterns were passed through the linearization and threshold finding procedures previously described. The increase in the number of peaks breaking through the threshold line, with increasing count time, is due to the decreasing relative error. All of the peaks other than the three lines due to silicon are below 1% relative intensity. These peaks have been identified as being due to minor impurity concentrations of SiC, Si_3N_4 and SiO_2. Figure 1-57 shows the theoretical limit of detectability as a function of count time for three different mixtures. For many materials detectability of 10 ppm is obtainable in a overnight run.

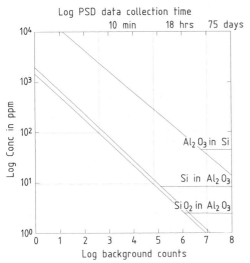

Figure 1-57. Theoretical limit of phase detectability for three mixtures.

1.10 Phase Identification by X-Ray Diffraction

We have seen that the set of diffracted d's and I's for any material is as characteristic as a fingerprint is for a human. We must now face the same problem the crime solvers have had to face – an observed set of fingerprints is useless unless you have an extensive collection of reference patterns for comparison. Our problem is even more complex in that unknown materials are commonly composed of more than one phase. When this occurs, the observed pattern contains all the diffraction peaks of each phase present. This is analogous to trying to unravel multiple finger prints superimposed on one another. We are fortunate in that usually powder diffraction patterns, particularly from high symmetry materials, are less complicated than a finger print.

1.10.1 The Powder Diffraction Data Base

Scientists have answered the need for a powder diffraction pattern database by creating an organization called the International Centre for Diffraction Data[1] (ICDD), which publishes the *Powder Diffraction File* (PDF) in annual installments. The ICDD is a nonprofit organization established by a number of international scientific organizations. The organization has historically sponsored the determination of powder patterns, through its association with the U.S. National Bureau of Standards (now called N.I.S.T.) and a number of other laboratories. However, most of the patterns published in the *Powder Diffraction File* are obtained from literature articles. They are reviewed by editors and published in book, and computer formats (e.g., CD-ROM). An example of such a pattern is shown in Fig. 1-58. The data base is published in sets with about 2000 new patterns released each year. Currently there are about 60 000 patterns in the file.

---

[1] 1601 Park Lane, Swarthmore, PA 19081, U.S.A.

A number of manual and computer searching methods have been developed over the years to accomplish phase identification. These methods place an emphasis on the accuracy of the intensities (Hanawalt method) or the d_{hkl} values (Fink method) or on the elements present (alphabetical search). Each year an alphabetical listing and a Hanawalt search manual are published for the updated PDF (by the ICDD). At non-regular intervals *Fink Search Manuals* and *Common Phase Manuals* are published. The *Common Phases* are a list of about 2500 of the most frequently encountered materials, its use often speeds phase identifications. In any search method, the ability to recognize a reference pattern in an unknown depends on the quality of the d's and I's in both the reference and unknown. The problem of phase identification is shown picturally in Fig. 1-59. The error in the d's and I's depend on a great number of factors. The experimental technique used to measure the pattern is one of the first quality indications to a user of the data base.

1.10.2 Phase Identification Strategies

The Hanawalt method involves sorting the patterns in the PDF according to the d-value of the 100% intensity line. This list is then broken into small intervals of d (with a ± 0.01 overlap in d between intervals) and each interval is sorted on the d-value of the second most intense line. Subsequent lines are listed in order of decreasing intensity.

Each phase is entered multiple times under the category of each of its most intense lines. The reason for the multiple entries is to help avoid the greatest weakness of this search method: distortion of the intensities due to preferred orientation. A sample of the Hanawalt index is shown in Fig. 1-60

1.10 Phase Identification by X-Ray Diffraction

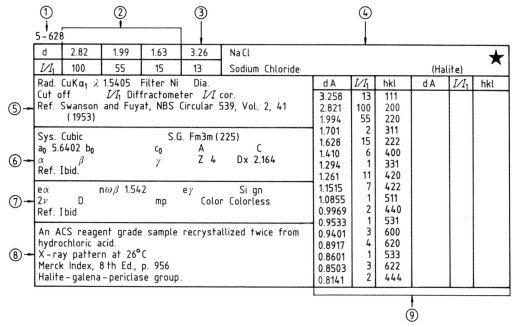

Figure 1-58. Example PDF card format: Standard 3 × 5 in. ICDD diffraction data card (card 628 from Set 5) for sodium chloride. Appearing on the card are: (1) set and file number, (2) three strongest lines, (3) lowest-angle line, (4) chemical formula and name of substance, (5) data on diffraction method used, (6) crystallographic data, (7) optical and other data, (8) data on specimen, and (9) diffraction pattern. Intensities are expressed as percentages of I_1, the intensity of the strongest line on the pattern. Most cards have a symbol in the upper right corner indicating the quality of the data: ★ (high quality), i (lines indexed, intensities fairly reliable), C (calculated pattern), and O (low reliability). A blank indicates undetermined quality.

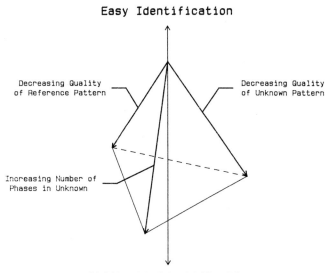

Aspects of Pattern Identification

Figure 1-59. The parameters affecting phase identification.

				2.84 – 2.80 (± 0.01)						File No.	I/Ic
i	2.83_x	2.00_3	1.64_3	1.42_1	1.27_1	1.07_1	2.32_1	1.16_1	$EuNbO_3$	26–1417	4.10
c	2.83_x	2.00_6	1.64_2	1.27_1	1.16_1	1.42_1	0.94_1	0.90_1	$Li_{0.4}Ag_{0.6}Br$	26– 858	
	2.83_x	2.00_6	1.64_6	1.07_6	4.62_4	2.42_4	1.42_4	1.27_4	K_2CaSiO_4	19– 943	
	2.83_x	2.00_9	1.63_9	1.42_7	1.27_7	2.31_6	4.90_5	4.01_5	$Sr(Mg_{0.33}Nb_{0.67})O_3$	17– 181	
i	2.83_x	2.00_4	1.62_4	2.79_3	1.63_3	1.77_3	1.27_3	4.00_2	$NdScO_3$	26–1275	2.90
	2.83_x	2.00_8	1.62_8	1.77_7	1.95_6	1.65_6	3.99_5	1.50_4	$PmScO_3$	33–1091	
	2.83_x	2.00_7	1.27_3	1.64_2	1.16_2	0.95_1	0.90_1	1.42_1	$AgSnSe_2$	33–1194	
	2.83_x	2.00_7	1.26_3	1.63_3	1.15_3	1.41_2	3.27_2	1.70_1	$ErSe$	18– 490	
	2.82_x	2.00_x	6.24_5	1.63_5	3.85_3	3.35_3	5.93_2	5.07_2	Yb_2Se_3	19–1434	
	2.82_x	2.00_8	4.67_6	4.52_6	3.98_6	2.38_4	1.82_4	3.78_2	Ca_3ReO_6	34–1328	
	2.82_x	2.00_5	2.23_3	1.82_2	1.41_1	1.26_1	1.15_1	1.28_1	$SiCl_4$	10– 220	
	2.82_x	2.00_9	1.74_9	7.12_8	3.31_8	2.13_8	3.51_7	3.10_7	NbS_2I_2	20– 811	
i	2.82_x	2.00_4	1.63_4	1.41_8	1.07_1	1.26_1	2.30_1	0.82_1	$Li_{0.25}SrNb_{0.75}O_3$	25–1383	
	2.82_x	2.00_8	1.63_8	1.41_8	1.27_8	1.07_8	2.30_5	1.98_5	Pb_2CoWO_6	17– 491	
	2.82_x	2.00_8	1.63_8	1.41_8	1.07_8	2.30_5	1.16_5	4.62_3	Pb_2CoWO_6	17– 494	
o	2.81_x	2.00_x	1.62_x	1.27_x	1.65_7	1.64_7	1.78_6	3.95_5	Ca_3TeO_6	22– 156	
o	2.81_x	2.00_x	1.54_x	6.32_8	4.48_8	2.24_8	2.11_8	1.88_8	$Pt(NH_3)_4PtCl_4$	2– 817	
	2.79_x	2.00_x	1.91_x	1.23_x	1.19_x	2.47_8	1.96_8	1.85_8	MnP	7– 384	
	2.84_x	1.99_6	3.39_5	3.45_5	2.25_4	3.61_3	3.27_3	2.93_3	$LiBi_3S_5$	33– 791	
	2.84_8	1.99_x	1.68_8	3.24_6	1.13_6	1.07_6	0.94_6	3.04_4	$CuS_{0.5}CN$	28– 402	
i	2.83_x	1.99_x	2.89_x	2.75_x	1.60_8	2.01_6	1.62_5	1.66_4	$YInO_3$	21–1449	
*	2.82_9	1.99_x	2.30_8	1.41_4	1.63_2	0.89_1	1.20_1	1.15_1	$KMgF_3$	18–1033	0.90
*	2.82_x	1.99_6	1.63_2	3.26_1	1.26_1	1.15_1	1.41_1	0.89_1	$NaCl/Halite, syn$	5– 628	4.40
c	2.82_x	1.99_8	1.26_3	1.63_2	1.15_2	0.94_1	0.89_1	1.41_1	$BePd$	18– 225	
i	2.81_x	1.99_4	3.98_3	2.79_3	1.62_3	2.30_3	1.63_2	4.02_2	$La_4Mn_4O_{11}$	35–1354	

Figure 1-60. Sample of the *Hanawalt Search Manual.*

and a flow chart of the search strategy is given in Fig. 1-61.

The Fink method is an alternative to the Hanawalt method when preferred orientation is suspected of distorting the relative intensities. The Fink method selects the eight strongest lines of each pattern in the PDF and rotates them to make entries in the search manual. The four most intense lines are used as entry points. The remaining seven lines are then entered in order of decreasing *d*-value. The deemphasis on the intensity allows materials with strong preferred orientation or with intensity distortions due to the use of a radiation other than Cu K_α, to be identified. In fact, this is the method of choice for identifying electron diffraction patterns or patterns obtained by energy dispersive techniques where the published X-ray intensities are of little value.

One of the first algorithms developed to take advantage of the large and rapid data handling capacity of the digital computer was developed at Pennsylvania by Johnson Jr. and Vand (1967). This approach compares each reference pattern in the PDF to the unknown pattern and computes a figure of merit for each match. The figure of merit is based on such quantities as the average error in matching the *d*'s and *I*'s. The absolute value of the merit figure is not important, since it will be computed for all reference patterns and the best

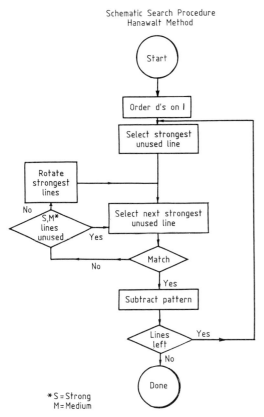

Figure 1-61. Flowchart of the Hanawalt Search Method.

fifty matches listed for evaluation. This first approach to computer phase identification had a number of shortcomings (Cherukuri et al., 1983). In addition, to the obvious problems, which will be caused by systematic d_{hkl} and I errors in the reference or unknown patterns, a number of other points can cause difficulties. If the unknown pattern contains a large number of lines then it becomes quite likely that some of the reference patterns with only a few lines will match well. Thus, these patterns will tend to rise to the top of the list. Even when a reference matches a large number of lines it still may contain chemical components known not to be present, so chemistry should always be used as a search criterion. However, when a good match with the wrong chemistry occurs, it most likely means that the unknown is isostructural with the reference and thus, still provides useful information.

The probability of a computer search producing correct candidate phases for matching is greatly enhanced by allowing the program to reject compounds based on their chemistry. True unknowns should always be examined by X-ray fluorescence first, to establish the chemical constraints to be used in the search. This advice applies equally well to the second generation type of computer search algorithms as well as the first.

The development of the full digital version of the *Powder Diffraction File* and its production on a CD-ROM has dramatically improved our abilities in phase identification and characterization. The latest version of the PDF contains nearly 60 000 phases and due to the high-density format of the CD-ROM, is readily accessible on a PC. The new digital version called PDF-2 contains all of the unit cell, indices, experimental conditions, etc. contained in the full database. A second generation of computer search algorithms (Snyder, 1981) have evolved into current search/match software which uses the CD-ROM database and performs identifications in about 60 seconds. Access to this database is essential to speed up data analysis, in that the instrumental developments mentioned earlier, have led to the ability to produce data much faster than it can be analyzed using second or third generation software or the even slower manual methods. The evaluation of the results of a search have moved to the fully visual in recent years. Figure 1-62 shows a screen from the phase identification section of Zorn's (1991) program SHOW.

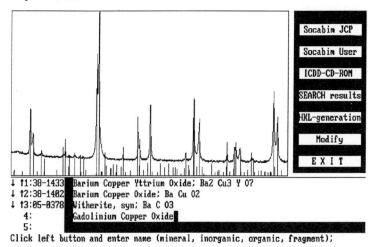

Figure 1-62. A screen from the fourth generation program, SHOW by G. Zorn.

1.10.3 The Crystal Data Database

The ICDD also publishes the *Crystal Data* database which has been built up over the years at the U.S. National Bureau of Standards (now called the National Institute of Science and Technology). This database contains the unit cell information on every material for which a report has been published. At the moment, it contains references to about 150 000 unit cells. Mighell and Himes (1986) have developed a computer algorithm which will take a unit cell derived, for example, from the powder pattern of an unknown, and compute all possibly related super- and subcells and then search the crystal data base for the phase or possible isomorphous phases.

As an example of the use of this database, we can look at a recent analysis of the crystal structure of BeH$_2$ (G. S. Smith et al., 1988). Due to the low symmetry and extremely low X-ray scattering ability of BeH$_2$, this structure defied analysis for many years. Using the very high-resolution diffractometer on beamline X7A at the BNL/NSLS synchrotron source mentioned previously, a very good quality pattern for this material was obtained. Computer indexing (G. S. Smith, 1987), of this pattern indicated an orthorhombic unit cell. A search of the *Crystal Data* database produced a reference to a high-pressure form of ice (i.e., OH$_2$). The very well resolved peaks allowed isolation of integrated intensities and the solution of this crystal structure based on this model.

1.10.4 The Elemental and Interplanar Spacings Index (EISI)

As mentioned above, the *Crystal Data* database contains about 150 000 references to published unit cells. As just described, this can be used with the modern computer indexing procedures to do phase identification and propose crystal structure models. Recently, using Eq. (1-20), powder patterns were computed from the unit cell information for all of the materials in *Crystal Data*. These calculated patterns have had the space group systematic extinct reflections removed, leaving a few *d*-values which might not occur in the observed pattern. These, occasional accidentally absent reflections are due to a particular structure factor having a low value, causing the intensity to be below the observational threshold. Since crystal structure information is not available for most of these phases, the intensities can not be computed. The new computed "*d*-value" patterns were added to the existing observed patterns in the PDF, producing a database with over 200 000 patterns. Because this database, based only on high *d*-values, is particularly suited to identifying patterns from electron diffraction, it has been called the electron diffraction database (EDD) or Max-*d* index. When the entries in the EDD are arranged by the elements present in each phase, as they would be determined by X-ray fluorescence, the resulting search manual is called the elemental and interplanar spacings index (EISI) and is also available from the International Centre for Diffraction Data.

As an example of the use of this new index, and an illustration that it is also useful for X-ray and neutron data, as well as electron diffraction, we can consider a recent application from the author's laboratory. We have been substituting various ions, including indium, into the high temperature superconducting material, $Bi_2Sr_2CaCu_2O_8$. One of the samples produced a phase whose powder pattern is not in the PDF. On carrying out the phase identifications using the EISI, a perfect match for the phase $In_{0.5}Sr_{0.5}O_3$ was observed, except that the database pattern contains two extra lines which are accidentally absent in the actual observed pattern. Thus, this index permits access to a much wider database for phase identification, even for XRD data.

1.11 Quantitative Phase Analysis

Quantitative phase analysis by X-ray powder diffraction dates back to 1925 when Navias (1925) quantitatively determined the amount of mullite in fired ceramics, and to 1936 when Clark and Reynolds (1936) reported an internal-standard method for mine dust analysis by film techniques. In 1948 Alexander and Klug (1948) presented the theoretical background for the absorption effects on diffraction intensities from a flat brickette of powder. Since then there have been numerous methods developed, based on their basic equations. Methods applicable to a wide range of phases and samples include the *method of standard additions* (also called the addition of analyte or *spiking method*) (Lennox, 1957), the *absorption diffraction method* (Alexander and Klug, 1948; S. T. Smith et al., 1979b) and the *internal-standard method* (Klug and Alexander, 1974). Formalisms have been established to permit inclusion of overlapping lines and chemical constraints in the analysis (Copeland and Bragg, 1958). For an up-to-date summary of modern quantitative analysis procedures see Snyder and Bish (1989) and Snyder and Hubbard (1984).

Quantitative analysis using any method is a difficult undertaking. It requires careful calibration of the instrument using carefully prepared standards and many repetitions during the set up-phase to establish the required technique. In general, quantitative analysis of a new phase system will require a minimum of a few days and often a week of setup time. After the technique and standards have been established, a well-designed computer algorithm, such as the NBS*QUANT84 system (Snyder et al., 1981b; Snyder and Hubbard, 1984) can produce routine analyses in as little as a few minutes to an hour per sample, depending on the precision desired.

Perhaps it is the difficult nature of quantitative analysis that has limited the number of carefully done literature examples. This, in turn, has resulted in a concentration of effort on a specific analysis technique, rather than on understanding the sample and instrument-dependent limitations of quantitative phase analysis by X-ray powder diffraction.

1.11.1 The Internal-Standard Method of Quantitative Analysis

The internal-standard method is the most general of any of the methods for quantitative phase analysis. In addition, this method lends itself most easily to generalization into the RIR (reference intensity ratio) method, and even further can be generalized into a system of linear equations which allow for the use of overlapped lines and chemical analysis constraints.

The intensity of a diffraction line i from a phase α, in the form of a flat plate, can be rewritten from Eq. (1-34) as

$$I_{i\alpha} = \frac{K_{i\alpha}}{\mu} \qquad (1\text{-}43)$$

where μ is the linear absorption coefficient of phase α and

$$K_{i\alpha} = \frac{I_0 \lambda^3 e^4}{32 \pi r m_e^2 c^4} \frac{M_i}{2 V_\alpha^2} |F_{i\alpha}|^2$$

$$\cdot \left(\frac{1 + \cos^2 2\theta_i \cos^2 2\theta_m}{\sin^2 \theta_i \cos \theta_i} \right) \qquad (1\text{-}44)$$

The intensity of a diffraction line i from phase α in a mixture of phases is given by

$$I_{i\alpha} = \frac{K_{i\alpha} X_\alpha}{\varrho_\alpha (\mu/\varrho)_m} \qquad (1\text{-}45)$$

where: X_α = weight fraction of phase α, ϱ_α = density of phase α, $(\mu/\varrho)_m$ = the mass absorption coefficient of the mixture, cf. Eq. (1-13).

The fundamental problem in quantitative analysis lies in the $(\mu/\varrho)_m$ term in Eq. (1-45). To solve for the weight fraction of phase α, we must be able to compute $(\mu/\varrho)_m$ and this, as can be seen from Eq. (1-13), requires a knowledge of the weight fractions of each phase. The internal-standard method is based on elimination of the absorption factor $(\mu/\varrho)_m$, by dividing two equations of type (1-45), giving

$$\frac{I_{i\alpha}}{I_{js}} = k \frac{X_\alpha}{X_s} \qquad (1\text{-}46)$$

which is linear in the weight fraction of phase α. The subscripts i and j refer to different hkl reflections. The k is the slope of the internal standard calibration curve: a plot of $(I_{i\alpha}/I_{js})$ vs. (X_α/X_s).

Equation (1-46) is linear and forms the basis of the internal-standard method of analysis. The addition of a known amount, (X_s), of an internal standard to a mixture of phases, which may include amorphous material, permits quantitative analysis of each of the components of the mixture by first establishing the values of k for each phase [i.e., the slope of the internal standard plot of $(I_{i\alpha}/I_{js}) X_s$ vs. X_α], from standards of known concentration.

Use of a pre-established calibration curve, defining the slope k, permits the weight fraction of any phase α in the original mixture to be computed. Note that k is a function of i, j, α, and s for a constant weight fraction of standard.

1.11.1.1 $I/I_{corundum}$

The slope k of the plot of $(I_{i\alpha}/I_{js}) X_s$ vs. X_a is a measure of the inherent diffracted intensities of the two phases. Visser and DeWolff (1964) were the first to propose that k values could be published as materials constants, if the concentration and the diffraction lines of the phases α and s were specified and s itself were agreed to by all researchers. Their proposal was to use corundum, as the universal internal standard, and to use the maximum or 100% lines of phase α and corundum, mixed in a 50% by weight mixture. This $I/I_{corundum}$ or I/I_c value has been widely accepted and now is published with patterns for standards in the *Powder Diffraction File*.

The simple 2-line procedure for measuring I/I_c is quick but suffers from several drawbacks. Preferred orientation commonly affects the observed intensities unless careful sample preparation methods (Cline and Snyder, 1985) are employed. Other problems include extinction (Cline and Snyder, 1987), inhomogeneity of mixing, and variable crystallinity of the sample due to its synthesis and history (Cline and Snyder, 1983; Gehringer et al., 1983). All of these effects conspire to make the published values of I/I_c subject to substantial error. For greater accuracy, multiple lines from both the sample (α) and the reference phase (corundum, in this case) should be used (Hubbard et al., 1976). Such an approach often reveals when preferred orientation is present and provides realistic measurement of the reproducibility in the measurement of I/I_c. In general, spray drying the samples is the best way to eliminate this serious source of error (S. T. Smith et al., 1979a, b; Cline and Snyder, 1985).

1.11.1.2 The Generalized Reference Intensity Ratio

The concept of the I/I_c value as a materials constant leads naturally to a broader definition, which permits the use of reference phases other than corundum, lines other than just the 100% relative intensity line, and arbitrary concentrations of the two phases α and s. The most general definition of the RIR has been given by Hubbard and Snyder (1988) as

$$RIR_{\alpha,s} = \frac{I_{i\alpha}\, I_{js}^{rel}\, X_s}{I_{js}\, I_{i\alpha}^{rel}\, X_\alpha} \qquad (1\text{-}47)$$

In Eq. (1-47) we see that the RIR is a function of α and s only, not of i, j, or X_s. It, of course, remains the slope of the calibration curve for phase α with internal standard s but now has been normalized so that it may be computed from any pair of diffraction lines in a calibration mixture. We also see that I/I_c values are simply RIR values where s is corundum. The RIR, when accurately measured, is a true constant and allows comparison of the absolute diffraction line intensities of one material to another (Chung, 1974a, b; Hubbard et al., 1976; Hubbard and Snyder, 1988). It also enables quantitative phase analysis in a number of convenient and useful ways.

1.11.1.3 Quantitative Analysis with RIR^s

The internal-standard method of quantitative analysis when stated in terms of an RIR comes from rearranging Eq. (1-47)

$$X_\alpha = \frac{I_{i\alpha}\, I_{js}^{rel}\, X_s}{I_{js}\, I_{i\alpha}^{rel}\, RIR_{\alpha,s}} \qquad (1\text{-}48)$$

The RIR value in Eq. (1-48) may be obtained by careful calibration, by determining the slope of the internal-standard plot, or it may be derived from other RIR values via

$$RIR_{\alpha,s} = \frac{RIR_{\alpha,s'}}{RIR_{s,s'}} \qquad (1\text{-}49)$$

where s′ is any common reference phase. When s′ is corundum, the RIR values are simply I/I_c values. Hence, we can combine I/I_c values for phases α and s to obtain the reference intensity ratio for phase α relative to phase s. Taking s to be an internal standard and substituting Eq. (1-49) into Eq. (1-48) and rearranging we have

$$X_\alpha = \frac{I_{i\alpha}}{I_{js}} \frac{I_{js}^{\text{rel}}}{I_{i\alpha}^{\text{rel}}} \frac{RIR_{s,c}}{RIR_{\alpha,c}} X_s \tag{1-50}$$

This equation is quite general and allows for the analysis of any crystalline phase in an unknown mixture, as long as the RIR is known, by the addition of an internal standard. However, if all four of the required constants ($I_{i\alpha}^{\text{rel}}$, I_{js}^{rel}, $RIR_{s,c}$ and $RIR_{\alpha,c}$) are taken from the literature the results should be considered as only semi-quantitative, since each of them may contain significant error.

Equation (1-50) is valid even for complex mixtures which contain unidentified phases, amorphous phases, or identified phases with unknown RIR^s.

1.11.1.4 Standardless Quantitative Analysis

The ratio of the weight fractions of any two phases whose RIR^s are known may always be computed by choosing one as X_α and the other as X_s. That is

$$\frac{X_\alpha}{X_\beta} = \frac{I_{i\alpha}}{I_{j\beta}} \frac{I_{j\beta}^{\text{rel}}}{I_{i\alpha}^{\text{rel}}} \frac{RIR_{\beta,s}}{RIR_{\alpha,s}} \tag{1-51}$$

If the RIR values for all n phases in a mixture are known, then $n-1$ equations of type (1-51) may be written. Karlak and Burnett (1966) and later Chung (1974a, b; 1975) pointed out that if no amorphous phases are present an additional equation holds

$$\sum_{m=1}^{n} X_m = 1. \tag{1-52}$$

Equation (1-52) permits analysis without adding any standard to the unknown specimen by allowing us to write a system of n equations to solve for the n weight fractions via

$$X_\alpha = \frac{I_{i\alpha}}{RIR_\alpha I_{i\alpha}^{\text{rel}}} \tag{1-53}$$

$$\cdot \left[1 \Big/ \left(\sum_{k=1}^{\text{number of phases}} \frac{I_{j,k}}{RIR_k I_{j,k}^{\text{rel}}} \right) \right]$$

Chung referred to the use of Eq. (1-52) as the matrix flushing or the adiabatic principle, but it should be called the normalized RIR method. It is important to note that the presence of any amorphous or unidentified crystalline phase invalidates the use of Eq. (1-53). Of course, any sample containing unidentified phases cannot be analyzed using the normalized RIR method in that the required RIR^s, will not be known.

When the RIR values are known from another source, for example, published I/I_c values, it may be tempting to use them along with the I^{rel} values on the PDF card to perform what some call a completely "standardless" quantitative analysis using Eq. (1-52). These I^{rel} and I/I_c values are seldom accurate enough to be used directly in quantitative analysis. The analyst should accurately determine the relative intensities and RIR values for an analysis by careful calibration measurements. In fact the word "standardless" is a misnomer. Standards are always required however, the RIR method allows you to use them from the literature for semi-quantitative analysis.

1.11.1.5 Constrained X-Ray Diffraction (XRD) Phase Analysis – Generalized Internal-Standard Method

A quantitative analysis combining both X-ray diffraction and chemical or thermal results, with a knowledge of the composition of the individual phases, can yield results of higher precision and accuracy than is generally possible with only one kind of observation. The analysis becomes more complex when several phases in a mixture have similar compositions and/or potential compositional variability, but it is possible with appropriate constraints during analysis to place limits on the actual compositions of the constituent phases.

The most general formulation of these ideas has been described by Copeland and Bragg (1958). Their internal-standard equations for multicomponent quantitative analysis, with possible line superposition and chemical constraints, take the form of a system of simultaneous linear equations. In terms of RIR values each equation is of the form

$$\frac{I_i}{I_{js}} = \left(\frac{I_{i1}^{rel}}{I_{js}^{rel}} RIR_{1,s}\right)\frac{X_1}{X_s} + \left(\frac{I_{i2}^{rel}}{I_{js}^{rel}} RIR_{2,s}\right) +$$
$$+ \ldots + \left(\frac{I_{ik}^{rkl}}{I_{js}^{rel}} RIR_{k,s}\right)\frac{X_k}{X_s} + \varepsilon \quad (1-54)$$

where

- I_i is the intensity of the ith line from the mixture of k phases which may contain contributions from one or more lines from one or more phases in the mixture,
- I_{js} is the intensity of a resolved line j of the internal standard s,
- X_k is the weight fraction of phase k in the mixture of sample plus internal standard,
- ε is a least-squares error term.

Line overlap is allowed for by having each intensity ratio (I_i/I_{js}) have contributions from multiple lines from multiple phases. As many terms, involving contributions to the intensity ratio, as needed are included in each linear equation. The Copeland-Bragg analysis results in n equations in n unknowns which may be solved via least squares.

Quantitative elemental data, obtained from X-ray fluorescence analysis for example, can be added to the system of equations, without increasing the number of unknowns, making it overdetermined. Coupling both the elemental and diffraction data will result in more accurate quantitative analysis of multicomponent mixtures. More importantly, estimates of the standard deviation of the results will be more accurate. Both the line overlap and chemical analysis features of the Copeland-Bragg formalism have been included in the AUTO and NBS*QUANT84 systems of Snyder and Hubbard (1981, 1982, 1984).

1.11.1.6 Full-Pattern Fitting

With the availability of automated X-ray powder diffractometers, digital diffraction data are now routinely available on computers and can be analyzed by a variety of numerical techniques. Complete digital diffraction patterns provide the opportunity to perform quantitative phase analysis using all data in a given pattern, rather than considering only a few of the strongest reflections. As the name implies, full-pattern methods involve fitting the entire diffraction pattern, often including the background, with a synthetic diffraction pattern. This synthetic diffraction pattern can either be calculated and fit dynamically from crystal structure data or can be produced from a combination of observed or calculated standard diffraction patterns.

The use of the whole pattern by D. K. Smith and his colleagues (D. K. Smith et al., 1987, 1988), involves collection of reference or standard data on pure materials using fixed instrumental conditions. These patterns of standards are processed to remove background and any artifacts, and the data may be smoothed. For standard materials that are unavailable in appropriate form, diffraction patterns can either be simulated from Powder Diffraction File data or calculated powder patterns. The next step in standardization is analogous to procedures used in conventional reference intensity ratio quantitative analyses. Data are collected on each of the samples of the standard that have been mixed with a known amount of corundum, α-Al_2O_3, in order to determine the RIR. These patterns yield the basic calibration data used in the quantitative analysis. The final step involves collection of data for the unknown samples to be analyzed, using conditions identical to those used in obtaining the standard data. Background and artifacts are also removed from unknown sample data.

Reference intensity ratios are obtained from the whole patterns by initially assigning the standard material a reference intensity ratio of 1.0. Then the values of "apparent" X_α and X_s are computed from Eq. (1-50). The ratio of the "apparent" weight fractions of α to s gives the correct RIR value. For example (D. K. Smith et al., 1987), a standard mixture of 50% ZnO and 50% corundum yields results of 57.8% ZnO and 42.2% corundum. The reference intensity ratio for ZnO is therefore 57.8/42.2 = 1.37.

The digital patterns of the unknown mixture are fit to the standards with a least-squares procedure minimizing the expression

$$\delta(2\theta) = I_{unk}(2\theta) - \sum_k X_k RIR_k I_k(2\theta) \quad (1\text{-}55)$$

where $I_{unk}(2\theta)$ and $I_k(2\theta)$ are the diffraction intensities at each 2θ interval for the unknown and each of the standard phases. Intensity ratios derived using this procedure are equivalent to peak-height RIR^s, rather than integrated-intensity RIR^s because the minimization is conducted on a step-by-step basis.

This whole pattern fitting procedure is simply a method for measuring RIR^s from whole pattern standards. Once this is accomplished the method becomes a conventional RIR quantitative analysis procedure, which requires standards and calibration. To make it "standardless" one must invoke the normalization assumption, i.e., Eq. (1-52).

1.11.1.7 Quantitative Phase Analysis Using Crystal Structure Constraints

Quantitative phase analysis using calculated patterns is a natural outgrowth of the *Rietveld method* (1969), originally conceived as a method of refining crystal structures using neutron powder diffraction data. Refinement is conducted by minimizing the sum of the weighted, squared differences between observed and calculated intensities at every step in a digital powder pattern. The Rietveld method requires a knowledge of the approximate crystal structure of all phases of interest (not necessarily all phases present) in a mixture (Hill and Howard, 1987; Bish and Howard, 1988).

The input data to a refinement are similar to those required to calculate a diffraction pattern i.e., space group symmetry, atomic positions, site occupancies, and lattice parameters. In a typical refinement, individual scale factors (related to the weight percents of each phase) and profile, background, and lattice parameters are varied. In favorable cases, the atomic positions and site occupancies can also be successfully varied. The method consists of fitting the complete experimental diffraction pattern with calculated profiles and backgrounds, and obtaining quantitative phase information from the scale factors for each phase in a mixture.

The $K_{i\alpha}$ of Eq. (1-44) can be divided into two terms. The first is

$$k = \left[\left(\frac{I_0 \lambda^3}{32 \pi r} \right) \left(\frac{e^4}{2 m_e^2 c^4} \right) \right] \quad (1\text{-}56)$$

which depends only on the experimental conditions and is independent of angle and sample effects. The second part

$$T_{i\alpha} = \frac{M_i}{V_\alpha^2}|F_{i\alpha}|^2 \left(\frac{1 + \cos^2 2\theta_i \cos^2 2\theta_m}{\sin^2 \theta_i \cos \theta_i}\right) \quad (1\text{-}57)$$

depends on the crystal structure and the specific reflection in question. Equation (1-43), for a pure phase, can now be written in terms of Eqs. (1-56) and (1-57) as

$$I_{i\alpha} = \frac{k T_{i\alpha}}{\mu_\alpha} \quad (1\text{-}58)$$

In a mixture, the intensity of reflection i from phase α is obtained from Eq. (1-45) as

$$I_{i\alpha} = k T_{i\alpha} \frac{X_\alpha}{\varrho_\alpha (\mu/\varrho)_m} \quad (1\text{-}59)$$

Here all of the structure sensitive parameters, which the Rietveld procedure optimizes, are in the T_α term.

In a multiple-phase mixture, other terms must be introduced to scale the computed powder pattern for each phase, before summing them together to fit the pattern of the mixture. In addition, terms to aid in the pattern fitting will be needed. The quantity minimized in Rietveld refinements is the conventional least squares residual

$$R = \sum_j w_j |I_j(o) - I_j(c)|^2 \quad (1\text{-}60)$$

where $I_j(o)$ and $I_j(c)$ are the intensity observed and calculated, respectively, at the jth step in the data and w_j is the weight. Thus, as is the case in whole pattern analysis, it is more appropriate to consider the intensity at a given 2θ step rather than for a given reflection. In a single phase powder pattern of phase α, the intensity at each step j, is determined by summing the contributions from background and all neighboring Bragg reflections as

$$I_j(c) = S \sum_i T_{i\alpha} G(\Delta\theta_{j,i\alpha}) P_i + I_{jb}(c) \quad (1\text{-}61)$$

where S is the conventional Rietveld scale factor, which puts the computed intensities on the same scale as those observed, P_i is a sometimes used preferred orientation function for the ith Bragg reflection, $G(\Delta\theta_{j,i\alpha})$ is the profile shape function, and $I_{jb}(c)$ is the background (Wiles and Young, 1981). The Rietveld scale factor, S, includes all of the constant terms in Eq. (1-56) along with the absorption coefficient from Eq. (1-58)

$$S = \frac{k}{\mu} \quad (1\text{-}62)$$

where S is the constant required to scale the observed and calculated patterns, and μ is the linear absorption coefficient for the sample. For a multi-phase mixture, Eq. (1-62) can be rewritten summing over the n phases in a mixture (e.g., Hill and Howard, 1987) as

$$I_{ic} = I_{ib} + \sum_n S_n \sum_j T_{jn} G_{ijn} \quad (1\text{-}63)$$

The scale factor for each phase can now be written as

$$S_\alpha = k \frac{X_\alpha}{\varrho_\alpha (\mu/\varrho)_m} \quad (1\text{-}64)$$

Therefore, in a Rietveld analysis of a multi-component mixture, the scale factors contain the desired weight fraction information as

$$X_\alpha = \frac{(\mu/\varrho)_m}{k} S_\alpha \varrho_\alpha \quad (1\text{-}65)$$

However, the sample mass absorption coefficient is not known and we are forced to apply the usual internal-standard analysis, calibrating the RIR^s of the phases to be analyzed, or applying the normalization assumption, constraining the sum of the weight fractions of the phases considered to unity. In fact, Hill and Howard (1987) recently showed the Rietveld scale factor acts in the role of a reference intensity ratio permitting conventional RIR analysis.

In order to apply the normalized RIR approach in the Rietveld method, we can think of Eq. (1-52) as

$$X_\alpha = \frac{X_\alpha}{X_\alpha + X_\beta + \ldots} \qquad (1\text{-}66)$$

Equation (1-66) can be solved for the weight fractions of each of the diffracting phases by substituting Eq. (1-65) into it, giving

$$X_\alpha = \frac{S_\alpha \varrho_\alpha}{\sum_n S_n \varrho_n} \qquad (1\text{-}67)$$

We see that this procedure is exactly analogous to the normalization assumption in which reference intensity ratios are measured prior to analysis. Instead of measuring reference intensity ratios to put all intensities on an absolute scale, the Rietveld method calculates what we should call normalized RIR_N values, in that they refer to the relative scale between the observed and calculated patterns

$$RIR_{N,\alpha} = S_\alpha \varrho_\alpha \qquad (1\text{-}68)$$

The conventional internal-standard method of analysis, applied to Rietveld quantitative analysis, requires that a known weight fraction of a crystalline internal standard be added to the unknown mixture or that one of the components be of known concentration. Thus, if X_s is known, then the normalized RIR can be used directly, to determine the weight fractions for other phases in the sample. For example, the weight fraction for the α phase is determined by

$$X_\alpha = \frac{S_\alpha \varrho_\alpha}{\left(\dfrac{RIR_{N,s}}{X_s}\right)} \qquad (1\text{-}69)$$

S_α is a refined parameter, and ϱ_α can be calculated from the refined composition and cell parameters of each phase. Therefore, the weight fraction of phase α, (X_α) can be easily determined. This internal-standard method does not constrain the sum of the weight fractions, as does the normalized RIR method.

The total weight fraction of any amorphous components, can also be determined with this method, by adjusting the Rietveld background polynomial to fit the broad amorphous scattering profile. The difference between the sum of the weight fractions of the crystalline components and 1.0 is the total weight fraction of the amorphous components. O'Connor and Raven (1988) used this method to conclude that their quartz contained an 18% amorphous component.

1.11.2 The Absolute Reference Intensity Ratio: RIR_A

Equations (1-64) shows that the Rietveld scale factors for all phases, including any amorphous material, contain the constant term $(\mu/\varrho)_m$. If this term could be factored out and refined independently, we could put the RIR^s onto an absolute scale by referring all of them to the scale of the calculated pattern. The problem is that a multiplicative factor cannot be independently refined using least squares. Only two other approaches remain. The first is to compute the overall $(\mu/\varrho)_m$ at the end of each cycle of least squares and then use it in computing the weight fractions to initiate the next cycle. The other is to break between cycles, of a nearly converged system, and run the simplex algorithm to determine the value of $(\mu/\varrho)_m$. The latter method suffers from removing the variable from the statistical environment of the least squares and if any correlations develop between this and any other parameter the procedure becomes invalid.

The absolute reference intensity ratio (Snyder, 1991a) can be defined as

$$RIR_{A,\alpha} = \frac{(\mu/\varrho)_m}{k} S_\alpha \varrho_\alpha \quad (1\text{-}70)$$

1.11.3 Absorption-Diffraction Method

In the internal-standard method, Eq. (1-45) was written twice, once for the unknown and again for the internal standard. The ratio of these gave Eq. (1-46) in which the absorption coefficient of the unknown sample cancelled out. In the absorption-diffraction method, we again write Eq. (1-45) twice: once for line i of phase α in the unknown and again for line i of a pure sample of phase α. The ratio here gives

$$\frac{I_{i\alpha}}{I^0_{i\alpha}} = \frac{(\mu/\varrho)_\alpha}{(\mu/\varrho)_m} X_\alpha \quad (1\text{-}71)$$

where $I^0_{i\alpha}$ is the intensity of line i in pure phase α. Equation (1-71) is the basis of the absorption-diffraction method for quantitative analysis. There are several methods for implementing Eq. (1-71). A few cases will be discussed in some detail. Each of these cases is dependent only on the validity of Eq. (1-71), and since we have made no assumptions in the derivation, this equation is rigorous.

It is more common than one might imagine that quantitative X-ray diffraction phase analysis is carried out on samples whose chemical composition is known. When this is the case, the mass absorption coefficients of the pure phase, to be analyzed, and of the mixture can be computed from Eq. (1-13). Thus, if $I_{i\alpha}$ and $I^0_{i\alpha}$ are measured from the unknown and a pure sample of α respectively, the weight fraction of the analyte (i.e., X_α) may be computed from Eq. (1-71).

Leroux et al. (1953) were the first to propose that quantitative analysis may be performed on unknowns by measuring the density and determining the linear absorption coefficients of the unknowns experimentally. This may be done using absorption experiments on specimens of different thickness and employing the mass absorption law given in Eq. (1-10). The difficulty with this approach is that the measurement of μ is extremely error prone. The low precision of this measurement dramatically limits the accuracy of a quantitative analysis.

It happens now and then that the absorption coefficient of the mixture and the phase to be analyzed are the same. In this case Eq. (1-71) reduces to

$$\frac{I_{i\alpha}}{I^0_{i\alpha}} = X_\alpha \quad (1\text{-}72)$$

since $(\mu/\varrho)_\alpha$ is exactly the same as $(\mu/\varrho)_m$. Common examples of this case are the analysis of cristobalite in a quartz and amorphous SiO_2 matrix or the analysis of the cubic, tetragonal, and monoclinic forms of ZrO_2 in pure ZrO_2 bodies. In cases where we have mixtures of polymorphs, the absorption-diffraction method of analysis becomes very attractive.

Another special case is that of a lightly loaded filter where the sample approximates a monolayer and presents a special case of Eq. (1-71). As long as all the particles lie alongside each other and do not shade each other from the X-ray beam, then all matrix absorption effects are eliminated. There can, of course, be no preferential absorption. Each crystallite of each phase diffracts as if it were a pure phase. Thus, the absorption coefficient for each phase is the same as for the pure phase. This causes Eq. (1-72) in case 3 to become operable. A plot of $I_{i\alpha}/I^0_{i\alpha}$ vs. X_α will be linear and immediately allows for quantitative analysis. This procedure is valid up to the limit of concentration of particles where crowding invalidates the assumption that $(\mu/\varrho)_\alpha = (\mu/\varrho)_m$. This method is commonly used, for example, in the analysis of respirable silica in air-filtered samples collected on silver membranes.

The last example of the use of the absorption-diffraction method is that of a binary mixture. In this case, Eq. (1-45) becomes

$$I_{i\alpha} = \frac{K_{i\alpha} X_\alpha}{\varrho_\alpha [X_\alpha (\mu/\varrho)_\alpha - (\mu/\varrho)_\beta] + (\mu/\varrho)_\beta} \quad (1\text{-}73)$$

Equation (1-45) for the pure phase α may be written as

$$I_{i\alpha}^0 = \frac{K_{i\alpha}}{\varrho_\alpha (\mu/\varrho)_\alpha} \quad (1\text{-}74)$$

Dividing Eq. (1-73) by Eq. (1-74) gives

$$\frac{I_{i\alpha}}{I_{i\alpha}^0} = \frac{X_\alpha (\mu/\varrho)_\alpha}{X_\alpha (\mu/\varrho)_\alpha + X_\beta (\mu/\varrho)_\beta} \quad (1\text{-}75)$$

Equation (1-75) can easily be used to compute the standard plot of $I_{i\alpha}/I_{i\alpha}^0$ vs. X_α for all possible compositions of mixtures of α and β.

1.11.4 Method of Standard Additions or Spiking Method

Lennox (1957) was the first to develop the spiking method, which has been widely adopted in X-ray fluorescence spectroscopy as the method of standard additions. This method, like the internal-standard method, is perfectly general applying to any phase α in a mixture. The only requirement is that the mixture also contains, among other phases, a phase β, which has a diffraction line unoverlapped by any line from α, which may be used as a reference. Phase β is not analyzed and does not even need to be identified.

In a sample containing α and β, the ratio of the intensities from a line from each phase can be obtained from Eq. (1-45) as

$$\frac{I_{i\alpha}}{I_{j\beta}} = \frac{K_{i\alpha} \varrho_\beta X_\alpha}{K_{j\beta} \varrho_\alpha X_\beta} \quad (1\text{-}76)$$

Using the method of standard additions, some of the pure phase α is added to the mixture containing the unknown concentration of α. After adding Y_α grams of the α phase, per gram of the unknown, the ratio $I_{i\alpha}/I_{j\beta}$ becomes

$$\frac{I_{i\alpha}}{I_{j\beta}} = \frac{K_{i\alpha} \varrho_\beta (X_\alpha + Y_\alpha)}{K_{j\beta} \varrho_\alpha X_\beta} \quad (1\text{-}77)$$

where

- X_α = the initial weight fraction of phase α,
- X_β = the initial weight fraction of phase β,
- Y_α = the number of grams of pure phase α added per gram of the original sample.

In general, the intensity ratio is given by

$$\frac{I_{i\alpha}}{I_{j\beta}} = K(X_\alpha + Y_\alpha) \quad (1\text{-}78)$$

where K is the slope of a plot of $I_{i\alpha}/I_{j\beta}$ versus Y_α, with Y_α in units of grams of α per gram of sample. Multiple additions are made to prepare a plot in which the negative x intercept is X_α, the desired concentration of the phase α in the original sample.

1.11.4.1 Amorphous Phase Analysis

The area under the amorphous scattering hump, in the approximate range between 15° and 30° 2θ, is related to the concentration of the amorphous phase present, in exactly the same manner as the intensity of a diffraction peak is related to concentration. If an unobscured section of the amorphous scattering region is integrated, the intensity may be treated according to any of the above methods. The only consideration which must be emphasized, is the points at which background is measured. Usually only a small error will result if a low-angle background in the vicinity of 10° 2θ, and a high-angle background in the range of 40° 2θ, are chosen. This technique was used by Hubbard to estimate the amount of amorphous silica in an NBS standard reference sample of quartz.

1.12 Indexing and Lattice Parameter Determination

Many of the patterns in the powder diffraction file give the hkl associated with each of the d-values. In this section we will discuss how these values may be obtained from powder diffraction patterns. In fact, many of the PDF cards have had their hkl's computed from lattice parameters obtained from single crystal X-ray diffraction measurements. However, current computer methods can routinely determine the indices of lines in patterns of unknowns if, and only if, the patterns are measured very accurately and corrected for systematic errors.

Indexing is the procedure used in assigning hkl values to the set of planes which give rise to each diffraction line. There are various procedures for indexing but in all procedures more difficulty is encountered as the crystal symmetry is decreased, i.e., cubic crystals are easy while triclinic are the most difficult. The procedure which is followed is dependent upon the information that is known about the crystal, that is, whether the unit cell dimensions and/or the crystal system is known.

1.12.1 Accuracy and Indexing

If the cell dimensions are known, then the pattern can be indexed by calculation, regardless of the crystal system. This is a result of the fact that d_{hkl} values are a geometric function of the size and shape of the unit cell. The relationship between d_{hkl} and the real unit cell is cumbersome and usually stated in a different form for each crystal system. However, the functional relation between the square of the reciprocal lattice vectors ($d^*_{hkl} = 1/d_{hkl}$) and the size and shape of the reciprocal unit cell, previously derived as Eq. (1-20), applies easily to all crystal systems.

For any indexing procedure to work, the data must be very accurate. The three most common errors result from:

– Uncorrected instrumental and sample related errors (e.g., diffractometer specimen displacement error). These can cause $\Delta 2\theta$ errors greater than 0.1°.
– Debye-Scherrer data containing absorption errors, in the low angle reflections. This causes shifts in d values which follow a $\cos\theta$ relation.
– The presence of one or more extraneous lines due to a second phase or the presence of a $K_{\alpha 2}$ or K_β peak. A single incorrect line will cause most indexing methods to fail, in that they all attempt to index all lines.

These problems may be eliminated by using diffractometer or focusing Guinier camera data, and correcting the observed 2θ's using the internal standard method. It can not be emphasized too strongly that only diffraction patterns with 2θ values of the highest possible accuracy should be used in the modern indexing procedures. The numerical procedures are particularly vulnerable to experimental error.

1.12.2 Figures of Merit

A figure of merit is a mathematical function which rates the quality of a match between any two things. In indexing powder patterns, we need some type of objective criterion to rate the different matches. Since d_{hkl} spacings are related to the intuitive, measured, 2θ values through a nonlinear sin function, we cannot rely on the users qualitative estimate of the correctness of a proposed cell. A number of authors prefer to use Q values, for both indexing algorithms and for figures of merit, rather than d_{hkl}^{*2} because the values are much larger. Q is defined as

$$Q_{hkl} = 10\,000\, d_{hkl}^{*2} = \frac{10^4}{d_{hkl}} \tag{1-79}$$

There are two figures of merit in common use today. Most of the popular computer indexing and lattice parameter refinement procedures use either one or both of them. The M_{20} figure of merit, proposed by de Wolff (1969), is defined as

$$M_{20} = \frac{Q_{20}}{2|\Delta Q|N_{20}} \tag{1-80}$$

where $|\Delta Q|$ is the average discrepancy between Q_{obs} and Q_{calc}, Q_{20} is the Q value of the 20th line and N_{20} is the number of possible, space group allowed, lines out to the value of Q_{20}. This figure of merit works well in determining the correctness of an indexing but has the disadvantage of having a different absolute value for each different crystal system.

Smith and Snyder (1979) have proposed another figure of merit called F_N which is defined as

$$F_N = \frac{N}{|\overline{\Delta 2\theta}|\, N_{\text{poss}}} \tag{1-81}$$

where N is the number of observed lines used in the calculation, N_{poss} is the number of possible, space group allowed, lines out to the 2θ of line N, and $|\overline{\Delta 2\theta}|$ is the average error between the observed and calculated 2θ values. This figure of merit is useful both in evaluating the quality of powder patterns and in establishing the correctness of an indexing. The advantage for indexing is that its value is independent of the crystal system, so a common scale can be applied to all crystal systems. Another advantage is that the reciprocal of F_N establishes an upper limit on the average $\Delta 2\theta$, which has some intuitive meaning.

1.12.3 Indexing Patterns with Unknown Lattice Parameters

The indexing of a unit cell with unknown lattice parameters is a formidable problem. The problem becomes more difficult as we go to lower symmetry because the number of unknowns we must determine increases. The d_{hkl}^{*2} equation can be rewritten as

$$Q_{hkl} = h^2 A + k^2 B + l^2 C + kl D + {} \\ + hl E + hk F \tag{1-82}$$

The problem of indexing can then be stated as having to find the six unknowns A through E. For example, we previously saw that in the cubic crystal class the d_{hkl}^{*2} equation reduces to

$$d_{hkl}^{*2} = (h^2 + k^2 + l^2)\, a^{*2} \tag{1-83}$$

With only one unknown lattice parameter the problem of indexing a cubic powder pattern is always solvable. However, it still requires solving one equation for two unknowns: $(h^2 + k^2 + l^2)$ and a^{*2}. Therefore, the simple mathematical procedures for solving, for example, two equations for two unknowns do not apply, in that we have only one equation. To supply the extra information needed to avoid this difficulty, we must take an indirect approach and apply a relationship between the lattice parameter and d, which can not be written in the form of an explicit equation. Thus, the method will be an algorithm rather than a simple mathematical formula evaluation.

The algorithm for cubic is based on our being able to further reduce the d_{hkl}^{*2} equation even further to

$$d_{hkl}^{*2} = n^2 a^{*2} \tag{1-84}$$

where $n^2 = h^2 + k^2 + l^2$. Notice that when one d_{hkl}^{*2} value is subtracted from any another

$$\Delta d^{*2} = d^{*2} - d^{*2\prime} = (n^2 - n^{2\prime}) a^{*2} \tag{1-85}$$

the result is an integral multiple of a^{*2}. Thus, if two reflections differ in n^2 by only 1, the result of subtracting their d^{*2} values is a^{*2}. Once a^{*2} is known we can calculate all possible d_{hkl} values from the d_{hkl}^{*2} equation.

Once the value of a^{*2} has been established, we need to consider how to obtain the "best" value. This is the subject of lattice parameter refinement described in the next section.

There are three fundamental approaches to indexing lower symmetry patterns. Since the problem gets progressively more difficult numerically as we decrease symmetry all three approaches employ computers. The three approaches are (1) intelligent algorithms, (2) searches of solution space and (3) searches of index ($h\,k\,l$) space.

The intelligent procedures try to determine the lattice parameters by applying known relationships from both geometry and crystallography and even previous experience with indexing problems. They are "intelligent" only in the sense of computer artificial intelligence and that is the context in which the word will be used. Two quite successful methods of this type have been developed by Visser (1969) and Werner (1964).

Search methods do a systematic search of solution space (i.e., look at all possible solutions within some boundaries). There are two types of index searching procedures: those that work in parameter space and those that examine index or $h\,k\,l$ space. Some of the search methods make assumptions about the regions of solution space that should be examined while others are exhaustive.

We should pause here to discuss the concept of a solution space. This space for an exhaustive parameter search, is envisioned as having up to six dimensions in a plane (corresponding to functions relating to a, b, c, α, β and γ) and the F_N figure of merit plotted perpendicular to this plane. For cubic problems this means a two dimensional problem: vary all possible values of a along the x axis with F_N plotted vertically. As described above, there are an infinite number of unit cells that can describe a lattice. In fact, there are also a great number of incorrect cells that will almost describe the lattice (within reasonable error limits). Each of these solutions will correspond to a maximum of F_N in solution space. Our assumption is that the highest maximum corresponds to the correct unit cell.

An exhaustive indexing procedure working in parameter space examines all possibe values of a, b, c, α, β and γ within some reasonable limits, usually adjustable by the user. The lattice parameters are divided into small increments and each possible set is used in an attempt to index the unknown pattern. For the triclinic case this means using a six dimensional space with an extremely large number of attempted indexings. The usual limit on exhaustive parameter space procedures is to only look at possible cells with less than a few thousand cubic angstroms in volume. Louër and Louër (1972) have established an extremely successful program based on this approach.

Exhaustive procedures in index or $h\,k\,l$ space examine all possible combinations of h, k and l which can be used to index each of the lines in the unknown pattern. Here

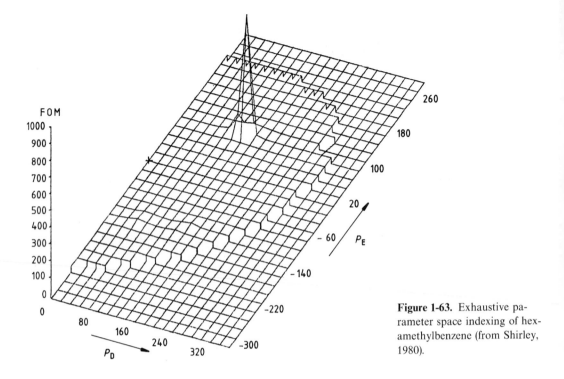

Figure 1-63. Exhaustive parameter space indexing of hexamethylbenzene (from Shirley, 1980).

our solution space is four dimensional for all symmetries: axis in integer units of h, k and l in a plane, F_N perpendicular to that plane. The advantage here is that hkl space is "naturally" quantized into integers, while in parameter space a reasonable quantum, or division factor, must be determined. If the spatial increment is too large the correct solution may be stepped over. A very powerful algorithm based on hkl space searching has been developed by Taupin (1973b).

Shirley (1980) has shown two extreme cases for an exhaustive parameter search. Figure 1-63 shows a very smooth solution space for the indexing of hexamethylbenzene. The cosine terms, D and E from Eq. (1-82), are in the plane with a sharpened figure of merit (FOM) vertical. Figure 1-64, on the other hand, shows the extremely lumpy or solution prone space for the indexing of Cu-phthalocyanine · HCl.

Before the days of powerful, fast PC's, exhaustive search procedures were only practical in relatively high symmetry systems and in a few exceptional cases (see Warren, 1969). For example the classic case of a manual orthorhombic indexing was performed by Jacob and Warren (1937) in the determination of the crystal structure of α-uranium. As the symmetry decreases to triclinic, the amount of computation required became impractical. However, today monoclinic and triclinic problems are routinely solved on researchers desk-top workstations. Usually the Niggli reduced cell, described in Sec. 1.6.5, is determined and then an analysis is performed to look for higher implied symmetry.

1.12.4 The Refinement of Lattice Parameters

Once a powder pattern is indexed we may routinely write from one to six d_{hkl}^{*2}

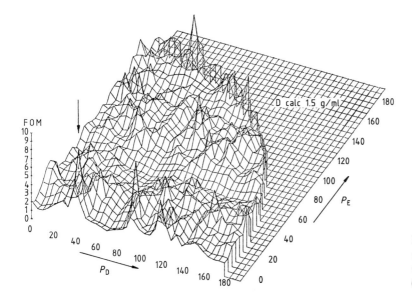

Figure 1-64. Exhaustive parameter space indexing of Cu-phthalocyanine · HCl (from Shirley, 1980).

equations with all terms known except the one to six lattice parameters. These equations may be solved simultaneously to give values for the lattice parameters, but these will not be the most precise values.

The precise and accurate measurement of lattice parameters, or unit cell dimensions, have been used for various purposes such as

- Determining thermal-expansion coefficients.
- Finding the true density of a material.
- Providing a direct measure of interatomic distances in the case of simple crystal structures.
- Studying interstitial and substitutional solid solution.
- Developing more satisfactory concepts of bonding energies.
- Developing phase-equilibrium diagrams.
- Determining stresses in materials such as steels.

In order to obtain accurate lattice parameters, the minimization and elimination of a number of errors is necessary. All sources of error, except the very accurately known wavelength, affect the accuracy of the lattice parameters through their effect on 2θ. Those errors that affect 2θ may be classified into two categories: (1) random errors and (2) systematic errors. Random errors are those made in locating and recording the center of the diffraction peak. These errors vary randomly with the angle 2θ. Random errors can be minimized by using back reflection ($2\theta > 90°$) lines, since precision in d-spacing or lattice constant is dependent upon $\sin\theta$ in the Bragg equation and not upon 2θ, the measured quantity. $\sin\theta$ changes very slow with 2θ near $180°$ and, therefore, a peak location error, $\Delta 2\theta$ will have a minimum effect upon the d-spacing. This is illustrated in Fig. 1-65. Systematic errors are those whose magnitude depends upon the position of the line, that is, upon the angle 2θ. In the cylindrical camera method, these systematic errors decrease as θ increases, vanishing completely when $\theta = 90°$. Therefore, in order to minimize these errors, high angle lines can be used. In the case of the diffractometer method, the sources of error cause the error in the line positions to vary in a compli-

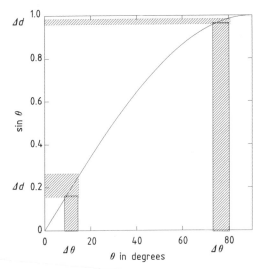

Figure 1-65. Variation of $\Delta \sin \theta$ with $\Delta\theta$.

cated manner with θ but must also minimize at high angle.

Systematic error is best eliminated, for any experimental techniques, by using an internal standard. With systematic error handled in this manner, the best approach for obtaining the most precise parameters from observed data, is to use the method of least squares, minimizing the differences between the observed d_{hkl} values and d_{hkl} values calculated from the d_{hkl}^{*2} equation, by varying the lattice parameters. This very powerful method will simultaneously produce estimates of the standard deviations of the lattice parameters. A general approach to this analysis has been incorporated into the powder pattern evaluation program NBS*LSQ by C. R. Hubbard.

1.13 Analytical Profile Fitting of X-Ray Powder Diffraction Patterns

The inherent asymmetry of many powder diffraction profiles has been a principal hindrance in extending the application of diffraction techniques beyond simple phase identification. The resurgence of developments in X-ray analytical procedures in recent years is primarily the result of the availability of computer automated powder diffractometers (Mallory and Snyder, 1979). The ability to collect digitized representations of the line profiles and apply numerical methods to their analysis, has led to a number of new and exciting applications (Snyder, 1991 b). The specimen shape contribution to the observed profile, when deconvolved from it, allows insights into the anisotropic strain and size of the crystallites as well as things like the ordering lengths of dopants or ions on lattice sites.

The most exciting of the profile fitting developments has been the continuing development of a whole pattern X-ray refinement method to obtain crystal structure information, begun by Rietveld (1969), who applied it to neutron diffraction. This method goes beyond the simple computation of intensities from Eq. (1-34) by distributing the calculated intensity over the kind of profile found using a particular instrument. The normally symmetric-Gaussian nature of neutron diffraction profiles has aided this application of this method to develop to its current, rather mature state. The two most common profile shape functions used in profile fitting are the Gaussian,

$$G(z) = I_0 \exp[-(\ln 2) z^2] \qquad (1\text{-}86)$$

and Lorentzian,

$$L(z) = \frac{I_0}{(1+z^2)^m} \qquad (1\text{-}87)$$

where

$$z = \frac{2\theta - 2\theta_0}{(\beta/2)}$$

$2\theta_0$ is the peak position, β is the full width at half maximum, I_0 is the peak intensity and $m = 1$. If the m term in the Lorentzian expression is made 1.5 (and z is multiplied by 0.5874), the profile shape function is called an intermediate Lorentzian. If m is made 2.0 (and z is multiplied by 0.4142), the profile function is called a modified Lorentzian. If the m term is used as an adjustable parameter and z is multiplied by $4(2^{1/m} - 1)$ then the function is called a Pearson VII. Figure 1-66 shows the relative shapes of these profile shape functions.

The application of the Rietveld method to X-ray patterns has been slower to develop, primarily because of the asymmetric and non-Gaussian nature of, and multiple spectral components in, most X-ray diffraction profiles (Snyder, 1992). These asymmetric profiles are not easily modeled and thus it is difficult to distribute the calculated intensity of the profile shape. Malmros and Thomas (1977) and Young et al. (1977) gave one of the first applications of this technique to X-ray data and the work of Wiles and Young (1981) marks the beginning of the much wider development of this method. However, the various successes fitting asymmetric profiles with symmetric functions, reported to date, have often been the result of defect disorder which tends to obliterate the inherent peak asymmetry and produce a more symmetric shape.

1.13.1 The Origin of the Profile Shape

There once was a time when a discussion of the shape of X-ray powder diffraction profiles could have been happily limited to the effects seen using a Bragg-Brentano geometry diffractometer with a sealed tube source. In fact, this has been done very well in the text by Klug and Alexander (1974). Today, a broader discussion is required to cover the common use of both pulsed and steady state neutron and of synchrotron and rotating anode X-ray sources, with incident and/or diffracted beam monochromators. Even the advances in X-ray detectors have an impact on this topic.

A diffraction line profile is the result of the *convolution* of a number of independent contributing shapes, some symmetric and some asymmetric. The process of convolution is one in which the product of two functions is integrated over all space. If can be represented as

$$h_{2\theta} = g_{2\theta} * f_{2\theta} = \int_{-\infty}^{+\infty} g_{2\theta'} f_{2\theta - 2\theta'} \, d(2\theta') \quad (1\text{-}88)$$

where $h_{2\theta}$ is the final observed profile and $g_{2\theta'}$ and $f_{2\theta}$ are shape functions contribut-

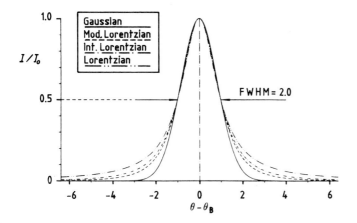

Figure 1-66. The Gaussian, modified-, intermediate- and Lorentzian profile shape functions.

ing to the resulting profile. Each point in the convolution is the result of summing the product of g and f over all possible values for $f_{2\theta}$. It is clear that, if this operation cannot be performed analytically, a lot of computer time will be used to evaluate it numerically. However, this may be reduced by the proper choice of functions, or by carrying it out in Fourier space, where convolution is mathematically more simple.

The components contributing to a diffraction profile can be divided into three categories.

1.13.1.1 Intrinsic Profile: S

The dynamical effects of an X-ray beam in a perfect crystal produce a reflection whose inherent width is called the Darwin width, after the author of the first dynamical treatment of diffraction. This inherent width is simply the result of the uncertainty principal ($\Delta p \Delta x = h$), in that the absorption coefficient of the specimen requires that the location of the photon in a crystal be restricted to a rather small volume. This means that Δp and in turn $\Delta \lambda$ ($\Delta p = h/\Delta \lambda$ by the de Broglie relation), must be finite, producing a finite width to a diffraction peak. The Darwin profile can usually be represented by a Lorentzian profile shape function (Parrish et al., 1976).

In addition to the inherent width, there are two physical sample effects which will broaden the profile shape function S, which the specimen contributes to the observed profile. The Scherrer crystallite size broadening is described by Eq. (1-21) and strain broadening of the specimen profile S, follows Eq. (1-22). Both of these specimen broadening effects generally effect the peak shape symmetrically.

1.13.1.2 Spectral Distribution: W

The most common X-ray source continues to be the sealed target X-ray tube. The inherent spectral profile, from the K transition in a Cu target X-ray tube, has a width of 1.18×10^{-4} Å and has been shown to be Lorentzian and not completely symmetric (see Frevel, 1987). The inherent width and asymmetry is usually overwhelmed by the fact that the various components of radiation in a polychromatic beam will each spread out as 2θ increases (i.e., with $\tan \theta$). This spectral dispersion is so great that this term dominates the convolution of diffraction lines at high angle, making them relatively symmetric and quite broad.

Monochromatization of an X-ray or neutron beam, using an incident beam monochromator, limits the breadth of the wavelength profile function W, to the Darwin width of the monochromator crystal and its intrinsic mosaicity. The use of both an incident and a diffracted beam monochromator in a parallel beam configuration, produces a broadening effect, in the diffraction profile, which depends on the distance, in 2θ, away from the point of optimum focus. (The point of optimum focus depends on the orientation of the two monochromators.) This distance in turn, depends on the crystal and Bragg reflection chosen for the monochromator and analyzer crystals (Cox et al., 1983, 1988). At the point of optimum focus, using perfect germanium crystals on a synchrotron source, the spectral profile is so narrow that the observed breadth seen is due primarily to the collimator divergence.

1.13.1.3 Instrumental Contributions: G

There are five possible profile contributions depending on the instrumental arrangement.

1. *The X-ray source image.* In a closed tube system this can be approximated with a symmetric Gaussian curve with a full width at half maximum (β) of 0.02° using a

take off angle of 3°. The effects of incident and diffracted beam monochromators are described by Cox et al. (1988) and do not introduce any asymmetry to the Gaussian shape. The long beam lengths on neutron and synchrotron ports allow for nearly perfect parallel optics. This, when coupled with both an incident and diffracted beam monochromator leaves the instrumental profile, at optimum focus, nearly a delta function (Cox et al., 1983). The use of focusing optics with an incident beam slit following the monochromator, while greatly increasing the intensity, introduces significant but symmetric broadening (Parrish et al., 1986).

2. *Flat specimen.* To maintain the Bragg-Brentano focusing condition the sample should be curved so that it follows the focusing circle. Since the focusing circle continuously changes radius with 2θ most experimental arrangements use a flat specimen, tangent to the focusing circle. This "out of focus" condition introduces a $\cot\theta$ dependence and produces a small asymmetry in the profile. It is particularly noticeable at low angle where the irradiated length of the sample is large. This term is not present on those neutron and synchrotron devices that use a cylindrical sample bathed in the beam.

3. *Axial divergence of the incident beam.* This also follows a $\cot\theta$ dependence and causes a substantial asymmetry in the profile particularly at low angles.

4. *Specimen transparency.* As the absorption coefficient of the sample decreases, the X-ray beam penetrates ever deeper, making the effective diffracting surface farther and farther off of the focusing circle. This produces a substantial asymmetric convolution term for low μ materials.

5. *Receiving slit.* For instruments using a receiving slit, another symmetric term contributes to the observed profile.

1.13.1.4 Observed Profile: *P*

Each of the terms described above conspire, via convolution, to produce a final diffraction profile that will range from the very asymmetric, in the case of sealed tube Bragg-Brentano systems, to quite symmetric, nearly Gaussian profiles, on neutron and some synchrotron instruments.

The observed diffraction profile, as stated by Jones (1938) and applied to powder diffraction systems by Taupin (1973a) and Parrish et al. (1976), is the result of the convolution of a specimen profile (*S*) and a combined function modeling the aberrations introduced by the diffractometer and wavelength dispersion. Taupin and Parrish grouped these terms together as $W*G$. The overall line profile can be expressed as

$$Profile = (W*G)*S + background \quad (1\text{-}89)$$

where the (*) represents the convolution operation. Since both *W* and *G* are fixed for a particular instrument/target system, ($W*G$) may be regarded as a single entity which we will refer to as the instrumental profile *P*. The specimen function *S* for a sample with no defect broadening has only the Darwin width, which can be approximated with a delta function (i.e., a line with height but no width). Using a delta function for *S* in Eq. (1-89) yields

$$Profile = P + background \quad (1\text{-}90)$$

Hence, for a pattern of an ideal sample, with background appropriately accounted for, the profiles are identical to the profiles of *P*. However, Parrish et al. (1976) and Howard and Snyder (1983) have indicated that the intensity ratio of components is affected by the setting of the monochromator, when one is present and hence, the *W* component of *P* must also be evaluated.

1.13.2 Modeling of Profiles

Khattak and Cox (1977) have shown fundamental problems in representing X-ray diffraction lines with either the Gaussian or simple Lorentzian functions. Of more recent interest are the Voigt (Langford, 1978; Cox et al., 1988), the pseudo-Voigt (see Young and Wiles, 1982) and the split-Pearson VII of Brown and Edmonds (1980). The Voigt function is the result of an analytical convolution of a Gaussian and a Lorentzian. It therefore ranges from pure Lorentzian to Gaussian type, depending on the ratio of both half widths. The function normalized to 1 has the form

$$V(x, \beta_l, \beta_g) = L(x/\beta_l) * G(x/\beta_g) \quad (1\text{-}91)$$

where L is the Lorentzian function with a full width at half maximum of β_l and G is the Gaussian function with half width of β_g and $x = \Delta 2\theta$. The Voigt-function can be calculated numerically using the complex error function.

The pseudo-Voigt conveniently allows the refinement of a mixing parameter determining the fraction of Lorentzian and Gaussian components needed to fit an observed profile. See Hastings et al. (1984) and Cox et al. (1988) for applications of this function. The pseudo-Voigt, although symmetric, allows for a flexible variation of the two most common profiles, ranging from the broad β Gaussian to the narrow β Lorentzian. However, Smith et al. (1987) have found an example of synchrotron profiles that are narrower than Lorentzian's and cannot be fit by the Voigt function.

The Pearson VII function also allows for the variation in shape from Lorentzian to Gaussian (Howard and Snyder, 1983) and is also symmetric. A split Pearson function, combined with a Gauss-Newton or a Marquardt least-squares optimization algorithm, gives excellent fits to asymmetric X-ray diffraction lines, obtained under a wide variety of conditions (Howard and Snyder, 1983, 1985, 1989). The approach here is to split each diffraction profile into a low angle and high angle part, by dividing at the profile maximum, and fitting a separate Pearson VII to each side. The Pearson VII function has the form

$$I = I_0/(1 + a x^2)^m \quad (1\text{-}92)$$

where $a = (2^{1/m} - 1)/(\beta/2)^2$ and m is the shape factor whose value determines the rate at which the tails fall and β is the full width at half maximum. The split-Pearson VII is illustrated in Fig. 1-67. The principal problem with this function is that its clumsy form prevents the analytical determination of convolutions, thus forcing extensive numerical approximations.

There are several demands on a profile shape function.

- The function must fit non-symmetric peaks.
- It should be mathematically as simple as possible, to make possible the calculation of all derivatives with respect to the variables.
- It must allow simple computation of the integral intensity.
- The convolution with a Lorentzian or a Gaussian function modeling S should be possible analytically.

Tomandl (Snyder, 1992), has proposed an asymmetric function which can be directly convoluted with the instrumental profile. Recently Hepp and Baerlocher (1988) have proposed a similar treatment for an asymmetry function in their "learned shape function" approach.

1.13.3 Description of Background

The description of the background in a powder diffraction pattern is critical to

1.13.4 Unconstrained Profile Fitting

The SHADOW algorithm, of Howard and Snyder (1985b), was developed as a general profile or pattern fitting program, which permits the use of a wide variety of shape functions and either a Gauss-Newton or Marquardt optimization algorithm. This was used in Fig. 1-68 to fit two peaks measured with an incident beam monochromator, to a simple unconstrained split Pearson function. The fit is quite good with the individual profiles looking like realistic diffraction lines. Figure 1-69 shows the same two peaks measured on a similar diffractometer but with different resolving power. Here we see that the low angle side of the split Pearson fitting the right peak has lifted in an unrealistic manner. In fact, there are a number of ways to place a group of profiles under an envelope, if the only criterion is to minimize the differences between the summed profiles and the observed envelope. All that any optimization

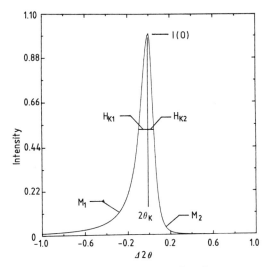

Figure 1-67. The Split-Pearson VII function uses two halves of the Pearson VII function with a common peak angle and intensity. The integral breadth β and the exponents m for each side are varied independently.

profile refinement because any background function must correlate strongly with the profile function. Two commonly used methods for describing background involve selecting points between peaks and interpolating between them (Rietveld, 1969), or refining the coefficients of a polynomial along with the profile parameters (Wiles and Young, 1981). While Sabine (1977) has described an analytical method for neutron powder diffraction by which the background scattering is analytically described and included in the refinement, there is no simple technique available for X-ray powder diffraction. In addition, no method proposed to date (Snyder, 1983a) will routinely produce an accurate description of X-ray background in complex patterns. The most serious problem in finding the true background occurs in low symmetry materials, where the powder pattern presents a continuum of peaks at high diffraction angle.

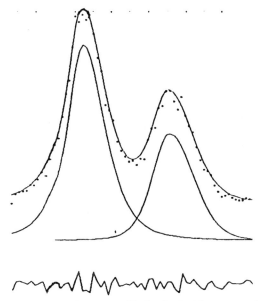

Figure 1-68. Correct profile fit of a doublet measured with only $K_{\alpha 1}$ radiation, using a split Pearson function.

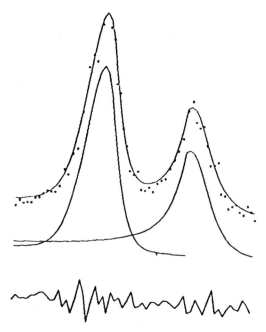

Figure 1-69. Incorrect profile fit of a doublet measured with only $K_{\alpha 1}$ radiation, using a split Pearson function.

procedure can do is to minimize this difference. Thus, it is clear that unconstrained profile fitting can always lead to failure to give a correct fit, unless we constrain the profiles used to be of the shape that are allowed for a particular experimental configuration. Any general procedure for profile fitting must then allow for the application of experimental constraints on the profile shape.

1.13.5 Establishing Profile Constraints: The P Calibration Curve

In order to establish constraints on the allowed shapes of the profiles from a particular instrument, Parrish and co-workers (1976) proposed a calibration procedure. The profiles of a standard, which shows no sample broadening, are fit with the split-Pearson VII function. Ideally, one would like a uniform particle size of about 5 μm, for an unstrained, high-symmetry material to use as a standard. In the Howard and Snyder approach, the 11 profiles of a standard sample of Si, observed with Cu K_α radiation are refined using the constrained split-Pearson function described above. Each profile is refined separately in a region with enough points to allow a precise description of the profile. The β values for each Pearson VII component, obtained from the refinement, are used to determine the value of the coefficients in the polynomial expression derived for neutron diffraction by Caglioti, Paoletti, Ricci (1958) as

$$(\beta_k)^2 \, 2\theta = U \tan^2(\theta_k) + V \tan(\theta_k) + W \qquad (1\text{-}93)$$

Two equations of this type were established by using a least-squares regression, analyzing for both the low angle and the high angle sides of the split-Pearson VII coefficients, from the fit to the standard's profiles. Similarly, a polynomial was used for the shape factors, m where

$$m = a'(2\theta_k)^2 + b'(2\theta_k) + c' \qquad (1\text{-}94)$$

Least squares regression of this function versus the two sets of shape factors completes the establishment of an analytical expression for evaluating P. Example polynomials for the high and low angle split Pearson parameters (m and β) from the calibration of a conventional diffractometer are illustrated in Fig. 1-70. This calibration procedure allows the evaluation of the instrumental P profile at any 2θ angle where a random specimen may have profiles that require fitting.

1.13.6 Modeling the Specimen Broadening Contribution to an X-Ray Profile

A powder diffraction pattern, for a material to be analyzed for size and strain, is determined. The observed peak positions

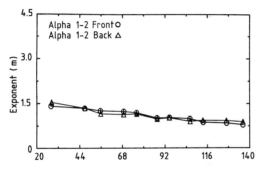

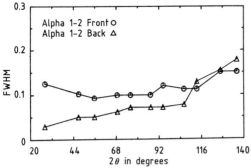

Figure 1-70. Variation of the split Pearson coefficients as a function of 2θ.

are used to generate P profiles from the instrumental function, which is expressed as the four polynomials in 2θ shown in Fig. 1-70. Lorentzian or Gaussian profile functions are generated to model the specimen broadening function S. It is generally accepted that S may be represented by a Lorentz function when profile broadening is caused by small crystallite size. When strain is responsible, S has often been assumed to be represented by a Gaussian function but sometimes a Lorentz profile shape is required.

In the absence of an analytical convolution function for the profile models, numerical techniques must be employed. The convoluted profile is obtained from

$$(P) * S_{(i)} = \sum_{j=-n}^{+n} (P)_{(i-j)} S_{(j)} \qquad (1\text{-}95)$$

The convolution gathers intensity contributions from all points in both P and S. The only approximation in the numerical convolution comes from the fact that the n limits are finite. The broader the full width of S, the greater the smearing of the profiles and loss of apparent resolution.

Since all non-specimen related broadening terms, like the $\tan \theta$ spectral broadening, have been accounted for in P, pure crystallite size and strain effects may be modeled by Eqs. (1-21) and (1-22). The slope of the curves β vs. $\lambda/\cos \theta_k$ and β vs. $4 \tan \theta_k$ give $1/\tau$ and ε respectively or, more simply, a Williamson-Hall plot (1953) of $B_k \cos \theta_k$ vs. $4\varepsilon \sin \theta_k$ gives a line of slope ε and y-intercept λ/τ. Program SHADOW uses these simple models to allow for the simultaneous determination and refinement of τ and/or ε. Figure 1-71 shows a measured region of two size broadened samples of Al_2O_3 along with their generated complete profiles and deconvoluted S profiles.

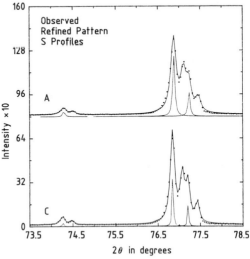

Figure 1-71. Both Linde A and C aluminas show significant line broadening. The inner curves are the deconvoluted S functions. The integral breadth of these S values show that the X-ray crystallite size of linde A is about half that of Linde C.

1.13.7 Fourier Methods for Size and Strain Analysis

We have explained both the origin and a method for analyzing size and strain in terms of the intuitive observed profiles. However, these easily understood procedures are relatively new. Before the days of inexpensive, powerful computers, the only way to carry out such an analysis was to simplify the computation by carrying it out in Fourier space. The traditional way of analyzing crystallite size and strain effects, was developed by Warren and Averbach (1969). In this method, the Fourier transformation of the diffraction peaks of a broadened sample and a standard material exhibiting no sample broadening, are computed. The deconvolution procedure is carried out in Fourier space. Since all intuition is lost on performing a Fourier transformation, this process can only be followed mathematically. The procedure presented above has the advantage of being understandable in terms of easily pictured profile shapes. If either procedure is carried out correctly the same information about size and strain will result.

1.13.8 Rietveld Analysis

With the P and S functions properly modeled for X-ray diffraction profiles, this method can now be applied to the pattern-fitting-structure-refinement procedure of Rietveld. The formalism developed above allows for the discarding of a number of empirical parameters in the Rietveld method, replacing them with τ and ε. The removal of all empirical parameters from the refinement, leaves in their place only parameters associated with known physical effects. All empirical parameters associated with the peak asymmetry are incorporated into the P instrumental function and are fixed during refinement.

When both strain and crystallite size effects are present, the integral breadth of the S profile is assumed to be a linear addition of the two components. That is, $\beta = \beta_\varepsilon + \beta_\tau$. The basis for this assumption is that the convolution of two Lorentzian functions yields another Lorentzian function. Thus,

$$L(\text{strain}) * L(\text{size}) * P = L(\text{combined}) * P \qquad (1\text{-}96)$$

The value for the full width of the new convoluted profile is simply a sum of the component β's. It is assumed that the order of convolution does not matter.

The X-ray version of the Rietveld procedure, as developed by Young, Mackie and von Dreele (1977) and later by Wiles and Young (1981), was modified by Howard and Snyder (1985) to incorporate the P deconvolution and modeling of S described above. The introduction of an analytical expression describing S broadening simplifies profile refinement. This constrained broadening reduces the number of parameters undergoing refinement while characterizing and quantifying the source of the broadening. However, the cost is a considerable increase in the execution time of the program. To eliminate the need for numerical convolution an analytically convolutable function is required like the Tomandl function.

1.14 Crystal Structure Analysis

All of the developments in hardware, software and application methods described in this article have produced a number of recent results which have a direct impact on the advancement of materials science. To conclude this article, we will survey some of the most exciting applications and how they have converged to produce exciting new results in the area of crystal structure analysis. These examples will also serve as illustrations of how all of the principles and techniques previously described can be applied to materials systems.

For example, the X-ray structure determination of the modulated structure of the $Bi_2Sr_2CaCu_2O_8$ superconductor was recently achieved by taking advantage of anomalous dispersion and obtaining diffraction patterns on each side of the absorption edge (Gao et al., 1989). Atomic resolution electron microscopy has made this structure dramatically clear. Figure 1-72 from Eibl (1990) dramatically shows the modulated wave misaligning the unit cells with respect to each other. If the wavelength of this modulation were a multiple of one of the lattice parameters, a superlattice would result. However, in this case the wavelength is not such a multiple so the modulation is said to be incommensurate.

Beyond the scanning transmission electron microscope, a wealth of new atomic resolution probes have developed in the last five years like, the scanning tunneling microscope which has spawned the atomic force microscope, friction force microscope, magnetic force microscope, electrostatic force microscope, chemical bonding force microscope, scanning thermal microscope, optical absorption microscope, scanning ion-conductance microscope, scanning acoustic microscope.... All have become tools of modern applied crystallography.

1.14.1 Structure of $YBa_2Cu_3O_{7-\delta}$

The crystal structure of the first superconducting material to break the liquid nitrogen barrier has probably been determined by more independent studies than any other. In our laboratory we isolated the diffraction peaks common to those

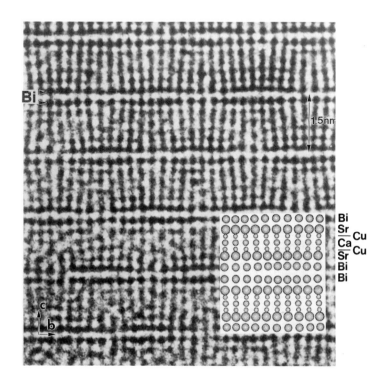

Figure 1-72. Lattice image of the modulated structure of the high temperature superconducting phase $Bi_2Sr_2CaCu_2O_8$ (from Eibl, 1990).

specimens (out of 40 compositions) which showed superconductivity. Microprobe analysis showed that the cations were in the approximate ratio of 1:2:3 for Y:Ba:Cu. Calibration of the XRD lines with a Si internal standard allowed each of the modern indexing programs, described in Sec. 1.12 to propose an orthorhombic unit cell of approximately $4 \times 4 \times 12$ Å^3. Use of the NBS*LATTICE program (Mighell and Himes, 1986), which computes super and sub cells related to the orthorhombic cell, and searches the *Crystal Data* database for matches, produced suggested matches for BaTiO$_3$ and a number of other cubic perovskites with unit cells near $4 \times 4 \times 4$ Å^3. In addition, it proposed a number of double celled perovskites and even one triple celled material. On placing the Y, Ba and Cu atoms into a triple celled perovskite structure, a Rietveld refinement led to convergence. This was far from the first structure analysis of this material but it occurred before the published reports of the structure arrived on our campus!

1.14.2 The Structures of the γ and η Transition Aluminas

As an example of the kind of analysis permitted by the new diffraction tools described in this article and currently available in the materials laboratory, we will conclude this article by describing a classical structural problem in an extremely important ceramic system (Zhou and Snyder, 1991). The transition aluminas refer to the group of partially dehydrated aluminium hydroxides, other than the anhydrous α-alumina (corundum). The diffuse character of their powder patterns reflects a high degree of structural disorder, but the similarity of the patterns indicates certain structural aspects that pervade all these phases.

The most disordered of these materials, γ and η, are catalytically active, and of high economic value, and have been studied for many years, but have resisted structure analysis.

1.14.2.1 Profile Analysis

The evolution of the profile fitting and Rietveld refinement procedures mentioned above allows new approaches to these previously intractable patterns. The deconvolution of the specimen broadening profile S from each of the profiles in the γ and η patterns immediately showed an unusual phenomenon (Zhou and Snyder, 1991). The patterns exhibit anomalous reflection broadening in two aspects: (1) irregular broadening among peaks of different hkl's [even from the same crystal zone – see the (111), (222), and (333) reflections in Fig. 1-73 and (2) the formation of an odd peak shape with broadened base and sharp top that can not be fit well by a single profile function. The differences between the γ and η patterns are in the degree of these anomalies.

Attributing the anomalous broadening to be indicative of the "crystallite size" or ordering lengths of the various sublattices that dominate the scattering for the different peaks, the following structural information can be extracted from profile fitting and pattern calculations, based on a spinel structure.

– The oxygen sublattice in both η and γ-alumina is fairly well ordered because the sharp (222) reflection is dominated by the scattering from the oxygen sublattice. The ordering domain of the oxygen sublattice estimated from the β value of the specimen convolute S, of the (222) reflection is 9.8(2) nm for η-alumina and 20(1) nm for γ-alumina. These numbers are comparable with the average crystal-

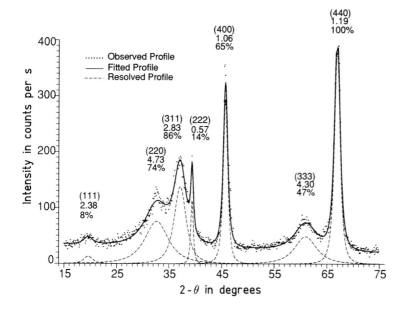

Figure 1-73. Profile fit of XRD pattern of γ-alumina. On top of peaks are hkl, β and relative intensity.

lite sizes of the precursor bayerite and boehmite. In fact, the η value agrees well with the average long dimension of TEM observed lamellae in the dehydroxylated tablets. Thus, it is the framework of close packed oxygens which supports the original crystallite shape and accounts for the pseudomorphosis of dehydroxylation.

- Both η and γ-alumina have octahedral coordinated Al as well as tetrahedral coordinated Al ions, since the (220) reflection is only due to scattering from the tetrahedral Al sublattice.
- The tetrahedral Al sublattices in both η and γ-alumina are very disordered because the (220) reflection is the most diffuse in the pattern. The ordering domain of the tetrahedral Al sublattice estimated from the β value of S, for the (220) reflection is 1.6(2) nm for η-alumina and 1.9(2) nm for γ-alumina. These dimensions also correspond to the TEM observed striations on the pseudomorphic tablets which result on heating.

The disparity in the size of the ordering domains of the different sublattices within a single spinel lattice, manifests itself in the formation of odd-shaped peaks (i.e., sharp top with broadened base) routinely observed on the XRD powder patterns of the transition aluminas. This disparity can be understood by considering the fact that the starting hydroxides have only octahedral Al ions and quasi cubic close-packed oxygen ions. Since the tetrahedral Al ions are the newcomers, from the dehydroxylation, they most reflect the diffusion induced disorder.

1.14.2.2 Rietveld Analysis

In the Rietveld refinement of powder diffraction patterns, the profiles are assumed to vary isotropically with diffraction angle. When this assumption is very poor the "two-step" method is used where the integrated intensities of the individual S profiles are treated as single crystal intensities and the traditional single crystal structure solving and refinement procedures can be employed. By contrast, this notation would refer to the Rietveld as the "one step" method. However, in the case of the transition aluminas, the extra statistical precision which comes from using all of the digital points in a pattern is essential due to the small number of reflections. The compromise here was to use a "one and one half step" procedure: namely, the anomalous β's from profile fitting were specified for each peak in the pattern and not refined during the Rietveld refinement. The Rietveld procedure was applied to neutron patterns collected on deuterated samples of γ and η-alumina, as well as on X-ray patterns. Structures were refined using traditional Fourier techniques by starting with the well determined oxygen atoms and iteratively refining the model and computing difference Fourier patterns to locate the other atoms in the structure.

The refinement results show that Al ions in η-alumina occupy only one octahedral site (16d), in $Fd3m$, one tetrahedral site (48f), and one *quasi-trihedral* site (32e). The quasi-trihedral Al ions are slightly displaced 0.0162 nm away from the oxygen plane toward the second octahedral site (16c) 0.0958 nm away from site (16c). In γ-alumina however, Al ions occupy only one octahedral site (16d), one tetrahedral site (8a), and one *quasi-octahedral* site (32e). The quasi-octahedral Al ions are 0.0373 nm away from the octahedral site (16c) and 0.0748 nm away from the nearest oxygen plane. The extraordinary presence of nearly pure three coordinated Al ions in η-alumina and of the quasi-octahedral Al ions in γ-alumina can be accounted for by considering them to exist in the surface of the crystallites. The size of these peaks as found in the difference Fourier map indicated a concentration of 13%. Using the crystallite size as determined in this study and confirmed by transmission electron microscopy, a simple computation predicts that 13% of the Al ions must lie in the surface of such small crystallites. These surface Al ions are clearly responsible for the strong Lewis-acidity (electron-acceptor property) observed on the surfaces of η and γ-alumina and must result in the catalytic activity of these phases. Thus the recent advances in applied crystallography, presented in this article, have converged in this case to produce the first structural model for catalytic activity in the most widely used industrial catalyst system.

The ability to determine the ordering lengths associated with the various sublattices in the many important disordered materials systems, offers the exciting prospect of finally understanding the structures of, and structure-property relationships in, these materials.

1.15 References

Alexander, L., Klug, H. P. (1948), *Anal. Chem. 20*, 886–894.
Apai, G., Hamilton, J. F., Stohr, J., Thompson, A. (1979), *Phys. Rev. Lett. 43*, 165–169.
Bayya, S. S., Kudesia, R., Snyder, R. L. (1991), *Superconductivity and Its Application,* Vol. 3: Kao, Y. H., Coppens, P., Kwok, H. S. (Eds.). Washington: Am. Phys. Soc., pp. 306–314.
Bearden, J. A. (1967), *Rev. Mod. Phys. 39*, 78.
Bellier, S. P., Doherty, R. D. (1977), *Acta Metallurgica 25*, 521.
Bish, D. L., Howard, S. A. (1988), *J. Appl. Cryst. 21*, 86–91.
Bragg, W. L. (1913), *Proc. Cambridge Phil. Soc. 17*, 43.
Bragg, W. H., Bragg, W. L. (1913), *Proc. Roy. Soc. London 88 A*, 428.
Brindley, G. W. (1945), *Phil. Mag. 36 (7)*, 347.
Brown, A., Edmonds, J. W. (1980), *Adv. in X-ray Anal. 23*, 361.
Buerger, M. (1964), *The Precession Method.* New York: Wiley.
Bunge, H.-J. (1982), *Texture Analysis in Materials Science.* London: Butterworths.
Caglioti, G., Paoletti, A., Ricci, F. P. (1958), *Nuclear Instruments and Methods 3*, 223–226.
Cahn, R. W. (1991), *Materials Science and Technology,* Vol. 15: Cahn, R. W., Haasen, P., Kramer, E. J. (Eds.). Weinheim: VCH, pp. 429–480.

Cava, R. J., Hewat, A. W., Hewat, E. A., Batlogg, B., Marezio, M., Rabe, K. M., Krajewski, J. J., Peck, W. F., Rupp, L. W. (1990), *Physica C 165*, 419–433.
Cherukuri, S., Snyder, R. L., Beard, D. (1983), *Adv. X-ray Anal. 26*, 99–105.
Chung, F. H. (1974a), *J. Appl. Cryst. 7*, 519–525.
Chung, F. H. (1974b), *J. Appl. Cryst. 7*, 526–531.
Chung, F. H. (1975), *J. Appl. Cryst. 8*, 17–19.
Clark, G. L., Reynolds, D. H. (1936), *Ind. Eng. Chem., Anal. Ed. 8*, 36–42.
Cline, J. P., Snyder, R. L. (1983), *Adv. in X-ray Anal. 26*, 111–118.
Cline, J. P., Snyder, R. L. (1985), *Advances in Material Characterization II:* Snyder, R. L., Condrate, R. A., Johnson, P. F. (Eds.). New York: Plenum Press, pp. 131–144.
Cline, J. P., Snyder, R. L. (1987), *Adv. in X-ray Anal. 30*, 447–456.
Coolidge, W. D. (1913), *Phys. Rev. 2*, 409.
Copeland, L. E., Bragg, R. H. (1958), *Anal. Chem. 30*, 196–206.
Cox, D. E., Hastings, J. B., Thomlinson, W., Prewitt, C. T. (1983), *Nucl. Instrum. Methods 208*, 273–278.
Cox, D. E., Hastings, J. B., Cardoso, L. P., Finger, L. W. (1986), *Mat. Sci. Forum 9*, 1.
Cox, D. E., Toby, B. H., Eddy, M. M. (1988), *Aust. J. Phys. 41*, 117–131.
Cromer, D. T., Waber, J. T. (1965), *Acta Cryst. 18*, 104.
Darwin, C. G. (1914), *Phil. Mag. 27*, 315.
Darwin, C. G. (1922), *Phil. Mag. 43*, 800.
Debye, P. (1913), *Ann. der Physik 43*, 49.
Debye, P. (1915), *Ann. der Physik 46*, 809.
Deslattes, R. D., Henins, A. (1973), *Phys. Rev. Lett. 31*, 972.
Delhez, R., de Keijser, Th. H., Mittemeijer, E. J., Langford, J. I. (1988), *Aust. J. Phys. 41*, 213–227.
de Wolff, P. M. (1968), *J. Appl. Cryst. 1*, 108–113.
Dragoo, A. L. (1986), *Powder Diffraction 1*, 294.
Eibl, O. (1990), *Physica C 168*, 215–238.
Ewald, P. P. (1921), *Z. Kristallogr. 56*, 129.
Frevel, L. K. (1987), *Powder Diffraction 2 (4)*, 237–241.
Friedrich, W., Knipping, P., M. v. Laue (1913), *Ann. der Physik 41*, 971.
Gao, Y., Sheu, H. S., Petricek, V., Restori, R., Coppens, P., Darovskikh, A., Phillips, J. C., Sleight, A. W., Subramanian, M. A. (1989), *Science 244*, 62.
Gaskell, P. H. (1991), *Materials Science and Technology*, Vol. 9: Cahn, R. W., Haasen, P., Kramer, E. J. (Eds.). Weinheim: VCH, pp. 175–278.
Gehringer, R. C., McCarthy, G. J., Garvey, R. G., Smith, D. K. (1983), *Adv. in X-ray Anal. 26*, 119–128.
Geiger, H., Müller W. (1928), *Phys. Z. 29*, 839.
Göbel, H. E. (1982a), *Adv. in X-ray Anal. 24*, 123–138.
Göbel, H. E. (1982b), *Adv. in X-ray Anal. 24*, 315–324.
Göbel, H. E. (1990), *Analytical Techiques for Semiconductor Materials and Characterization*, Vol. 90-11: Kolbesen, B. O., McCaughan, D. V., Vandervorst, W. (Eds.). Pennington, NJ: The Electrochemical Society Inc. pp. 238–260.
Hargittai, I. (1990), *Quasicrystals, Networks, and Molecules of Fivefold Symmetry*. New York: VCH.
Hastings, J. B., Thomlinson, W., Cox, D. E. (1984), *J. Appl. Cryst. 17*, 85–95.
Hart, M. Parrish, W. (1989), *Mater. Res. Soc. Symp. Proc. 143*, 185–195.
Hepp, A., Baerlocher, Ch. (1988), *Aust. J. Phys. 41*, 229–236.
Hermann, C. (1931), *Z. Kristallogr. 76*, 559.
Hill, R. J., Howard, C. J. (1987), *J. Appl. Cryst. 20*, 467–474.
Howard, S. A., Snyder, R. L. (1983), *Adv. in X-ray Anal. 26*, 73–81.
Howard, S. A., Snyder, R. L. (1985), *Advances in Material Characterization II:* Snyder, R. L., Condrate, R. A., Johnson, P. F. (Eds.). New York: Plenum Press, p. 43–56.
Howard, S. A., Snyder, R. L. (1989), *J. Appl. Cryst. 22*, 238–243.
Huang, T. C., Cohen, P. C., Eaglesham, D. J. (Eds.) (1991), *Advances in Surface and Thin Film Diffraction*. Warrendale: Materials Research Society.
Hubbard, C. R., Snyder, R. L. (1988), *J. Powder Diffraction 3*, 74.
Hubbard, C. R., Evans, E. H., Smith, D. K. (1976), *J. Appl. Cryst. 9*, 169–174.
Hubbard, C. R., Robbins, C. R., Snyder, R. L. (1983), *Adv. X-ray Anal. 26*, 149–157.
Hubbard, C. R., Robbins, C. R., Wong-ng, W. (1987), *Standard Reference Material Silicon 640b*. Available from National Institute of Science and Technology Office of Standard Reference Materials, Gaithersburg, MD 20899, USA.
Inokuti, Y., Maeda, C., Ito, Y. (1987), *Trans. Iron and Steel Inst. of Japan 27*, 139, 302.
International Centre for Diffraction Data – JCPDS, 1601 Park Lane, Swarthmore, PA 19081.
International Tables for Crystallography (1983), Vol. A: *Space-Group Symmetry:* Hahn, T. (Ed.). Dordrecht: D. Reidel.
Jacob, C. W., Warren, B. E. (1937), *J. Chem. Soc. 59*, 2588.
Johnson, G. G., Vand, V. (1967), *Ind. Eng. Chem. 59*, 19–31.
Jones, F. W. (1938), *Proc. R. Soc. London Ser. A 166*, 16–43.
Karlak, F., Burnett, D. S. (1966), *Anal. Chem. 38*, 1741–1745.
Khattak, C. P., Cox, D. E. (1977), *J. Appl. Cryst. 10*, 405–411.
Kinney, J. H., Stock, S. R., Nichols, M. C., Bonse, U., Breunig, T. M., Saroyan, R. A., Nusshardt, R., Johnson, Q. C., Busch, F., Antolovich, S. D. (1990), *J. Mat. Res. 5 [5]*, 1123–1129.
Klug, H. P., Alexander, L. E. (1974), *X-ray Diffraction Procedures*, 2nd ed. New York: John Wiley and Sons.

Kossel, W. (1936), *Ann. der Physik* 27, 694.
Ladell, J., Zagofsky, A., Pearlman, S. (1970), *J. Appl. Cryst.* 8, 499–506.
Langford, J. I. (1978), *J. Appl. Cryst.* 11, 10–14.
Langford, J. I., Delhez, R., de Keijser, Th. H., Mittemeijer, E. J. (1988), *Aust. J. Phys.* 41, 173–187.
Lawton, S. L., Jacobson, R. A. (1965), *The Reduced Cell and Its Crystallographic Applications*, U.S.A.E.C., *Ames Laboratory Report IS-1141*. Ames, IA: Iowa State University.
LeGalley, D. P. (1935), *Rev. Sci. Instr.* 6, 279.
Lengeler, B. (1990a), *Adv. Mater.* 2, 123–131.
Lengeler, B. (1990b), *Photoemission and Absorption Spectroscopy:* Campana, M., Rosci, R. (Eds.). New York: North Holland, p. 157–202.
Lennox, D. H. (1957), *Anal. Chem.* 29, 767–772.
Leroux, J., Lennox, D. H., Kay, K. (1953), *Anal. Chem.* 25, 740–748.
Lim, G., Parrish, W., Ortiz, C., Bellotto, M., Hart, M. (1987), *J. Mater. Res.* 2, 471.
Louër, D. Louër, M. (1972), *J. Appl. Cryst.* 5, 271–275.
Mallory, C. L., Snyder, R. L. (1979), *Adv. X-Ray Anal.* 22, 121–132.
Mallory, C. L., Snyder, R. L. (1980), in: *Accuracy in Powder Diffraction, National Bureau of Standards Special Publication 567*. Gaithersburg: Nat. Bur. Stand., p. 93.
Malmros, G., Thomas, J. O. (1977), *J. Appl. Cryst.* 10, 7–11.
Mauguin, C. (1931), *Z. Kristallogr.* 76, 542.
Mighell, A. D. (1976), *J. Appl. Cryst.* 9, 491–498.
Mighell, A. D., Himes, V. L. (1986), *Acta Cryst.* A42, 101–105.
Navias, A. L. (1925), *J. Amer. Ceram. Soc.* 8, 296–302.
Newsam, J. M., Yang, C. Z., King, H. E. J., Jones, R. H., Xie, D. (1991), *J. Phys. Chem. Solids*, in press.
O'Connor, B. H., Raven, M. D. (1988), *J. Powder Diffraction* 3, 2–6.
Parrish, W., Gordo, S. (1945), *Am. Minerol.* 30 (5-6), 326.
Parrish, W., Huang, T. C., Ayers, G. L. (1976), *Am. Cryst. Assoc. Monograph* 12, 55–73.
Parrish, W., Hart, M., Huang, T. C. (1986), *J. Appl. Cryst.* 20, 79–83.
Patterson, A. L. (1935), *Z. Kristallogr.* A90, 517.
Rietveld, H. M. (1969), *J. Appl. Cryst.* 2, 65–71.
Rodriguez, M. A., Matheis, D. P., Bayya, S. S., Simmins, J. J., Snyder, R. L., Cox, D. E. (1990), *J. Mater. Res.* 5 (9), 1799–1801.
Sabine, T. M. (1977), *J. Appl. Cryst.*, 10, 277–280.
Savitzky, A., Golay, M. (1964), *Anal. Chem.* 36 (8), 1627–1639.
Scherrer, P. (1918), *Nachr. Ges. Wiss. Göttingen*, 98–100.
Schönflies, A. (1891), *Krystallsysteme und Krystallstruktur*. Leipzig.
Shechtman, D., Blech, I., Gratias, D., Cahn, J. W. (1984), *Phys. Rev. Let.* 53, 1951.
Shirley, R. (1980), in: *Accuracy in Powder Diffraction, National Bureau of Standards Special Publication 567*. Gaithersburg: Nat. Bur. Stand., pp. 361–382.
Smith, D. K., Nichols, M. C., Zolensky, M. E. (1982), *POWD10. A FORTRAN IV program for calculating X-ray powder diffraction patterns – version 10*. University Park, Pa: The Pennsylvania State University.
Smith, D. K., Johnson, Jr., G. G., Scheible, A., Wims, A. M., Johnson, J. L., Ullmann, G. (1987), *Powder Diffraction* 2, 73–77.
Smith, D. K., Johnson, Jr., G. G., Wims, A. M. (1988), *Aust. J. Phys.* 41, 311–321.
Smith, G. S., Snyder, R. L. (1979), *J. Appl. Cryst.* 12, 60–65.
Smith, G. S., Johnson, Q. C., Cox, D. E., Snyder, R. L., Smith, D. K., Zalkin, A. (1987), *Adv. in X-Ray Anal.* 30, 383–388.
Smith, G. S., Johnson, Q. C., Smith, D. K., Cox, D. E., Snyder, R. L., Zhou, R. S., Zalkin, A. (1988), *Solid State Commun.* 67, 491–494.
Smith, S. T., Snyder, R. L., Brownell, W. E. (1979a), *Adv. X-ray Anal.* 22, 77–88.
Smith, S. T., Snyder, R. L., Brownell, W. E. (1979b), *Adv. X-ray Anal.* 22, 181–191.
Snyder, R. L. (1981), *Adv. X-ray Anal.* 24, 83–90.
Snyder, R. L. (1983a), *Advances in Material Characterization*, Rossington, D. R., Condrate, R. A., Snyder, R. L. (Eds.). New York: Plenum Press, pp. 449–464.
Snyder, R. L. (1983b), *Adv. X-ray Anal.* 26, 1–11.
Snyder, R. L. (1991a), *The Use of Reference Intensity Ratios in Quantitative Analysis, X-ray and Neutron Structure Analysis in Materials Science II*, in press.
Snyder, R. L. (1991b), *Applied Crystallography in Advanced Ceramics, EPDIC1 Proceedings of the First European Powder Diffraction Conference:* Delhez, R., Mittermeijer, E. J. (Eds.), Zürich: Trans Tech Publications.
Snyder, R. L. (1992), *Profile Analysis*, in: *Rietveld Analysis:* Young, R. A. (Ed.), in press.
Snyder, R. L., Bish, D. (1989), *Modern Powder Diffraction: Reviews in Mineralogy*, Vol. 20: Bish, D. L., Post, J. E. (Eds.). Washington, DC: Mineralogical Society of America, pp. 101–145.
Snyder, R. L., Carr, W. L. (1974), *Interfaces of Glass and Ceramics:* Frechette, V. D. (Ed.). New York: Plenum, pp. 85–99.
Snyder, R. L., Hubbard, C. R. (1984), *NBS*QUANT84: A System for Quantitative Analysis by Automated X-ray Powder Diffraction*, NBS Special Publication.
Snyder, R. L., Hubbard, C. R., Panagiotopoulos, N. C. (1981a), *AUTO: A Real Time Diffractometer Control System*. Gaithersburg: U.S. National Bureau of Standards, NBSIR 81-2229.
Snyder, R. L., Hubbard, C. R., Panagiotopoulos, N. C. (1981b), *Adv. X-ray Anal.* 25, 245–260.
Snyder, R. L., Rodriguez, M. A., Chen, B. J., Göbel, H. E., Zorn, G., Seebacher, F. B. (1992), *Adv. X-ray Anal.*, in press.

Steurer, W. (1990), *Z. Kristallogr. 190,* 179.
Tanner, B. K. (1976), *X-ray Diffraction Topography*. Oxford: Pergamon Press.
Taupin, D. (1973a), *J. Appl. Cryst. 6,* 266–273.
Taupin, D. (1973b), *J. Appl. Cryst. 6,* 380–385.
Teo, B. K. (1986), *EXAFS: Basic Principles and Data Analysis*. Heidelberg: Springer-Verlag.
Teo, B. K., Lee, P. A., Simons, A. L. Eisenberger, P., Kineaid, B. M. (1977), *J. Am. Chem. Soc. 99,* 3854–3857.
Thomson, J. J. (1906), *Conduction of Electricity through Gases*, 2nd ed. London: Cambridge University Press.
Visser, J. W. (1969), *J. Appl. Cryst. 2,* 89–95.
Visser, J. W., de Wolff, P. M. (1964), "Absolute Intensities". *Report 641.109*. Delft, Netherlands: Technisch Physischer Dienst.
Waller, I. (1928), *Z. Phys. 51,* 213.
Warren, B. E. (1969), *X-ray Diffraction*. Reading, MA: Addison-Wesley.
Warren, B. E., Averbach, B. L. (1950), *J. Appl. Phys. 21,* 595.
Warren, B. E., Gingrich, N. S. (1934), *Phys. Rev. 46,* 368.
Werner, P.-E. (1964), *Z. Kristallogr. 120,* 375–387.
Wiles, D. B., Young, R. A. (1981), *J. Appl. Cryst. 14,* 149–151.
Williamson, G. K., Hall, W. H. (1953), *Acta Metall 1,* 22–31.
Wright, A. (1992), "Neutron and X-ray Amorphology": *Analytical Techniques in Glass Science*, El-Bayoumi, O. H., Simmons, C. J. (Eds.). Columbus, OH: Amer. Ceram. Soc.
Young, R. A., Wiles, D. B. (1982), *J. Appl. Cryst. 15,* 430–438.
Young, R. A., Mackie, P. E., VonDreele, R. B. (1977), *J. Appl. Cryst. 10,* 262–269.
Zachariasen, W. H. (1945), *Theory of X-ray Diffraction in Crystals*. New York: Dover Publications, Inc.
Zhao, J., Ramanathan, M., Montano, P. A., Shenoy, G. K., Schulze, W. (1989), *Mat. Res. Soc. Symp. Proc. 143,* 151–156.
Zhang, Z., Ye, H. Q., Kuo, K. H. (1985), *Philos. Mag. A 52,* L49.
Zhou, R. S., Snyder, R. L. (1991), *Acta Cryst. B 47,* 617–636.
Zorn, G. (1991), *Programs DIFF and SHOW*. München: Siemens, Private communication.
Zorn, G., Wersing, W., Göbel, H. E. (1985), *Japanese J. of Appl. Phys. 24 (2),* 721–723.
Zorn, G., Hellstern, W., Göbel, H., Schultz, L. (1987), *Adv. X-ray Anal. 30,* 483–491.

General Reading

For more tutorial information on X-ray diffraction and applications:
Cullity, B. D. (1978), *Elements of X-ray Diffraction,* 4th ed. Reading, MA: Addison-Wesley.
This is an excellent introductory treatment fully covering the fundamentals of X-ray diffraction.

For more depth on any particular topic and a good historical literature review:
Klug, H. P., Alexander, L. E. (1974), *X-ray Diffraction Procedures,* 2nd ed. New York: Wiley and Sons.
This book gives a comprehensive development of all of the methods of powder diffraction with a complete literature survey up through the early 1970's.

For a modern treatment of the techniques of applied crystallography:
Bish, D. L., Post, J. E. (Eds.) (1989), *Reviews in Mineralogy,* Vol. 20, *Modern Powder Diffraction*. Washington, DC: Mineralogical Society of America.
This is an excellent, fully up-to-date treatment of current applications of powder diffraction. It compliments the first two books listed in that they do not cover the modern applications of X-ray diffraction which this article concentrated on.

For an introduction to single crystal techniques:
Stout, G. H., Jensen, L. H. (1968), *X-ray Structure Determination*. New York: The Macmillan Co.
This is an excellent introduction to single crystal structure analysis techniques.

The "bible" of space group theory is:
Hahn, T. (Ed.) (1983), *International Tables for Crystallography,* Vol. A: *Space-Group Symmetry*. Dordrecht, Holland: D. Reidel.

An excellent introduction to classical crystallography is:
Phillips, F. C. (1963), *An Introduction to Crystallography,* 3rd ed. New York: Wiley.

A good review of the new area of quasi-crystals may be found in:
Hargittai, I. (Ed.) (1990), *Quasicrystals, Networks, and Molecules of Fivefold Symmetry*.

A good presentation of EXAFS techniques may be found in:
Teo, B. K. (1986), *EXAFS: Basic Principles and Data Analysis*. Heidelberg: Springer-Verlag.

2 Application of Synchrotron X-Radiation to Problems in Materials Science

Andrea R. Gerson[1], Peter J. Halfpenny, Stefania Pizzini[2], Radoljub Ristić, Kevin J. Roberts[3], David B. Sheen, and John N. Sherwood

Department of Pure and Applied Chemistry, University of Strathclyde, Glasgow, U.K.
[1] Department of Chemistry, King's College, University of London, U.K.
[2] LURE, Bâtiment 205D, Centre Universitaire Paris-Sud, Orsay-Cedex, France
[3] and also at SERC Daresbury Laboratory, Warrington, U.K.

2.1	**Introduction**	107
2.1.1	Synchrotron Storage Rings	107
2.1.2	Characteristics of Synchrotron Radiation	109
2.1.2.1	Radiation from Bending Magnets	109
2.1.2.1	Radiation from Insertion Devices	109
2.1.2.3	Other Characteristics	109
2.1.3	Summary	110
2.2	**X-Ray Absorption Spectroscopy**	110
2.2.1	Basic Principles and Essential Theory	111
2.2.2	Methods of Detection	112
2.2.3	Structure of Cu-Rich Precipitates in α-Fe	113
2.2.4	Amorphisation of $Ni_{50}Mo_{50}$ by Mechanical Alloying	115
2.2.5	Structure of the Surface Oxide on GaAs (100)	115
2.2.6	Surface Corrosion of Borosilicate Glasses	117
2.2.7	Oxidation of Polished Stainless Steels	118
2.2.8	Changes in the Atomic Environment Around Zn in K_2ZnCl_4 during Phase Transformations	120
2.2.9	Dynamic Investigation of the Reduction of Nickel Formate	121
2.2.10	Role of the Sulfidation Process in the Operation of Pt/Re Catalysts	122
2.3	**Structural Studies Using X-Ray Diffraction**	123
2.3.1	Introduction	123
2.3.2	High Resolution Powder Diffraction	124
2.3.2.1	Introduction	124
2.3.2.2	High Resolution Refinement of the Lattice Parameters of Tungsten	124
2.3.2.3	Determination of the Structure of Cimetidine Using Rietveld Refinement	125
2.3.2.4	Influence of Growth Environment on the Structure of Long Chain Hydrocarbons	125
2.3.3	Energy Dispersive Powder Diffraction	128
2.3.3.1	Introduction	128
2.3.3.2	In Situ Characterisation of Solid-State Chemical Transformations	129
2.3.3.3	Structural Phase Transformations at High Pressure	130

2.3.3.4	Particle Formation in a Liquid Environment	133
2.3.4	Laue Diffraction	134
2.3.4.1	Introduction	134
2.3.4.2	Instrumentation and Experimentation	135
2.3.4.3	Effect of Radiation Damage	135
2.3.4.4	Assessment of the Quality of Organic Crystals	137
2.3.4.5	Solid-State Polymerisation of PTS	137
2.3.4.6	Influence of Strain on the Growth Rate of Secondary Nucleated Particles	139
2.3.4.7	Probing Structural Changes During Phase Transformations	140
2.3.4.8	Application to Crystal Structure Determination	141
2.4	**Diffraction and Scattering from Real Surfaces and Interfaces**	142
2.4.1	Introduction	142
2.4.2	Information Content Available from Surface Diffraction Data	142
2.4.3	Instrumentation	143
2.4.4	Lattice Coherency in Group III–V Quantum-Well Structures	144
2.4.5	Ge/Si (001) Strained-Layer Semiconductor Interfaces	144
2.4.6	In Situ Structure of Au (100) Under Electrochemical Potential Control	146
2.4.7	Crystal/Solution Interface Structure of ADP (100) During Growth	148
2.4.8	Thin Polyphenylene Films Prepared from Different Precursors	149
2.4.9	Photodissolution of Silver on Chalcogenide Glasses	149
2.4.10	Freely Supported Films of Octylcyanobiphenyl	151
2.5.	**X-Ray Standing Wave Spectroscopy**	152
2.5.1	Introduction	152
2.5.2	Surface Coordination of Adsorbed Br on Si (111)	152
2.5.3	Structural Environment Around Cu^{2+} Ionic Habit Modifiers in Ammonium Sulfate	153
2.6	**Characterisation of Lattice Defects Using X-Ray Topography**	155
2.6.1	Introduction	155
2.6.2	White Radiation Synchrotron Topography	155
2.6.2.1	Introduction	155
2.6.2.2	Instrumentation	155
2.6.2.3	Advantages of the Broad Spectral Range	156
2.6.2.4	Influence of Beam Flux, Dimensions and Divergence	158
2.6.2.5	Experiments Involving the Time Structure	161
2.6.3	Double Crystal Topography	162
2.6.3.1	Introduction	162
2.6.3.2	Instrumentation	162
2.6.3.3	Applications	163
2.7	**Conclusions and Future Perspective**	164
2.8	**Acknowledgements**	165
2.9	**References**	165

2.1 Introduction

2.1.1 Synchrotron Storage Rings

When charged particles travelling at relativistic energies are constrained to follow a curved trajectory, as for example when they pass through a magnetic field, they emit electromagnetic radiation. The total emitted intensity of the radiation is proportional to Γ^4, where

$$\Gamma = E_p/(mc^2) \qquad (2\text{-}1)$$

E_p and m are the energy and mass of the particle and c is the velocity of light. From this expression it follows that, for a certain energy, the highest intensity will be emitted by the lightest moving particle, and hence that electrons and positrons will be the most efficient emitters. This type of radiation is produced in all circular particle accelerators (*synchrotrons*) and thus it is usually referred to as *synchrotron radiation*.

The emitted radiation shows not only a high intensity compared with conventional sources but also covers a broad spectral range from the infra-red to the hard X-ray region. The considerable potential of this type of source as demonstrated by early experiments using radiation emitted by synchrotrons built and operated primarily for nuclear physics research has led to the development of *synchrotron storage rings* dedicated to the production of synchrotron radiation. These sources and their use are now so widespread that many countries in the world either have built or are building sources. Those which are not, often have ready access to such facilities.

The general form of a storage ring and the basic mode of its operation is shown schematically in Fig. 2-1. Electrons or positrons developed from a suitable source are first accelerated to energies around 60 MeV in a linear accelerator. From here they are injected into a booster synchrotron in which the energies are increased to 600 MeV levels by acceleration in the circular orbit. These high energy particles are then injected in bunches into the main storage ring until integrated particle currents of circa 200–300 mA are attained. The storage ring consists of straight sections linked by curved portions around which are situated the electromagnets (bending magnets) which cause the essential curvature of the beam.

In the storage ring the bunches of particles are still further accelerated to relativistic (2–8 GeV) energies. This stored current of the eletron or positron bunches continues to circulate under the accelerating field.

The current, and hence beam intensity, decays over a long period (10–20 hours) as a result of the loss of energy due to collisions of the particles with residual gas

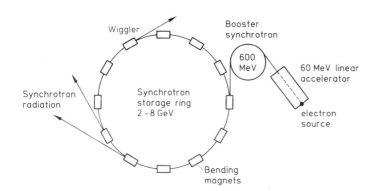

Figure 2-1. Schematic representation of the general form of an electron synchrotron storage ring and associated facilities.

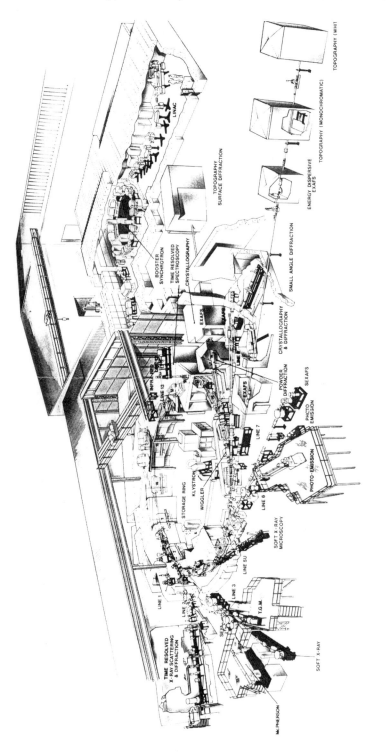

Figure 2-2. General view of the experimental hall of the SERC UK Synchrotron Radiation Source facility at Daresbury, U.K.

molecules and the storage ring walls. During this period, utilisable radiation is emitted as the electrons pass through the bending magnets. The radiation is emitted in the forward direction tangentially to the particle beam. It is extracted for use principally through some type of tangential beam port into a beam line, fully evacuated to the ring vacuum for the softer radiation or through a beryllium window into air for the harder radiation ($\lambda < 0.3$ nm).

The beam lines and their associated experimental hutches are accommodated in an experimental hall. A general view of such a facility is shown in Fig. 2-2.

2.1.2 Characteristics of Synchrotron Radiation

2.1.2.1 Radiation from Bending Magnets

The typical spectral distribution of synchrotron radiation is shown in Fig. 2-3. The horizontal scale is defined by a critical wavelength λ_c and the vertical scale of intensity by the electron current and the energy of the beam. λ_c is given by the formula:

$$\lambda_c = 5.6\, R_s/E^3 = 1.86/(BE^3) \text{ (mm)} \qquad (2\text{-}2)$$

where the energy of the beam, E, is expressed in GeV, R_s the radius of the ring in metres and B the bending magnet strength in Tesla. From this we see that in order to obtain a spectrum of energy extending from the infra-red to the hard X-ray regions then $\lambda_c \leq 0.1$ nm and hence E should be 2–5 GeV and $R_s \approx 10$–20 m.

2.1.2.2 Radiation from Insertion Devices

In any particular system the energy range and beam intensity can be extended and modified by the use of insertion devices known as wigglers and undulators.

(a) *Wigglers.* A wiggler consists of a series of superconducting magnets (3–5)

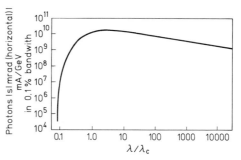

Figure 2-3. Synchrotron radiation spectrum emitted by an electron moving in a curved trajectory.

placed in a straight section of the ring. The purpose of this is to cause the electron to execute a path of shorter local radius of curvature than in the dipole bending magnets. The result [Eq. (2-2)] is to increase the critical energy and to shift the spectrum to higher energies by a parallel amount, thus providing even harder radiation than from the bending magnets. The radiation is again emitted tangentially to the smaller curve and can be extracted via a window and beam-pipe.

(b) *Undulators.* Undulators, which are again inserted in a linear section of the storage ring are multipole wigglers which develop the acceleration by causing either a transverse sinusoidal or helical magnetic field. The result, under some geometrical operating circumstances, is to produce a line spectrum of harmonics of increased power density and spectral brilliance.

Further details on the design, operation and properties of synchrotron storage rings and insertion devices will be found in Kunz (1979), Winick and Doniach (1980), Stuhrmann (1982) and Koch (1983).

2.1.2.3 Other Characteristics

In addition to this broad spectral range the beam shows the following general properties:

(a) *Low divergence.* For $\lambda = \lambda_c$ the vertical divergence of the beam $\Theta \approx 1/\Gamma$ where

$\Gamma = E_p/(m_0 c^2)$ and $m_0 c^2$ is the rest mass energy of the electron. When E is of the order of 1 GeV, $\Theta \leq 0.1$ mrad (0.006°, 20″).

(b) *Defined polarisation*. The radiation is 100% polarised with the electric vector parallel to the orbital plane. Above and below the orbital plane the radiation is elliptically polarised. The degree of ellipticity depends on the angle of viewing.

(c) *A time structure*. The electron bunches can be injected with bunch lengths typically 0.05–1 ns and at intervals which can range from a few nanoseconds to a few microseconds.

(d) *High intensity and brightness*. The intensity of the emitted radiation is directly proportional to the current circulating in the synchrotron. It is typically $\approx 1.6 \times 10^{10} E$ (GeV) photons/[s · mrad (horizontal) · mA · 0.1% (bandwidth)]. Brightness depends on the dimensions of the radiating area as viewed along the beam. Typically this is $> 10^{10}$ photons/ [mm^2 · s · mrad^2 · 0.1% (bandwidth)] at $\lambda = 1$ Å.

(e) *Large beam size*. The emitted beam has a relatively large cross-section, 1–2 cm diameter, depending on the synchrotron source.

2.1.3 Summary

In operational terms synchrotron X-ray sources offer the advantages of a broad spectrum or tunable monochromatic source of high and uniform spectral brilliance some 100–10000 times brighter than traditional sources depending on the particular use. The well-defined polarisation offers prospects for improving signal to noise ratio in some applications. The time structure coupled with the intensity allows dynamic experiments on a wide time range. The low beam divergence allows for improved resolution. In combination these characteristics offer a wide range of prospects for novel and extended experiments using both diffraction and spectroscopic techniques.

In the following pages we present a number of examples of the various ways in which synchrotron X-radiation can be used to examine problems in materials science. In each case we detail how the characteristics referred to above can be used to advantage. These examples are taken predominantly from our own studies of crystal growth, surface and interface science, and solid-state chemistry. They and the bibliography should allow extrapolation to a wider range of problems and experiments.

Background on X-ray diffraction generally will be found in this Volume, Chap. 4.

2.2 X-Ray Absorption Spectroscopy

In recent years X-ray absorption spectroscopy (XAS) has achieved an increasing importance in materials science due to its ability to obtain information on the local structure of selected atomic species in complicated systems which may not be suitable for conventional X-ray crystallographic analysis, e.g. amorphous and multi-component materials, alloys, surfaces, etc. The development of XAS as a tool for structural determination started effectively in the late 1970s when synchrotron radiation sources became widely available. XAS benefits from the high intensity, the collimation and the broad spectral range of synchrotron radiation. Since X-ray fluxes emitted by synchrotron sources are typically 3–4 orders of magnitude greater than those obtained with laboratory sources, the data acquisition times for absorption edge spectra are reduced from the order of weeks to minutes. The collimation of the

X-ray synchrotron beam is such that energy resolutions as good as 10^{-4} can be obtained. Finally the continuous synchrotron radiation spectrum (Fig. 2-3) allows L and K-edges of most elements to be accessed (Winick and Doniach 1980).

2.2.1 Basic Principles and Essential Theory

A typical X-ray absorption spectrum is given in Fig. 2-4 which shows the K-edge spectrum of an Fe foil and illustrates the modulation of the absorption cross section for the photoexcitation of a 1 s core shell electron. The strong oscillations which extend beyond the edge for about 30–40 eV, the so called X-ray absorption near-edge structure (XANES), involve the multiple scattering of the excited photoelectrons and are determined by the geometrical arrangement of atoms in a local cluster around the absorbing atom (Bianconi, 1988; Durham, 1988).

The structure observed in the absorption spectrum beyond the XANES region is known as EXAFS (extended X-ray absorption fine structure). The oscillations are a final state electron effect, arising from the interference between the outgoing photoelectron wave ejected in the absorption process and the wave backscattered from the neighbours of the absorbing atom (Sayers et al., 1971; Lee et al., 1981; Hayes and Boyce, 1982; Köningsberger, 1988). These techniques are also treated in Section 1.9.

The primary objective of EXAFS investigations is to determine the local atomic environment of the absorbing atom, by analysing the measured structure. The EXAFS function is defined as:

$$\chi(k) = (\mu - \mu_0)/\mu_0 \qquad (2\text{-}3)$$

where k is the photoelectron wave vector and μ_0 is the background absorption i.e. the absorption coefficient of an isolated atom. In the so called "plane wave approximation" the EXAFS function can be expressed (Stern, 1974) as:

$$\chi(k) = -\sum_i \{[N_i/(k\,r_i^2)]\,F_i(k)\,S_0^2(k)$$
$$\cdot \exp[-r_i/\lambda_d(k)]$$
$$\cdot \exp(-2k^2\sigma_i^2)\sin(2kr_i + \Phi_i)\} \qquad (2\text{-}4)$$

which applies to the photoexcitation of a 1 s electron (K-edge) in a polycrystalline sample. This equation expresses the EXAFS function as a sum of contributions, one for each "shell" of neighbouring atoms. Each atomic shell contains N_i atoms of type i at a radial distance r_i from the absorbing atom. A shell centred at r_i contributes with a sinusoidal function of wavelength $\approx \pi/r_i$ in k space. The phase of each sinusoidal function is related to the distance r_i of the neighbouring atoms from the central atom, and to the phase shift Φ_i introduced by the potentials of central and backscattering atoms. The envelope of each sinusoidal is determined by the number of atoms in the shell and by the amplitude of the backscattering factor F_i. The damping terms in Eq. (2-4) take into account multibody effects, S_0^2, inelastic scattering, $\exp(-r_i/\lambda_d)$,

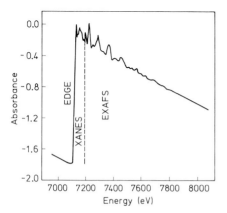

Figure 2-4. Fe K edge X-ray absorption spectrum of a Fe foil illustrating the main regions of interest in an absorption spectrum (after Pizzini, 1990 d).

and thermal and structural disorder, $\exp(-2\sigma^2 k^2)$.

A simple Fourier transform of the EXAFS data essentially provides a radial distribution function centred on the absorbing central atom. For more detailed analysis we can model the EXAFS data [i.e., using the kind of relationship given in Eq. (2-4)] to provide information on the distance, the number and the type of the neighbours of the absorbing atom. Since only elastically scattered electrons can interfere, and the elastic mean free path of electrons is short, EXAFS only contains information on the *local* structure of the absorbing atom. Since crystallinity is not a prerequisite for this technique, amorphous materials can also be investigated. The atom-specific nature of the EXAFS process makes it suitable for the study of complex (e.g., multicomponent) systems or for the determination of the environment of dilute species.

2.2.2 Methods of Detection

X-ray absorption spectra can be measured directly in a transmission experiment by monitoring the attenuation of the X-ray beam incident on the sample as a function of photon energy, or indirectly by measuring the energy dependence of physical processes which depend on absorption, such as fluorescence emission, photoelectron yield or X-ray reflectivity (e.g., Heald, 1988).

In a transmission experiment the absorption of the sample is measured by monitoring the incoming (I_0) and the transmitted (I_t) flux, which are related by the expression:

$$I_t = I_0 \exp(-\mu t) \qquad (2\text{-}5)$$

where μ is the linear absorption coefficient of the sample and t is the sample thickness.

When the EXAFS signal is only a small fraction of the total absorption (typically less than 5%) it is more advantageous to measure processes which are specific to the absorbing atom of interest, such as fluorescence emission. In the case of dilute analytes, the fluorescence intensity is proportional to the absorption coefficient μ and to a good approximation is given by:

$$I_f \approx \mu I_0 \qquad (2\text{-}6)$$

A schematic diagram of the experimental equipment for XAS measurements in transmission and in fluorescence modes is shown in Fig. 2-5.

Information obtained from the transmission and fluorescence measurements reflects the local structure of the *bulk* material. By detecting the photoelectrons (which have a short escape depth in a solid material) emitted as a consequence of the absorption process, surface structure can also be specifically investigated. Alternatively, surface sensitivity can be enhanced by working in glancing incidence geometry. Since the refractive index of X-rays is slightly less than unity (James, 1958), an incident beam at a glancing angle under-

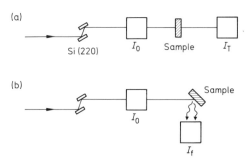

Figure 2-5. Sketch of the experimental set-up for an X-ray absorption spectroscopy measurement. The white synchrotron beam is monochromated typically by a Si(220) channel cut crystal and recorded with ion chambers (I_0 and I_T) and a fluorescence detector (I_f). (a) Transmission detection geometry; (b) fluorescence detection geometry.

goes total external reflection where the X-rays penetrate only 3–5 nm into the material. EXAFS spectra recorded under such conditions (ReflEXAFS) provides a highly effective structural probe for surface studies.

It can be shown (Parratt, 1954; Bosio et al., 1984) that the X-ray absorption coefficient μ of a condensed medium can be obtained from the energy dependence of the reflectivity coefficient $R(E)$, measured for a fixed incident glancing angle ϕ close to the critical angle. The absorption coefficient so obtained is specific to the surface layer probed by the X-ray beam. If the measurement is carried out for photon energies beyond an absorption edge, the reflectivity spectrum exhibits a structure, which can be related to the EXAFS spectrum (Martens and Rabe, 1980).

If the atomic species of interest is diluted at the surface, fluorescence detection, still at glancing incidence, gives a more favourable signal to noise ratio. A sketch of the equipment for glancing angle XAS measurements is shown in Fig. 2-6. Since the typical critical angles for X-rays are of the order of a few mrad, an appropriate precision goniometer system has to be used. The instrumentation used for glancing angle EXAFS at the Daresbury SRS has been described by Pizzini et al. (1989).

An overview of the application of EXAFS to the characterisation of materials systems has been given by Gurman (1982). In this review we consider some specific case studies of structural characterisation of dilute metallic alloys, amorphous metals, semiconductor surfaces, corrosion, phase transformations, solid-state reactivity and heterogeneous catalysis.

2.2.3 Structure of Cu-Rich Precipitates in α-Fe

Understanding the role of irradiation-induced second phase precipitation in the embrittlement of steel is of crucial importance in defining the long term stability of nuclear reactor pressure vessels. Through its ability to determine the local order around the precipitated impurities, EXAFS can provide important structural information which cannot be obtained using other techniques.

α-Fe based alloys containing 1.3 at.% Cu were investigated using fluorescence detection above the Cu K-edge (Pizzini et al., 1990a and 1990b). The precipitation of small Cu-rich clusters was studied in a series of thermally aged model alloys, well characterised with transmission electron microscopy (TEM) (Buswell and Brock 1987; Phythian and English, 1991) and small angle neutron scattering (SANS) measurements (Beaven et al., 1986; Lucas and Odette, 1986; Buswell et al., 1986). These studies show that at peak hardness condition (2 h of thermal ageing) small precipitates of mean diameter 2.5 nm are formed. Interpretation of SANS magnetic ratio and positron annihilation data also

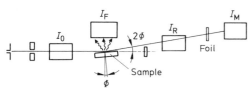

Figure 2-6. Sketch of the experimental set-up for a glancing angle XAS experiment (after Pizzini et al., 1989): the X-ray beam, collimated by entrance slits, hits the sample at an angle ϕ. The incident and reflected beams are monitored by ionisation chambers. For measurements on diluted species, a fluorescence detector is positioned above the sample. A third detector (ionisation chamber, scintillation counter as solid-state detector) can be used to measure the transmission spectrum of a model compound to achieve energy calibration.

indicate a possible association of copper in the precipitates with Fe and/or vacancies. Although the TEM measurements suggest that the Cu-rich precipitates might have a b.c.c. structure, the structural properties of the clusters and their evolution on ageing can be better investigated using X-ray absorption spectroscopy.

The local structure around Cu in Fe-Cu alloys was investigated in the "as quenched" condition and after 2 h, 10 h and 760 h of thermal ageing at 550 °C. The fluorescence signal was collected using a multi-element solid-state detector which provided an energy resolution of about 300 eV at 10 keV and allowed the Cu K emission lines to be discriminated from the intense Fe K background emission. The results of the EXAFS experiments are summarised in Fig. 2-7, where the EXAFS spectra for the thermally aged alloys (data acquisition time $\approx$ 5 h) are compared with the transmission EXAFS spectra of metallic Fe and Cu. An analysis of the EXAFS data indicates that in the "as quenched" material Cu is in a typically b.c.c. environment. This is consistent with Cu being in solid solution in the Fe matrix, i.e., Cu being substitutional in the host b.c.c. lattice. Analysis of the data, least-squares fitted using calculated phase shifts (Gurman, 1988) shows a reduction of the Cu-Fe distance (0.246 nm) with respect to the Fe-Fe distance in Fe (0.248 nm). After 2 h of thermal ageing the EXAFS spectra still retain the structure observed for the data of the "as quenched" samples. Previous SANS and TEM data (Buswell et al., 1988) showed that at this stage of the thermal ageing about 50% of the Cu is retained in the matrix, while the remaining copper precipitates in small clusters with an average diameter of about 2–3 nm. Therefore about a half of the Cu atoms probed by EXAFS are expected to be coordinated to Fe atoms in the α-Fe matrix and to exhibit a b.c.c.-like radial distribution function. The remaining Cu atoms are coordinated to Cu atoms in the small clusters. The maintenance of a b.c.c.-like EXAFS spectrum strongly suggests that at "peak hardness" condition the precipitates have b.c.c. structure. Least-squares fitting analysis of the EXAFS data confirms these qualitative indications. Although Cu and Fe neighbouring atoms cannot clearly be distinguished by EXAFS, because of the closeness of their atomic numbers, least-squares fitting of the experimental data seem to indicate that Cu is largely coordinated to Cu nearest neighbours, pointing to the formation of Cu rich clusters.

The results indicate that as the precipitates grow with increasing annealing time, their structure changes from b.c.c. to f.c.c. The EXAFS spectra recorded for the 760 h-aged samples more closely resemble those typical of an f.c.c. structure than a b.c.c. structure. This is consistent with previous observations by TEM of f.c.c. precipitates produced in over-aged materials (greater than 2 h at 550 °C).

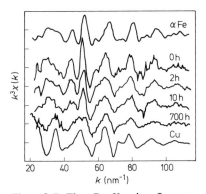

Figure 2-7. The Cu K edge fluorescence EXAFS spectra of the Fe-Cu alloy thermally aged for 0 h, 2 h, 10 h and 700 h are compared with the transmission EXAFS spectra of α-Fe and Cu foils (after Pizzini et al., 1990).

2.2.4 Amorphisation of $Ni_{50}Mo_{50}$ by Mechanical Alloying

The formation of novel amorphous metallic alloys (Samwer, 1988) obtained by the mechanical reaction of the pure constituent elements is currently of significant interest in materials science (Weeber and Bakker, 1988). It has been shown (Schultz, 1987) that glassy binary metals with unexplored composition ranges can be created by mechanical alloying. $Ni_{50}Mo_{50}$ alloys, obtained by mixing together and milling equimolar quantities of pure crystalline elemental powders of Ni and Mo, were investigated using transmission EXAFS at both Ni K- and Mo K-edges (Cocco et al., 1989 and 1990).

The Ni and Mo K-edge EXAFS spectra of the samples as a function of milling time are shown in Fig. 2-8. These spectra show a significant reduction of EXAFS amplitude with milling time, due to the gradual loss of long range order associated with the onset of amorphisation.

The environment around the Ni atoms changes significantly during milling with the f.c.c. lattice becoming more disordered after 2 h milling and rapidly losing its crystallity between 2 and 10 h. The Fourier transforms of the EXAFS data (Cocco et al., 1989, 1990, 1992) show the loss of the 4th neighbour shell after 5 h milling time. At 10 h milling time the 2nd and 3rd shells are also smeared-out and the 1st shell shifted to a shorter bond length. The latter effect is commonly observed in glassy alloys (Teo et al., 1983) and in this case is indicative of Ni-Mo interactions.

The Mo K-EXAFS data reveal that, after milling, the local environment of Mo is more crystalline than that of Ni. An amplitude reduction of the EXAFS oscillations is still observed, but the overall b.c.c. structure remains, even after 10 h milling. The

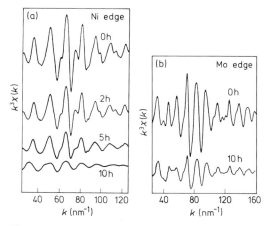

Figure 2-8. The EXAFS spectra of the Ni-Mo alloys, recorded as a function of milling time (after Cocco et al., 1990). (a) Ni K edge; (b) Mo K edge.

Fourier transform shows (Cocco et al., 1990) that the 4 b.c.c. shells essentially remain. X-ray diffraction measurements (Cocco et al., 1989) reveal no change in the Mo b.c.c. lattice parameter with milling and confirm the low solubility of Ni in Mo (Hansen, 1958). However, as Mo is up to 20 wt.% soluble in Ni, one can presume that amorphisation follows a substantial insertion of Mo in the Ni lattice. The stress and defect generation associated with this is probably reflected in the loss of long range order which takes place after mechanical alloying (Johnson, 1986; Dubois, 1988).

2.2.5 Structure of the Surface Oxide on GaAs (100)

The characterisation of surface oxides of group III-V semiconductors is important in the definition of device fabrication techniques. In this study the surface oxide structure on a commercial GaAs (100) wafer was investigated using X-ray absorption spectroscopy in glancing angle geometry and reflectivity detection (Barrett et al.,

1990; Pizzini et al., 1990c). ReflEXAFS spectra were recorded above the As K and Ga K absorption edges, so that the local structure around both atomic sites could be obtained separately. The spectra were recorded for two incident angles (ϕ_1 and ϕ_2) below the critical angles, corresponding to X-ray penetration depths of about 2 nm and 3 nm. The structural parameters obtained from least-squares fits of the ReflEXAFS data (Table 2-1) indicate that the surface of the GaAs (100) wafer is partially oxidised. The ratio of the coordination numbers of bulk-like structures in the surface layer, to those typical of GaAs, indicates that the oxide thickness is 0.7–0.9 nm thick.

The results summarised in Table 2-1 show that Ga and As atoms have different oxygen environments in the surface oxide. For both Ga and As atoms there is an oxygen neighbour shell at ≈ 0.17 nm, close to that found for the tetrahedral coordination in $GaAsO_4$. However, Ga atoms have an additional oxygen neighbour shell at ≈ 0.195 nm, which is typical of the octahedral coordination of Ga_2O_3. Thus while As atoms seem to be exclusively in a tetrahedral environment at the surface, Ga seems to be present in both tetrahedral and octahedral co-ordinations.

The two matched cation shells (Ga or As) at the same distances (≈ 0.28 nm and ≈ 0.31 nm for both Ga and As central atoms) are probably associated with the two different oxygen coordinations for Ga. The shell at ≈ 0.28 nm possibly arises from the coordination between cations when both are in a tetrahedral environment whilst the shell at ≈ 0.31 nm is related to the coordination between octahedral Ga and a tetrahedral cation (As or Ga). The matching of these second shell distances for both Ga and As central atoms shows the oxide to be a single component which can thus be modelled as a *microscopically* random mixture of tetrahedral and octahedral sites linked via oxygen bridges, with Ga occupying both and As only the former 4-fold site. A sketch of the structural model based on these results is shown in Fig. 2-9a.

Comparison between the Ga and As environments as a function of incident angle (i.e., of depth) shows no significant variation for As concentration while it indicates a distinct increase in Ga coordination at the surface. This trend indicates a greater association of Ga atoms with O in the outermost oxide layer and a corresponding As depletion.

A structural model (Pizzini et al., 1990c) of this oxide on GaAs (100) based on a continuous random network (Polk and Boudreaux, 1973; Greaves and Davies, 1974) with oxygens attached to the dangling bonds at the surface, has been built up from a mixture of octahedral and tetrahedral Ga's, tetrahedral As's and bridging O's. The surface oxide so constructed results in an open structure with an atomic density much smaller than that of any of

Table 2-1. Coordination numbers and shell distances extracted from least-squares fits of the As K and Ga K-edge ReflEXAFS spectra of the GaAs(100) spectra (angle ϕ_1) (after Barrett et al., 1990).

r(nm)	N	r(nm)	N	r(nm)	N
Ga K-edge					
Ga-O		Ga-O-M		Ga-As	
0.172	0.6	0.288	3.1	0.246	3.0
0.195	2.2	0.315	3.4		
As K-edge					
As-O		As-O-M		As-Ga	
0.168	0.9	0.284	1.1	0.243	3.1
		0.309	1.0		
Bulk GaAs					
				Ga-As	
				0.2446	

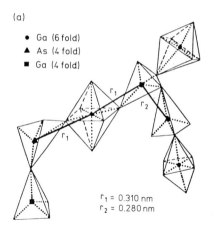

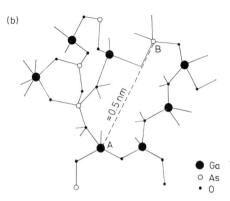

Figure 2-9. (a) Sketch of the local coordination of Ga and As atoms in the oxide coating on GaAs (100): 6-fold and 4-fold coordinated Ga atoms are linked to 4-fold coordinated As atoms via oxygen bridges (after Barrett et al., 1990); (b) sketch of the section of a typical fissure in the oxide on GaAs (100), obtained from a three-dimensional molecular model of the oxide (after Pizzini, 1990d).

the crystalling oxides of Ga and/or As. The presence of 2-fold coordinated oxygens and the variable cation coordination combine to encourage the formation of micro-voids (typical size $\approx 0.5-0.8$ nm) whose internal surfaces are oxygen rich (Fig. 2-9).

2.2.6 Surface Corrosion of Borosilicate Glasses

A knowledge of the leaching mechanism of metallic ions from glass surfaces is impor-

2.2 X-Ray Absorption Spectroscopy

tant to the understanding of materials requirements for the vitrification processes associated with nuclear waste management. Detailed studies have been carried out of the leaching of Fe (Sacchi et al., 1989) and U (Greaves et al., 1989) through the glass matrix. In the study of the corrosion-induced transport of Fe^{3+} ions in borosilicate glasses loaded with Fe_2O_3 (Sacchi et al., 1989) fluorescence X-ray absorption spectroscopy was used to follow the changes in the ionic environment during the leaching process. The measurements were carried out using a glancing angle geometry commensurate with a penetration depth of about 1 μm.

The results of this study show that after leaching in H_2O at 373 K, with the sample surface still in the "wet" state, dramatic changes in the edge structures and in the EXAFS spectra are observed. These changes are seen both as a function of depth and of the extent of leaching (Fig. 2-10) and give some insight into the transport properties of Fe^{3+} ions through the glass network during corrosion.

In contrast to most minerals where ferric sites are octahedral, in glasses Fe^{3+} is commonly tetrahedrally coordinated (Binsted et al., 1986). This configurational change can be readily recognised in EXAFS by a shortening of the Fe-O distance. The two different geometries can also be distinguished in the edge structure from the magnitude of the "white line" (peak A in Fig. 2-10) which is considerable for octahedrally coordinated Fe^{3+}, but much reduced when the metal occupies tetrahedral sites. A converse behaviour is exhibited by the pre-edge feature (peak B in Fig. 2-10) which is larger for tetrahedrally coordinated Fe^{3+} in oxide glasses (Binsted et al., 1986). Both types of ferric coordination can be identified at the surface of borosilicate glass (Sacchi et al., 1989).

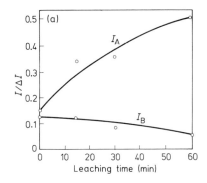

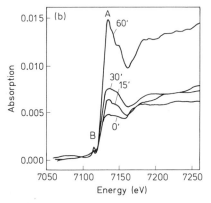

Figure 2-10. The changing near-edge structure of Fe K edge measured in glancing angle fluorescence mode for borosilicate glass as a function of leaching time (after Sacchi et al., 1989). The inset shows the increase in the white line (peak A) and the decrease in the pre-edge feature (peak B) with advancing corrosion. Both are normalised to the step height I.

The increase in the absorption step height as leaching progresses demonstrates that ferric ions migrate to the glass surface, quadrupling in concentration over a period of ≈ 1 h. Initially the white line (A) is weak and the pre-edge feature strong, indicative of extensive tetrahedral bonding. After 1 h of leaching the converse is true. pointing to the developement of a substantial fraction of octahedrally coordinated ferric ions. The same picture emerges from the EXAFS data which show the presence of a single Fe-O distance at first, followed by the appearance of a longer distance as corrosion continues. From what is known about the bulk structure of Fe^{3+} in aegerine glass and in the mineral, the glancing angle XAS behaviour of the borosilicate glass surface strongly suggests that leaching promotes dissolution of Fe from the glass network and its transport towards the corrosion interface where it precipitates in a form similar to that of a ferric silicate.

2.2.7 Oxidation of Polished Stainless Steels

The detection of the early stages of corrosion together with the structural characterisation of any surface phases formed can provide a useful insight into our understanding of fundamental and applied aspects of surface reactivity. X-ray absorption spectroscopy was used to characterise changes in the environment around Fe in the surfaces of polished polycrystalline stainless steel samples which had been oxidised at 1000 °C for up to 4 min (Barrett et al., 1989a and 1989b). Angle dependent reflectivity curves were recorded for a fixed photon energy beyond the Fe K edge. XAS spectra were recorded above the Fe-K edge for an incident angle below the critical angle, corresponding to an X-ray penetration depth of about 3.5 nm. The results, summarised in Fig. 2-11 demonstrate that self-consistent information can be achieved by combining the near edge, edge height, chemical shift and reflectivity data. Figure 2-11a shows that with increasing oxidation time the slope of the angle-dependent reflectivity curves dramatically decreases. This behaviour reflects the increased absorption of the X-ray beam (Parratt, 1954) and is evidence for the migration of Fe ions to the surface layer.

The variation of the absorption edge height as a function of oxidation time (Fig. 2-11b) indicates an increase in the

Fe^{3+} content near the surface, which probably relates to Fe diffusion through the protective Cr_2O_3 layer normally present on stainless steel. For steel measured in transmission the edge position coincides with that of metallic Fe. The increase in chemical shift with corrosion time (Fig. 2-11c) indicates that the oxidation state is changing as corrosion proceeds. The edge position is shifted by 1 eV for 1 min corrosion time and after 4 min the shift is 5 eV, corresponding to that observed for Fe_2O_3.

In Fig. 2-11d the near-edge spectra for the uncorroded, 1 min, 3 min and 4 min corroded steel surfaces are compared to bulk steel, Fe_3O_4 and Fe_2O_3 model spectra. The bulk steel has a near edge structure similar to that of metallic iron (Grunes, 1983). A white line, typical of oxide spectra, appears as corrosion proceeds, as does the pre-edge feature $\approx$ 10 eV below the edge. The uncorroded steel surface shows evidence for an oxide-like white line. After 1 min the pre-edge structure resembles that found for Fe_3O_4. As corrosion proceeds, Fe_2O_3 becomes the more dominant phase, as indicated by the splitting of the pre-edge and main peaks. After 3 min the near edge structure is virtually identical to that of Fe_2O_3.

The EXAFS results (Barrett et al., 1989a and 1989b) confirm these indications. In agreement with the near-edge data, the polished steel surface is metal-like. The dominant structure is that of the bulk steel although there is some reduction of the first shell (f.c.c.) amplitude. After 4 min corrosion time the f.c.c crystal structure of the metallic steel has disappeared and the Fourier transform is virtually identical to octahedral Fe_2O_3.

These results indicate that the polished surface, although retaining its metallic character, shows signs of initial oxidation. As corrosion proceeds the surface content

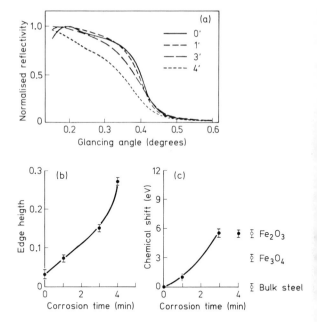

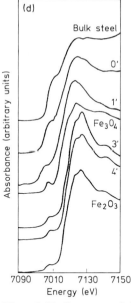

Figure 2-11. Summary of the measurements on corroded stainless steels (after Barrett et al., 1989a). (a) Normalised reflectivity curves as a function of oxidation time. (b) The step height at the Fe K edge as a function of corrosion time. (c) The chemical shift as a function of corrosion time. (d) Near-edge spectra of the Fe K edge for steel surfaces corroded for: (B) 0 min, (C) 1 min, (E) 3 min, (F) 4 min; compared to the near-edge spectra for: (A) the bulk steel, (D) Fe_3O_4 and (G) Fe_2O_3.

of Fe is enriched and the oxidation state increases. The first oxide phase to be formed on oxidation is probably similar to Fe_3O_4 but further reaction yields preferential oxidation to Fe(III) in the form of Fe_2O_3.

2.2.8 Changes in the Atomic Environment Around Zn in K_2ZnCl_4 during Phase Transformations

The formation of modulated and incommensurate crystalline phases is often accompanied by very subtle changes in atomic structure which may not be detectable using conventional analysis. EXAFS can be used to probe such changes and has been applied in situ to analyse changes to the local environment around Zn atoms in potassium tetrachlorozincate K_2ZnCl_4 (KZC). KZC crystallises with a ferroelectric (Gesi, 1978) structure (orthorhombic $Pna2_1$, Mikhail and Peters, 1979) which is commensurate with a wave vector with the magnitude $q = a^*/3$ and which undergoes phase transitions at temperatures close to 127 °C and 282 °C (Gesi, 1978; Gesi and Iizumi, 1979; Jacobi, 1972; Itoh et al., 1980; Milia et al., 1983; Quilichini et al., 1982). Above the lock-in transition at ≈ 127 °C the structure is incommensurately modulated along the pseudo-hexagonal a-axis with a wave vector of magnitude $q = 1/3 \cdot (1 - \delta) q^*$, where the deviation parameter δ increases with temperature (Kucharczyk et al., 1982) until close to 284 °C at which the structure transforms to a paraelectric phase with the space group $Pnam$.

The environment around the Zn atoms in the room temperature structure forms a distorted tetrahedron (Mikhail and Peters, 1979) where the Zn-Cl distances range from 0.2241 to 0.2289 nm. The structural changes around the Zn atoms in the KZC structure associated with the 127 °C and 282 °C phase transitions have been studied using EXAFS. The samples were mounted in a compact furnace (Bhat et al., 1990 a) and transmission EXAFS spectra were recorded as a function of temperature (Bhat et al., 1990 b).

Figure 2-12 shows the Fourier transforms of the EXAFS data as a function of temperature. In Table 2-2 the structural parameters obtained from least-squares fits of the EXAFS data are reported. The expected increase in disorder associated with the phase transition can be seen in the

Table 2-2. EXAFS characterisation of the tetrahedral distortion in KZC using a two shell model (after Bhat et al., 1990 b).

T (°C)	N	r (nm)	A (nm²)	N	r (nm)	A (nm²)
25	2.0	0.221	0.00007	2.1	0.229	0.00007
127	2.0	0.220	0.00012	2.0	0.227	0.00008
212	1.4	0.216	0.00009	2.7	0.228	0.00008
286	1.4	0.218	0.00018	2.7	0.227	0.00013
(25	2.0	0.221	0.00007	2.0	0.229	0.00007)

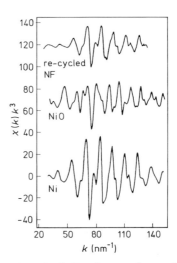

Figure 2-12. Fourier transforms of the EXAFS data recorded for KZC as a function of temperature (after Bhat et al., 1990 b).

modelled Debye-Waller factors and is noticeable by a reduction in amplitude in the Fourier transforms. The change in tetrahedral distortion as a function of temperature is also evidenced in Table 2-2. Although the Zn radial distribution function (RDF) is the summation of the 12 closely distributed Zn-Cl bond lengths, a 2-shell model provides a useful insight to the change in the relative weighting of the neighbouring shells with temperature. Below 212 °C the data can be modelled by 2 shells equally weighted (2:2) at Zn-Cl distances of about 0.220 nm and 0.229 nm whereas above 212 °C the weighting changes to about 1.5:2.5 respectively.

Upon recycling to room temperature the structure returns close to that observed before transformation. Thus, although X-ray topography (see Sec. 2.6) shows that these phase transformations do induce changes in macro-structure, i.e., the generation of defects (El Korashy, 1988), this study indicates that these changes are not directly correlated with changes in the local atomic structure.

2.2.9 Dynamic Investigation of the Reduction of Nickel Formate

It has been demonstrated (Frahm, 1989a and 1989b) that high quality X-ray absorption spectra can be obtained using the quick-scanning or QEXAFS technique. QEXAFS uses a continuous scan of the monochromator axis and time integration of the data rather than the conventional step/count approach. The improvement in data acquisition time is significant and opens the possibility of the application of the technique to the characterisation of structural changes in chemical systems, particularly in cases where the kinetics of transformation may be comparatively slow (≈ 1 min).

The applicability of the technique to studies of thermally-induced solid-state decomposition is here demonstrated by examining the decomposition of nickel formate dihydrate [$Ni(COOH)_2 \cdot 2H_2O$, hereinafter NF]. The sample was mounted in a small transmission micro-furnace (Bhat et al., 1990a) which can operate in the temperature range 25–320 °C. QEXAFS spectra were recorded as a function of temperature (Edwards et al., 1990).

Differential scanning calorimetry of NF reveals three noticable features at temperatures of 170 °C, 210 °C and 270 °C. These have been ascribed to transitions from the hydrated to anhydrous formate (A), from anhydrous formate to oxide (B) and from oxide to metal (C) (Vecher et al., 1985; Iglesia and Boudart, 1984; Waiti et al., 1974). The reaction to the metallic state progresses by a nucleation and growth mechanism (Waiti et al., 1974) and this is evidenced from the isothermal ($T = 257$ °C) Fourier transformed QEXAFS data (Fig. 2-13) taken in the metastable region close to transition C.

The nature of the transformed phase is clearly shown in the room temperature EXAFS data shown in Fig. 2-14. It can be seen that the re-cycled material resembles the metal rather than the oxide structure from which it has transformed. The main difference is the significant amplitude reduction observed in the recycled formate which is indicative of the formation of rather small metal clusters. The data were modelled using curved wave theory (Gurman, 1988). Modelling the refined data against hypothetical radial distributions around Ni based on spherical clusters of various radii (Table 2-3) yields a particle size of around 0.5 nm. This is in good agreement with measurements of particle size by diffraction line broadening (Waiti et al., 1974).

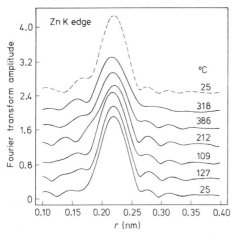

Figure 2-13. Ni K edge Fourier transformed QEXAFS data of nickel formate dihydrate (100 ms integration time and scan speed 20 mdeg s^{-1}) recorded in situ and isothermally in the metastable region close to transition C showing the slow thermal reduction of NiO to metallic Ni (after Edwards et al., 1990).

2.2.10 Role of the Sulfidation Process in the Operation of Pt/Re Catalysts

The preparation of reforming catalysts involves a number of classic but poorly understood basic steps such as impregnation, calcination and reduction followed by modifying steps such as sulfiding. The utility of EXAFS in the characterisation and optimising the sulfidation stage for a Pt/Re catalyst has been demonstrated by Oldman (1986). Figure 2-15 shows Pt L$_3$ absorption edges for differential oxidation states. At the edge, absorption occurs via a fully allowed pseudo-atomic transition, 2p → 5d, with the intensity depending on the number of unoccupied valence levels. An excellent correlation exists between the normalised white line amplitude and the oxidation state.

The edge data, the Fourier transforms (Fig. 2-16) and least-squares fitting of the EXAFS data show that a poor catalyst performance is associated with the com-

Table 2-3. Ni f.c.c. radial distributions calculated as a function of cluster size compared to the refined data for the cycled formate (after Edwards et al., 1990).

Cluster rad. (nm)	Atoms in cluster	1st shell	2nd shell	3rd shell	4th shell
0.3	13	5.54	1.85	3.69	0.92
0.5	55	7.85	3.27	9.60	4.15
0.8	201	9.43	3.94	14.57	6.86
∞	∞	12.00	6.00	24.00	12.00
This data		5.8	3.1	10.9	4.7

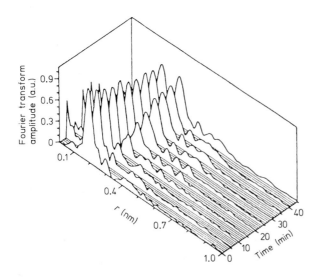

Figure 2-14. The Ni K edge k^3-weighted EXAFS data recorded at room temperature showing the comparison between thermally cycled nickel formate dihydrate, NiO and Ni metal (after Edwards et al., 1990).

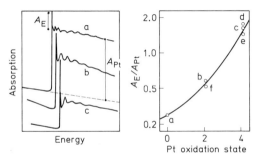

Figure 2-15. L$_3$ edge XAS for some model materials showing the correlation between normalised white line intensity (A_E/A_{Pt}) and oxidation state (after Oldman, 1986): (a) Pt foil, (b) PtCl$_2$, (c) PtO$_2 \cdot$ 2H$_2$O, (d) Na$_2$Pt(OH)$_6$, (e) H$_2$PtCl$_6$, (f) PtS.

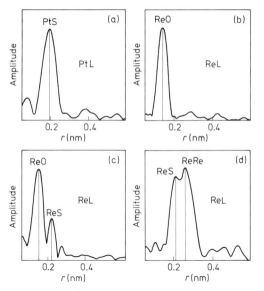

Figure 2-16. k^3 weighted Fourier transforms without phase shifts for 2.5% Pt/2.5% Re on alumina (a) Pt L$_3$ edge (b) Re L$_3$ edge both with original sulfiding procedure followed by air exposure. (c) Re L$_3$ edge revised procedure and air exposed, (d) Re L$_3$ edge reduced in H$_2$ at 450°C (after Oldman, 1986).

plete conversion of Pt to PtS (4S atoms at 2.31 Å) while Re is not sulfided at all. A revised sulfiding procedure shows partial conversion of both Pt and Re. PtS forms a coating on Pt preventing air oxidation, whilst Re is clearly not protected in this way. Under in situ reduction conditions at 450°C PtS is completely reduced to Pt f.c.c. crystallites while ReS$_2$ remains.

2.3 Structural Studies Using X-Ray Diffraction

2.3.1 Introduction

The combination of high flux, wide spectral range (tunability) and low natural divergence makes synchrotron radiation an optimum source for the investigation of long range order in crystalline solids. Over the past decade these properties have been utilised for studies using both white-beam and monochromatic diffraction techniques in the examination of crystalline powders, single crystals and the surfaces of ordered interfaces.

Conventional X-ray analysis is routinely used by crystallographers for the definition of the 3-D molecular structure of small and medium size molecules. For most crystal structures synchrotron radiation offers no significant advantages over laboratory based techniques. Thus the application of synchrotron radiation to single crystal studies has focussed on research areas where the specific and unique properties of this photon source can be utilised for the solution of specific problems. Such areas of application fall into a number of groups; studies of macromolecular biological systems (e.g., proteins and viruses), high resolution powder diffraction and energy dispersive diffraction, dynamical diffraction studies using nearly-perfect single crystals, X-ray standing wave spectroscopy and diffraction from surfaces. Applications to studies of biological systems are beyond the scope of this review and the reader is best referred to a number of reviews (e.g., Helliwell, 1984 and 1991) on this subject.

Studies of surfaces and interfaces (see Sec. 2.4), X-ray standing waves (see Sec. 2.5) and the topographic assessment of nearly perfect crystals (see Sec. 2.6) are described elsewhere in this review.

2.3.2 High Resolution Powder Diffraction

2.3.2.1 Introduction

Conventional powder diffraction analysis using Bragg/Brentano geometry is prone to peak displacement and deformation due to sample misalignment or beam focussing errors. The use of synchrotron radiation sources enable excellent angular resolution to be achieved (Cox et al., 1983; Hastings et al., 1984). Diffractometers designed specifically for use with synchrotron radiation have been developed by Hart, Parrish and co-workers (Hart and Parrish, 1989; Parrish et al., 1986; Hart and Parrish, 1986; Lim et al., 1987).

Figure 2-17 shows a schematic view of the powder diffractometer available on station 8.3 (Cernik et al., 1990 b) at the Daresbury SRS. The wavelength is selected by a Si(111) monochromator which operates in the wavelength range 0.07–0.25 nm. The incident beam cross-section is defined by horizontal and vertical entrance slits. The sample is contained in a flat-plate sample stage which can rotate the specimen about an axis orthogonal to the diffractometer axis and the axis of the synchrotron orbit. The angular resolution of the diffractometer is defined by a set of parallel stainless steel foils which are housed in a vacuum vessel to reduce air scatter. These foils are attached to the 2Θ arm. The diffracted radiation is monitored by an Ar/Xe proportional detector mounted after the collimator at the end of the 2Θ arm.

2.3.2.2 High Resolution Refinement of the Lattice Parameters of Tungsten

The high resolution attainable with the synchrotron powder diffraction technique has been demonstrated through the work of Hart et al. (1990) who demonstrated the accuracy achievable when measuring the lattice parameters of tungsten. To do this four sets of data using two wavelengths (0.1 nm and 0.125 nm) were collected on station 8.3 at the Daresbury SRS. A Si internal (SRM640B obtained from the Na-

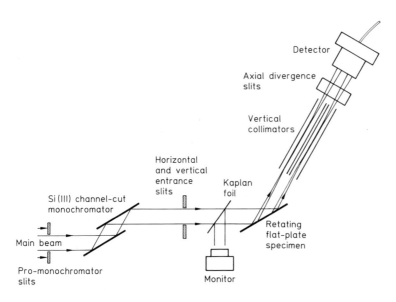

Figure 2-17. The parallel-beam powder diffractometer on station 8.3 at the SRS (after Cernik et al., 1990 b) which comprises Si (111) monochromator, slits, a rotating flat-plate sample holder, two-circle optically-encoded diffractometer, diffracted-beam collimator and an Ar/Xe proportional detector.

tional Institute of Standards and Technology) standard was mixed with the samples in the ratio 3:1 Si:W. Data was collected between 60° and 100° for 2Θ. Peak positions were determined to within 0.1 mdeg for 2Θ by fitting the experimental data to a pseudo-Voight function (Fig. 2-18). All fitted functions were perfectly symmetrical with halfwidths between 0.05° and 0.07° for 2Θ. The mean deviation between calculated and measured 2Θ values was at most 0.0005° to give a lattice parameter accuracy of parts per million. The lattice parameter for tungsten was determined to be 0.3165629(19) nm.

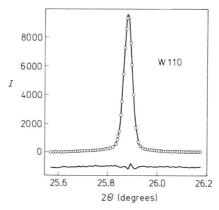

Figure 2-18. High resolution powder diffraction study of tungsten (after Hart et al., 1990). 110 Bragg reflection profile; circles: experimental data; full line: theoretical fit using a pseudo-Voigt function; bottom line: difference between theory and experiment.

2.3.2.3 Determination of the Structure of Cimetidine Using Rietveld Refinement

The use of conventional single crystal methods for the determination of molecular structure is often limited by the difficulties associated with preparing crystals of adequate quality. Cimetidine $[Cd_3(OH)_5(NO_3)]$ is a powerful histamine antagonist of pharmaceutical interest. Its molecular structure has been determined using ab inito methods (Cernik et al., 1990a). Data were collected between 6° and 105° for 2Θ using $\Theta/(2\Theta)$ geometry. Atom positions for non-hydrogen atoms were obtained by direct methods followed by iterative cycles of least squares refinement and Fourier synthesis. These positions were then refined using Rietveld analysis (Rietveld, 1969; Murray et al., 1990, see Section 1.14.8). After convergence was achieved a difference Fourier synthesis showed positive electron density at the expected positions for the hydrogen atoms. These positions were then included in the model structure. Figure 2-19a shows the collected diffraction pattern together with the simulated pattern based on the Rietveld method. The resulting crystal structure (Fig. 2-19b) was found to be monoclinic with four molecules to the unit cell and a space group of $P2_1/n$. That the structure of a low symmetry organic system including unconstrained hydrogen positions can be solved by ab initio methods is an extreme test of the reliability of the synchrotron radiation powder method.

2.3.2.4 Influence of Growth Environment on the Structure of Long Chain Hydrocarbons

The crystal structure of homologous mixtures of n-alkanes is of considerable importance in view of the deleterious influence of wax crystallisation in diesel fuel. The potential influence of the crystallisation environment on the structure of the precipitated wax was investigated (Gerson et al., 1990; Cunningham et al., 1991 a) for the model system of mixtures of n-eicosane ($C_{20}H_{42}$) and n-docosane ($C_{22}H_{46}$) recrystallised from the melt and from n-dodecane ($C_{12}H_{26}$). X-ray diffraction patterns obtained from samples containing multiple phases of low symmetry crystal systems

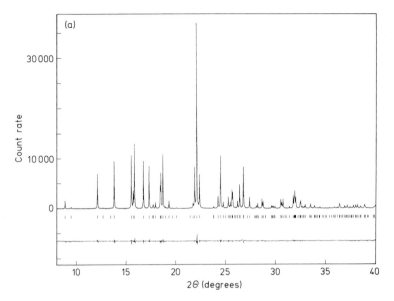

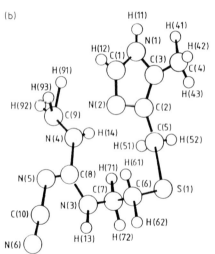

Figure 2-19. The crystal structure of Cimetidine; (a) showing experimental data overlaied with Rietveld refinement, the vertical lines indicate the positions of all expected diffraction peaks and the bottom curve shows the difference between experiment and theory; (b) model showing the refined molecular structure (after Cernik et al., 1990a).

are extremely complex and to index these patterns good quality high resolution data are needed. An indication of the complexity of the patterns produced is shown in Fig. 2-20a which presents the central section of the diffraction pattern recorded from a sample comprising 18% $C_{20}H_{42}$, 82% $C_{22}H_{46}$ crystallising from $C_{12}H_{26}$. The number of phases present in such a mixture is clearly indicated by the 3 sets of (001) peaks present in Fig. 2-20b.

The phase system of the mixed homologues can be clearly seen from the refined unit cell parameters and lattice plane spacings summarised in Table 2-4. The structures of the samples crystallised from the melt confirm the observations of Lüth et al. (1974); the pure and mixed homologues crystallising with triclinic (one molecule in the unit cell, $z = 1$) and orthorhombic ($z = 4$) structures respectively. The orthorhombic phase observed in the mixed systems is thought to be of a related structure to that known for the single odd n-alkanes, the main difference being in the degree of disorder of the molecules (Lüth et al., 1974). A different behaviour was observed when the same mixtures were crystallised from n-dodecane where a rather complex phase behaviour occurs involving a number of crystalline structures including a new polymorphic form which phase-separates from the mixtures over the

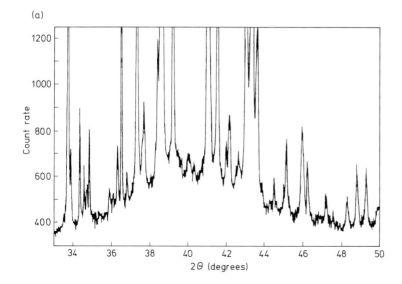

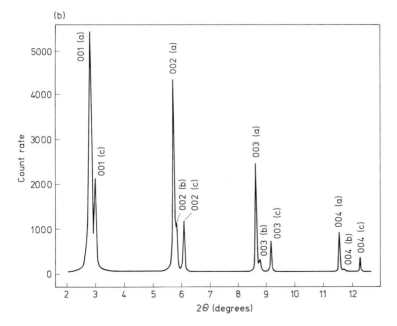

Figure 2-20. Synchrotron radiation diffraction patterns showing phase separation in a mixture comprising 18% $C_{20}H_{42}$ and 82% $C_{22}H_{46}$ crystallised from $C_{12}H_{26}$: (a) wide angle diffraction pattern; (b) the low angle diffraction pattern (after Cunningham et al., 1990b).

composition range from ≈ 45 to 90% *n*-docosane.

Crystallographically the orthorhombic and triclinic ($z = 2$) phases of mixed homologues may be able to be considered to exhibit positional disordered crystal structures with the *n*-eicosane and *n*-docosane molecules randomly distributed over the available crystal sites. Further disordering may also be present due to partial molecular rotation. Molecular packing difficulties arising in this disordered orthorhombic structure result in unit cells less dense than those observed for single *n*-alkane systems.

The use of synchrotron X-rays to study phase transitions under high pressure is discussed by Ruoff (1991).

Table 2-4. Unit cell parameters for homologous mixtures of $C_{20}H_{42}$ and $C_{22}H_{46}$ recrystallised from the melt and from dodecane solution, (*) is the average molecular volume based on the refined lattice parameters (after Cunningham et al., 1991 b).

C20:C22	a (nm)	b (nm)	c (nm)	α (°)	β (°)	γ (°)	* (nm³)	z	Growth method
100:0	0.428	0.482	2.552	91.00	93.54	107.39	0.501	1	melt/soln
99:1	0.429	0.483	2.555	91.42	93.49	107.41	0.503	1	soln
	0.499	0.769	5.797	90.00	90.00	90.00	0.536	4	
84:16	0.502	0.769	5.633	90.00	90.00	90.00	0.443	4	melt
64:36	0.504	0.767	5.747	90.00	90.00	90.00	0.550	4	soln
50:50	0.502	0.765	5.825	90.00	90.00	90.00	0.556	4	melt
40:60	0.504	0.762	5.852	90.00	90.00	90.00	0.561	4	soln
28:72	0.498	0.753	5.922	90.00	90.00	90.00	0.559	4	soln
	0.464	0.698	2.802	84.28	93.22	97.50	0.447	2	
	third phase, longest lattice plane spacing = 2.934 nm								
24:76	0.501	0.752	5.959	90.00	90.00	90.00	0.561	4	soln
	0.464	0.699	2.818	84.72	92.99	97.39	0.451	2	
18.82	0.495	0.753	5.925	90.00	90.00	90.00	0.558	4	soln
	0.464	0.696	2.802	84.72	93.16	97.46	0.446	2	
	third phase, longest lattice plane spacing = 2.921 nm								
14:86	0.498	0.750	5.923	90.00	90.00	90.00	0.553	4	soln
	0.464	0.703	2.802	84.51	93.11	97.29	0.449	2	
	third phase, longest lattice plane spacing = 2.920 nm								
8:92	0.463	0.707	2.803	84.66	93.18	97.03	0.453	2	soln
1:99	0.496	0.750	5.861	90.00	90.00	90.00	0.545	4	son
	0.429	0.481	2.791	91.77	91.96	107.09	0.550	1	
0:100	0.428	0.483	2.793	91.59	92.39	107.83	0.548	1	melt/soln

2.3.3 Energy Dispersive Powder Diffraction

2.3.3.1 Introduction

Monochromatic synchrotron radiation techniques suffer from the obvious disadvantage that only a fraction ($\approx 0.01\%$) of the available wavelength bandwidth provided by a synchrotron source is used. This disadvantage can be overcome by using energy dispersive techniques. Energy dispersive X-ray diffraction (EDXRD) is defined from the alternative formulation of Bragg's law in terms of energy (E);

$$E \text{ (keV)} \approx 6.2/(d \sin \Theta) \qquad (2\text{-}7)$$

in which each set (d) of lattice planes satisfies Bragg's law for a discrete energy and the synchrotron white radiation source is used with an energy sensitive detector to provide a diffraction spectrum as a function of energy at a fixed detector angle. The fixed detector scattering geometry means that diffraction patterns can be recorded in a few minutes. Also since it is a one beam in, one beam out technique small aperture environmental cells for in-situ studies can be used.

Figure 2-21 shows a schematic view of the dedicated instrument for EDXRD studies, station 9.7 at the Daresbury SRS

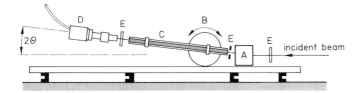

Figure 2-21. Energy dispersive X-ray diffraction facility on station 9.7 of the Daresbury SRS; sample area (A), detector axis (B), collimator (C), detector (D) and slits (E) (after Cunningham et al., 1990b).

(Clark, 1989; Clark and Miller, 1990). The axial divergence and therefore the resolution is controlled by a set of Soller slits placed prior to the solid-state detector but is limited by the electronics of the detector system. These are composed of 25 metal foils 50 cm long each separated by 0.1 mm. Slits and detector are both mounted on a 2Θ rotation arm. Horizontal and vertical slits reduce the incident beam size to within the counting limits of the solid-state detector.

2.3.3.2 In Situ Characterisation of Solid-State Chemical Transformations

The fixed diffraction geometry of the EDXRD technique is particularly suitable for in situ studies of chemical transformations in the solid state (e.g., Anwar et al., 1990). Applications have been made to a number of systems including studies of the calcination of zirconium hydroxide (Mamott et al., 1991) and the hydration of cement (Barnes et al., 1991).

Zirconium hydroxide, an important chemical precursor used in the synthesis of zirconia, can be precipitated from a zirconium salt by means of base addition or the reaction of a slurry of an insoluble zirconium salt with a base (Houchin, 1987). One of the main structural units thought to be associated with zirconium hydroxide are tetramer units which join together during precipitation by a polymerisation reaction (Clearfield, 1964). Subsequent processing of the precipitated powder involves calcination where the amorphous phase is converted to a crystalline form and conversion where this phase transforms to the desired monoclinic phase. Figure 2-22 shows energy dispersive data from a detailed study by Mamott et al. (1991) who carried out an in-situ investigation of how a change in pH of the liquid surrounding the zirconium hydroxide precipitate affected the dynamic parameters of the synthesis process and the kinetics of conversion from the tetragonal to monoclinic zirconia. Their study shows that the kinetics of the initial precipitation stage directly affects the kinetics of both the calcination and conversion reactions. In particular their data appear to confirm the mediating role played by OH^- ions which seem to control the balance between the hydration and dehydration stages of this reaction. They find that an increase in pH during precipitation generates excess OH^- ions which supress dehydration, enhancing the polymerisation rate and resulting in a more amorphous phase.

Despite its universal role in the construction industry there is surprising little information concerning structural changes associated with the hydration of cement. The action of hydraulic cement commences with a mixing of a finely ground (unhydrated) cement powder with water in a typical ratio of 0.5–0.4. The setting time of a cement depends on its nature and on the conditions but can be as short as a few minutes. The hydration of a fast setting cement, based on hydrates of calcium trisulpho-aluminate (often referred to as

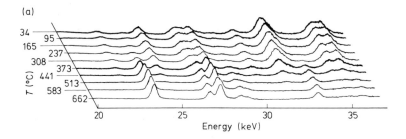

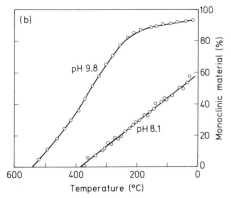

Figure 2-22. Dynamic energy dispersive diffraction data showing the conversion stage in the preparation of zirconia (after Mamott et al., 1991). (a) Part of the diffraction pattern taken during the cooling sequence, the transformation from the tetragonal to monoclinic phases at $T \approx 373$ K is clearly shown. (b) Plot of the growth of the monoclinic phase for samples prepared at pH values of 9.8 and 8.1 as determined from an assessment of the relative peak areas of the two phases.

Ettringite) has been investigated (Fig. 2-23) in-situ by Barnes et al. (1992). Figure 2-23a shows a plot of the diffraction pattern obtained as a function of hydration time. From an analysis of the peak positions of the 110, 112 and 114 diffraction profiles it can be seen that there are significant changes in the a and c parameters in the hexagonal Ettringite unit cell as a function of hydration with, for example, the a parameter increasing by about 0.01 nm ($\approx 1\%$) during hydration. Barnes et al. (1992) ascribe the lattice dilation to the initially formed Ettringite being deficient in sulphate due to the relatively low solubility of the anhydrous calcium sulfate. In the initial stage this is substituted by the more plentiful carbonate phases. The time dependent evolution of the hydration reaction is shown in Fig. 2-23b which reveals that the development of the Ettringite phase begins immediately after mixing and continues to grow in a sigmoidal manner without any obvious induction time. Further experiments show that the maximum phase formation takes, typically, 1–2 days to occur.

2.3.3.3 Structural Phase Transformations at High Pressure

Energy dispersive diffraction using synchrotron radiation is well suited to high pressure studies since it permits the examination of minute volumes of samples in high pressure cells. Moderate pressure can be achieved in Drickamer cells and exceptionally high pressures in diamond anvil cells.

Hauserman and coworkers (Hauserman et al., 1989 and 1991) have used a Drickamer-type high pressure cell to measure the compressibility of TiB_2. Such measurements demand the use of high photon energy (i.e., >50 KeV) part of the synchrotron radiation spectra as the X-rays need to penetrate a mixture of sample and pressure calibration material surrounded by several millimeters of pressure transmitting material. TiB_2 crystallises in a hexagonal crystal system with a space group $P6/mm$. Hauserman et al. (1989 and 1990) typically

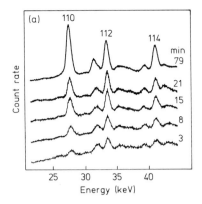

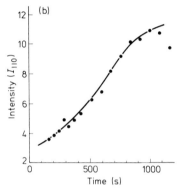

Figure 2-23. Time dependant energy dispersive diffraction data revealing the development of Ettringite phases during the in situ hydration of cement (after Barnes et al., 1991). (a) Diffraction patterns showing the 110, 112 and 114 Bragg reflections. (b) Plot of the evolution of the intensity of the 110 Bragg peak as a function of hydration time.

used 4 or 5 copper reflections to determine the pressure and 6 to 8 reflections from TiB_2 to calculate the changes in both the a and c lattice parameters. Figure 2-24a shows some of the data recorded at pressures of 25.2 (± 0.1) and 123.5 (± 0.6) kbar. Measurements of the lattice constants as a function of pressure revealed (Fig. 2-24 b) the bulk modulus K_0, of TiB_2 to be 4.1 (± 0.2) Mbar and showed TiB_2 to be one of the least compressible materials known (diamond has $K_{0'} \approx 4.4$ Mbar (Mc-Skimin and Bond, 1957) and thus a potentially useful industrial material.

In contrast to the moderate pressures in Drickamer cells very high pressures (up to ≈ 1 Mbar) can be obtained using diamond anvil cells. Diamond is an ideal window material due to its transparency not only to X-rays but also to optical frequencies. The optical transparency of the cell windom material can be applied by using the fluorescent lines of Ruby, which shift linearly with pressure, to calibrate the cell pressure.

Interest in the solid-state properties of mixed metal oxides has increased significantly since the discovery by Bednorz and Müller (1986) of superconductivity at temperatures up to 30 K in the La-Ba-Cu-O system. Chu et al. (1987) have shown that under pressures of up to 2 kbar the maximum temperature for superconductivity is increased to 52.5 K. This work has led to an increase in effort in both theoretical and experimental studies towards gaining an understanding of the mechanism responsible for high temperature superconductivity. Akhtar et al. (1988) studied the structure of La_2CuO_4, prepared by solid-state reaction (Longo and Raccah, 1973) at high pressure using a diamond avil and energy dispersive diffraction. The data, summarised in Fig. 2-25 show that up to ≈ 150 mbar the lattice constants decrease linearly with pressure whilst the near constancy of the c/a axial ratio demonstrates that the compressibility of this material is isotropic. Akhtar et al. (1988) correlate their data with band structure calculations (Stocks et al., 1988) which appear to suggest that the inter-ionic bonding for this material is dominated by the 3-d network of La-O interactions. Such behaviour contrasts with the known 2-d magnetic properties (Shirane et al., 1987) which originate from the $Cu-O_2$ layers.

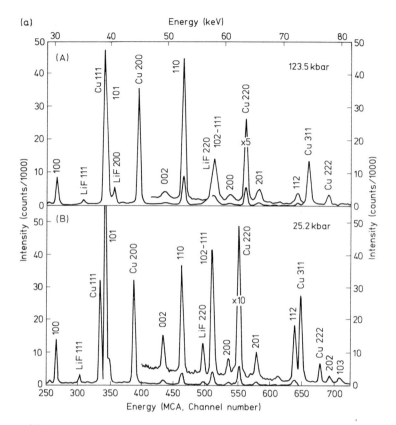
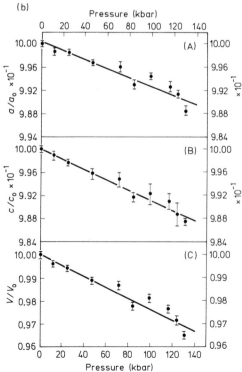

Figure 2-24. Studies of the compressibility of TiB$_2$ using energy dispersive diffraction and a Drickamer-type high pressure cell (after Hausermann et al., 1991). (a) Diffraction patterns taken at 123.5 and 25.2 kbar, the reflections labelled in black are of the added copper pressure calibration sample, some LiF contaminant peaks are also discernable; (b) reduced a and c lattice parameters (A and B) together with unit cell volumes (C) as a function of pressure.

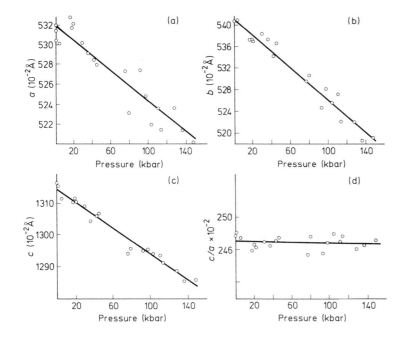

Figure 2-25. Energy dispersive data showing the isotropic compressibility of La_2CuO_4 using a diamond anvil cell (after Akhtar et al., 1988), $2\Theta = 9.81°$, open circles data points and full lines show the computed fit; (a)–(c) effect of pressure on the a, b and c lattice parameters respectively; (d) effect of pressure on the c/a axial ratio.

2.3.3.4 Particle Formation in a Liquid Environment

Very little information is available on the structural nature of crystalline particles as they form within solution and in particular the relationship between structure and growth kinetics. In principle X-rays should prove to be an effective probe for detecting crystal nucleation and growth. The much shorter wavelength of X-rays compared to visible radiation should allow the detection of crystal nuclei at much smaller sizes than when using conventional methods such as optical turbidity (e.g., Gerson et al., 1991 b). Using the high energy of synchrotron radiation the EDXRD technique has been applied (Gerson et al. 1990; Cunningham et al., 1990 and 1991) to collect powder diffraction patterns in situ on n-alkanes during crystallisation from solution. However, the general applicability of this technique has been limited to some extent by problems associated with solution heating.

Recent work by Doyle et al. (1991), summarised in Fig. 2-26, used a variation of the EDXRD technique in which a fast scanning monochromator is used to provide the wavelength range needed for the energy dispersion. The diffraction patterns Fig. 2-26a taken during the crystallisation of n-eicosane from a saturated solution of n-dodecane show the formation of well-resolved Bragg peaks demonstrating that in situ diffraction data can be obtained from n-eicosane crystals in the solution environment. Scans of diffraction intensity at constant energy Fig. 2-26b used to detect the onset of crystallisation revealed three distinct regions; firstly between 20 and 18 °C there is little change in the scattered intensity, indicating the absence of crystalline material. Between 18 and 14 °C the intensity increases approximately linearly as crystallisation occurs. Finally, below 12 °C the intensity becomes constant indicating effectively complete crystallisation of the material. The dotted line in Fig. 2-26b represents the equivalent experiment carried out in the laboratory using an automated turbidometric apparatus (Gerson et al., 1991 a). Whilst the behaviour in the two experiments is similar the obvious discontinuity in the initial slope of the data collected using X-rays provides potential direct evidence of significant ordering in the liquid phase prior to crystallisation.

2.3.4 Laue Diffraction

2.3.4.1 Introduction

Since the publication of the first Laue diffraction patterns in 1912 (Friedrich et al., 1912) this technique has been used routinely by metallurgists and mineralogists to define crystal orientation and symmetry. The general applications of the technique are well documented in the book of Amoros et al. (1975).

The Laue method uses a stationary crystal and a polychromatic X-ray beam. Synchrotron radiation is therefore an ideal source for this purpose and offers many advantages over traditional sources for routine use. The very broad spectral range provides diffraction patterns with a very high information content (many thousands of reflections) (Cruickshank et al., 1987) and the highly collimated beam offers a much superior resolution. Of greatest importance, however, is that the high intensity of the beam allows these patterns to be recorded from very small crystals ($\approx 100\ \mu m^3$) (Andrews et al., 1988) in very short periods of time (0.1–10 s). This allows the technique to be used for the assessment of crystal structure and quality at the very earliest stages of materials preparation when only small crystals are available. It also offers an alternative prospect of dynamic and in situ examinations of structural changes during processing.

A further and great advantage of the synchrotron radiation Laue technique is that it can be used for molecular structure determination. In the early days, this prospect was discounted by Bragg (1949 and 1975) who argued that this was precluded

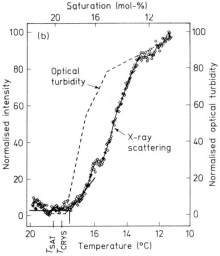

Figure 2-26. In situ data taken using the wavelength scanning EDXRD technique (after Doyle et al., 1991) with a detector angle 2Θ of 4° during the crystallisation of a saturated solution of 20% n-eicosane in n-dodecane; (a) a series of diffraction peaks at varying temperatures, (b) plot (◇◇◇◇◇) of the diffracted beam intensity for the diffraction peak labelled A in (a) as a function of cooling the solution at 0.5 °C/minute. Turbidometric data (– – –) are overlaied for comparison.

because several orders of reflection could be superimposed on one Laue diffraction spot, thus preventing the evaluation of an electron density map. Cruickshank et al. (1987) have since shown that with synchrotron radiation the overlapping orders problem is not limiting. This has opened the way, in combination with the above properties, to molecular structure determinations using small crystals in short time scales (Andrews et al., 1988) and hence with labile materials. Dynamic changes in molecular structure can also be followed even in complicated systems such as catalysis in crystals of phosphorylase-b (Hajdu et al., 1987), and phase transitions between different molecular forms of insulin (Reynolds et al., 1988).

2.3.4.2 Instrumentation and Experimentation

The typical Laue diffraction system shown in Fig. 2-27 is that used on the Daresbury synchrotron radiation source. The front end consists of a water cooled tungsten carbide aperture followed by a set of foil attenuators to reduce radiation damage by the softer X-rays in the spectrum and an incident beam monitor.

This is followed by a collimation system comprising a fast shutter giving exposures down to a few milliseconds and a set of pinhole collimators. These provide a timed and defined beam to the crystal situated on a goniometer which provides two orthogonal axes of rotation. Data recording is carried out on high speed films such as Reflex 25. Exposure times are the orders of fractions of a second for normal size (≈ 1 mm) crystals and still only a few seconds or minutes for very small crystals.

2.3.4.3 Radiation Damage

With this method, as with all other synchrotron radiation methods which use the full spectral range of the synchrotron beam, the problems of radiation damage are ever present (Roberts et al., 1982; Bhat et al., 1982, 1984 and 1988). Radiation damage may be minimised or removed by correct use of the attenuator positioned in the front end of the system. With materials such as sodium chlorate, which we find to be ultrasensitive, the introduction of an aluminium absorber (200–1000 µm) removes sufficient of the softer radiation to allow reliable experiments for a reasonable experimental period; at least 10–100 times the experimental recording period. For less sensitive inorganic and organic materials, thinner absorbers yield satisfactory results. The most obvious consequence of radiation damage is the gradual development of a radial asterism around the diffraction spots. Assessment of the rate of development of this can give information on the rate of structural damage which can be correlated with other factors such as the formation of colour centres (Halfpenny et al., 1991). The assessment of this prob-

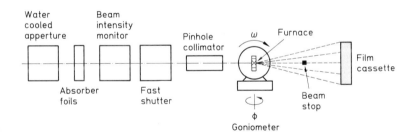

Figure 2-27. Schematic representation of the Laue diffraction camera at the Daresbury Synchrotron Radiation Source.

Table 2-5. The mosaic spread of some organic single crystals determined by synchrotron radiation Laue diffraction (after Roberts et al., 1983).

Material	Formula	Crystal system	Growth technique	Sample size (mm) h	v	t	Accumulated dose (Mrad)	Mosaic spread	Remarks
benzoic acid	C_6H_5-COOH	monocl.	melt	23	21	6.0	0.5	~24'	uniformly strained
anthracene	$C_{14}H_{10}$	monocl.	melt	18	8	2.0	1.7	~1.6°	grossly strained, sample colours
naphthalene	$C_{10}H_8$	monocl.	melt	12	6	2.0	5.2	~2°	grossly strained
p-terphenyl	$C_6H_5-C_6H_4-C_6H_4$	monocl.	melt	31	10	4.0	1.3	~7°	grossly strained
stilbene	$C_6H_5-(CH_2)_2-C_6H_5$	monocl.	melt	21	11	3.0	0.4	~2.5'	uniformly strained
terephthalic acid	$COOH-C_6H_4-COOH$	tricli.	solution	0.1	0.1	0.1	0.1	~27"	grossly strained
TTF–TCNQ[a]	$C_6S_4H_4 \cdot C_{12}N_4H_4$	monocl.	diffusion	10	3	0.1	9.9	~70"	no asterism, some resolution of micro-defects
benzophenone	$C_6H_5-CO-C_6H_5$	orthor.	solution	22	15	1.5	58.7	~70"	micro-defect resolution, some colouring of the sample
benzophenone	$C_6H_5-CO-C_6H_5$	orthor	melt	8	8	2.0	5.9	~50'	uniformly strained
benzil	$C_6H_5-(CO)_2-C_6H_5$	trigon.	melt	7	7	2.0	11.1	~6'	uniformly strained, sample colours
fluorenone	$C_6H_4-CO-C_6H_4$	orthor.	solution	15	12	1.5	47.8	~70"	no asterism, some resolution of micro-defects
carbazole	$C_6H_4-NH-C_6H_4$	orthor.	melt	13	5	1.0	12.9	~70"	no asterism, surface damage obscures bulk defects
adamantane	$(CH)_4(CH_2)_6$	cubic	solution	3	3	3.0	9.8	~70"	resolution of microdefects, region in sample misaligned by ~20'
adamantane	$(CH)_4(CH_2)_6$	cubic	melt	8	8	2.0	7.0	2.7°	grossly strained
urea	$CONH_2$	tetrag.	solution	4	3	0.5	0.6	~70"	no asterism, some resolution of micro-defects
urea	$CONH_2$	tetrag.	prilled	2.5	2.5	2.5	0.3	~1'	several crystallites (<10), low asterism for individual grains
urea/NaCl	$CONH_2 \cdot NaCl$	orthor.	solution	13	6	1.0	1.0	~10'	some localised surface strain
pentaerythritol	$C-(CH_2OH)_4$	tetrag.	solution	12	8	2.0	2.4	~8'	low asterism, some resolution of macro-structure
oxalic acid dihydrate	$COOH-COOH \cdot 2H_2O$	orthor.	solution	21	5	3.0	0.3	~11'	thick sample, low asterism, some resolution of surface defects

Compound	Formula	Crystal system	Growth method					Comment
n-eicosane	$CH_3-(CH_2)_{18}-CH_3$	tricli.	melt	5	5	2.0	5.2	$-2°$	grossly strained, significant radiation damage
PTS[b]	$R=CH_3-C_6H_5-SO_2-O-CH_2-$	monocl.	solution	5	3	2.0	3.5	$-10'$	low localised strain
TCDU[b]	$R=C_6H_5-NH-CO-O-(CH_2)_4-$	monocl.	solution[c]	3	3	1.0	5.9	$-70'$	uniformly strained
DDEU[b]	$R=C_2H_5-NH-CO-O-(CH_2)_3-$	monocl.	solution	6	2	0.1	1.7	$-1.9°$	grossly strained

[a] Tetrathiofulvalene-tetracyanoquinodimethane.
[b] Single-crystal diacetylene polymers; general formula of monomer $R-C\equiv C-C\equiv C-R$.
[c] The monomer crystals were grown from solution, before polymerisation.

lem is an essential preliminary to all such synchrotron experiments since many do involve the existence or development of strain in samples.

2.3.4.4 Assessment of the Quality of Organic Crystals

The Laue method is a useful and simple technique for the assessment of crystal quality. Beyond the limit of X-ray topography with synchrotron radiation ($<500''$ mosaic spread) (see Sec. 2.6) there are few techniques available to the crystal grower which can be used to assess the quality of large crystals produced when developing a new growth technique. Such information is essential in order to assess the influence of changing growth parameters on increasing perfection. Small beam Laue diffraction can be used to search the volume of a crystal to produce information on the local perfection. The larger synchrotron beam (≈ 2 cm diameter at the Daresbury SRS) allows the complete assessment of crystals of this size in one exposure of around a few minutes duration. The overall perfection of the crystal is demonstrated by the quality of the image. A net and well-defined image is typical of a crystal of high quality which is potentially useful for more detailed X-ray topography. An increase in mosaic spread is evidenced by an asterism of the image streaking radially across the plate. Analysis of the degree of asterism (Amoros et al., 1975) can be used to calculate mosaic spread. Details of a typical study (Roberts et al., 1983) of a range of organic crystals are given in Table 2-5.

2.3.4.5 Solid-State Polymerisation of PTS

The monomer 2,4-hexadiynediol-bis-(p-toluene sulfonate) (PTS) undergoes polymerisation in the solid state by a single crystal to single crystal transformation. The

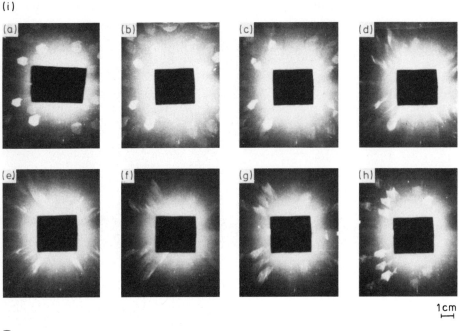

Figure 2-28. (i) Selected Laue patterns for PTS from a series taken on Polaroid film recording the progress of radiolytic polymerisation after cumulative radiation doses of; (a) 77 Mrad, (b) 185 Mrad, (c) 259 Mrad, (d) 458 Mrad, (e) 551 Mrad, (f) 619 Mrad, (g) 657 Mrad, (h) 716 Mrad (after Dudley et al., 1991). (ii) The kinetics of polymerisation of PTS expressed as a change in relative spot area (x) (324-reflection) versus accumulated time in the beam A. Unfiltered beam at 295 K: B. Unfiltered beam at 150 K: C. Filtered beam (715 mm Al) at 295 K.

reaction can be initiated radiolytically, photochemically and thermally. The progress of reaction can be followed by the use of either full beam or narrow beam "spot" Laue patterns. Figure 2-28a shows a range of Laue patterns from a typical experiment. The patterns were recorded on polaroid film and represent a selection from a total of 60 taken over a 30 minute period. The original and final patterns which represent those of the monomer and polymer respectively are distinctive and confirm the perfection of each. It will be noted that as the reaction initiates and proceeds strain develops due to the formation of polymer in the crystal. This causes an asterism. As the reaction proceeds to completion this strain relaxes and the pattern gradually develops to that of the polymer. This experiment is a radiolytic experiment carried out using the synchrotron beam as both the initiator and probe of the reaction.

Analysis of the rate of development of the asterism allows the definition of the kinetics of the process. Figure 2-28b shows that the influence of decreasing temperature on the induction time to nucleation of the reaction was less than that obained by filtering the softer X-ray components from the beam. By using the latter method, the induction time could be increased sufficiently to allow the study of thermal and UV polymerisation well before rapid radiolytic polymerisation was initiated. Since each crystal image is a topograph (see Sec. 2.6) this also allowed the study of the influence of lattice perfection on the polymerisation process (Dudley et al., 1985, 1986, 1990, and 1991).

2.3.4.6 Influence of Strain on the Growth Rate of Secondary Nucleated Particles

A major problem in the performance of industrial crystallisers is the growth rate dispersion of secondary nuclei (attrition fragments). Many of these fragments show a zero growth rate. Studies of the structural perfection of micro crystals (5–25 mm diameter) using Laue diffraction techniques have shown that the growth rate of the crystals at constant supersaturation is proportional to the mosaic spread of the crystal (Fig. 2-29). This influence has been found for sodium chlorate (Ristić et al., 1988), potash alum (Ristić et al., 1991), sodium nitrate and Rochelle Salt (Mitrovic et al., 1990).

These results parallel similar observations on other small crystals of more complicated materials, such as proteins and zeolites, for which a cessation of growth is found when they have attained a certain size (Andrews et al., 1987). Those which cease to grow were found to have the largest mosaic spread.

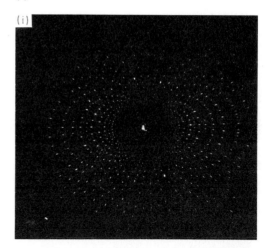

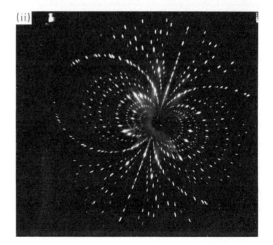

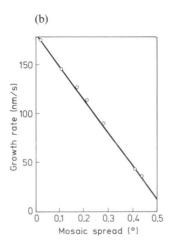

Figure 2-29. (a) Laue photographs of two different Rochelle salt crystals taken using the SRS "white" beam: wavelength range approximately 3–25 nm, collimator 0.2 mm, crystal film distance 50 mm. (i) Asterism obtained for a crystal with growth rate 114.2 nm s^{-1}; (ii) stronger asterism for a crystal with growth rate 52.3 nm s^{-1}. (b) The variation of growth rate for Rochelle salt crystals as a function of mosaic spread (after Mitrovic et al., 1990).

2.3.4.7 Probing Structural Changes During Phase Transformations

Although the Laue topographic technique (Tuomi et al., 1974) has been used for investigations of solid-state transformation (Bordas et al., 1975, Aleshko-Ozhevsij 1982 and 1983; Zarka, 1983; Gillespie et al., 1989), there have been surprisingly few applications of the use of synchrotron radiation Laue patterns to study phase transitions (El Korashy et al., 1987) or lattice pseudosymmetry (Docherty et al., 1988) in small molecular systems.

The applicability of this technique to solid-state phase transitions has recently been demonstrated (Bhat et al., 1990) for the three systems: potassium tetrachlorozincate (KZC), pentaerythritol and caesium periodate. These three systems were chosen to include a typical range, mechanistically, of the types of phase transition currently of interest and hence to demonstrate the broad applicability of synchrotron Laue techniques to the study of solid-state transitions.

As an example of this work we show (Fig. 2-30) four Laue photographs taken of

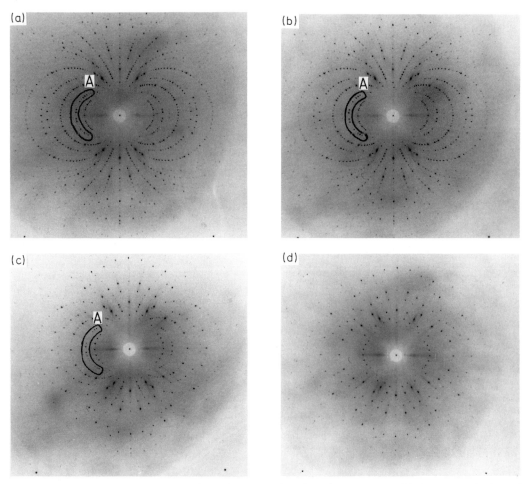

Figure 2-30. Sequence of Laue patterns taken of KZC as a function of temperature with the beam axis along the [010] lattice direction (a) 60.3 °C; (b) 139.8 °C; (c) 278.9 °C; (d) 328.0 °C (after Bhat et al., 1990a).

KZC as a function of temperature. The data were recorded for the room temperature ferroelectric phase (a), at the beginning of the incommensurate phase (b), towards the end of the incommensurate phase (c) and in the paraelectric phase (d). Zones of satellite spots (labelled A) are clearly visible for the ferroelectric phase (a). These superlattice reflections change in relative intensities [(b) (c)] within the incommensurate phase and disappear completely in the paraelectric phase (d).

2.3.4.8 Application to Crystal Structure Determination

As Laue diffraction patterns do not suffer from the problems associated with overlapping reflections, and due to the excellent properties of synchrotron radiation, synchrotron Laue patterns can be used for crystallographic and molecular structure determination using microcrystals of considerably smaller size than are necessary for traditional methods of examination. In fact, specimens usually regarded as powder particles are potentially useful.

Combined with unit cell parameters, which still have to be obtained from a monochromatic pattern, Laue data have been used to calculate difference Fourier maps of proteins (Hajdu et al., 1987) and to solve small crystal structures by the Patterson method (Harding et al., 1988).

Film can be recorded in one second or less giving the intensities of 2000 or more reflections.

Typical of the complexity of the type of compounds which have been assessed are $C_{25}H_{20}N_2O_2$ ($R = 0.05$) (Helliwell et al., 1989), $Mo_5S_2O_{23}(NEt_4)_4PhCN$ ($R = 0.10$) (McGinn, 1990), $Rh_6(CO)_{14}Ph_2PCH_2PPh_2$ ($R = 0.16$) (Clucas et al., 1988), $FeRhCl(CO)_5Ph_2PC(=CH_2)PPh_2$ ($R = 0.14$) (Harding et al., 1988). The accuracy of

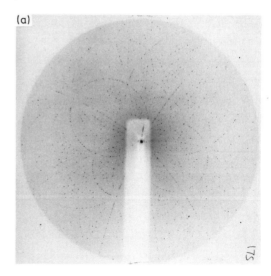

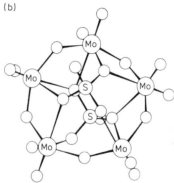

Figure 2-31. (a) Laue diffraction photograph of a small crystal of tetraethyl pentamolybdo-phosphate, taken on station 9.7 on the SRS Wiggler beam line at Daresbury Laboratory, U.K. Crystal size $100 \times 140 \times 15$ mm^3, crystal to film distance 48 mm, exposure time 0.2 s. (b) The molecular structure derived from analysis of these data. (Reproduction by permission of Drs. M. M. Harding and S. J. McGinn, Liverpool University.)

the determination in each case is defined by the refinement factor R. The crystal sizes were of the order of $100 \times 150 \times 15$ μm^3. Figure 2-31 shows a typical example of the Laue pattern and derived structure for one of these materials.

Some examples of synchrotron Laue studies of phase transitions under high pressure, for instance structural studies of solid hydrogen at room temperature at 25 GPa, are cited in Ruoff (1991).

2.4 Diffraction and Scattering from Real Surfaces and Interfaces

2.4.1 Introduction

A comparison of the scattering lengths of X-rays (about 10 µm) and electrons (about 1 nm) shows that in a normal X-ray experiment the radiation scattered from the bulk material will dominate over that from a surface monolayer by about 6–7 orders of magnitude. This inherently low signal, specific to the surface regions, can be overcome by using synchrotron radiation. This has the additional advantage that the high energy and intensity of the radiation enables the structure of interfaces buried under dense overlayers to be examined non-destructively in a variety of environments in addition to high and ultra-high vacuum.

2.4.2 Information Content Available from Surface Diffraction Data

Surface diffraction can be used to assess and characterise the structural order in both "in-plane" and "normal" directions with respect to the interface under examination. In-plane structure can be deduced from an equatorial diffraction scan about an axis parallel to the surface normal. The depth sensitivity of such a measurement can be varied by changing the incident glancing angle and is dramatically enhanced under conditions of total external reflection where the X-rays are confined to the surface regions of the sample (≈ 0.2 nm) as a damped evanescent wave. Measurement of the width of Bragg surface diffraction features provides information on the coherence length or domain size of the interfacial structure. For solid-solid interfaces, scans in reciprocal space with respect to the underlying lattice of the substrate provide direct evidence of reduction in symmetry, as might be expected if the surface reconstructs. They also show whether the layer is incommensurate or not. Such equatorial scans together with scans of the reciprocal lattice normal to the interface, along the Bragg truncation rods, indicate the structural registry of substrate and overlayer. The intensity of Bragg diffracted surface reflections can be analysed using conventional crystallographic techniques. An inherent advantage of this technique is that the data can be readily analysed using kinematic scattering theory. Dynamic theory is not implicitly required.

Diffraction scans can be complemented by high resolution measurements of X-ray reflectivity, made by varying the angle of incidence of the X-rays on the sample surface. Measurements in the angular range $(0.5-10)\phi_c$, where ϕ_c is the critical angle, allow the atomic structure and structural variation normal to the surface to be determined. Structural information can be obtained directly from the interference of Kiessig fringes observed in reflectivity profiles (Kiessig, 1931; Koenig and Carron, 1967; Isherwood, 1977). This is shown schematically in Fig. 2-32 for a three layer system (air/thin-layer/substrate). A scan of reflectivity versus incidence angle firstly shows the total external reflection from the top surface followed by a series of Kiessig fringes due to interference between the specularly scattered X-rays from the "top" and "buried" interfaces. From the angular positions of the interference fringes the layer thickness (t) and the mean electron density of the film can be calculated (Koenig and Carron, 1967).

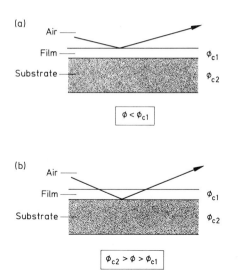

Figure 2-32. Reflection geometry in a three layer (i.e., air/film/substrate) system in which the substrate is more dense than the film. ϕ – angle of incidence, ϕ_{c1} – critical angle for the film and ϕ_{c2} – critical angle for the substrate. (Note: $\phi_{c1} < \phi_{c2}$).

An analysis of fringe positions is only optimal where surface layers are chemically homogeneous as a function of depth. In systems exhibiting a graded interface the complete reflectivity profile needs to be fitted to a model which sums the reflectivity contributions from each sub-layer of the surface. Using such a model the reflectivity from the surface can be calculated from a classical Fresnel recursion relationship (e.g., Parratt, 1954; Wainfan et al., 1959).

2.4.3 Instrumentation

The experimental recording of surface diffraction and X-ray reflectivity data requires the use of a 5-circle diffraction geometry. Figure 2-33 shows a schematic view of the surface diffraction facility available on station 9.4 of the Daresbury SRS.

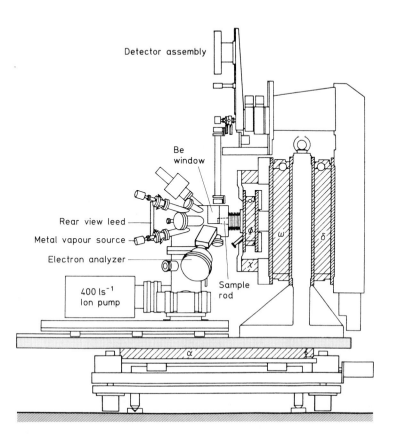

Figure 2-33. Schematic view of the 5-circle goniometer for surface X-ray diffraction used on station 9.4(b). This is shown with the UHV surface science analysis chamber (after Norris, Moore and Harris, unpublished).

This comprises a high resolution 5-circle diffractometer together with monochromator and a 2-circle analyser stage. This diffractometer has the capacity to hold extremely large and massive environmental stages. The experimental requirements for X-ray reflectivity measurement alone are less severe and 2-circle goniometry (see Sec. 2.3) can also be used.

2.4.4 Lattice Coherency in Group III–V Quantum-Well Structures

The definition of materials characterisation procedures for the preparation of semiconductor multilayers is of crucial importance for the development of novel low dimensional structures for applications in integrated opto-electronics. Surface diffraction is being increasingly applied to the characterisation of lattice coherency between the substrate and its heteroepitaxial overlayer (Ryan et al., 1987; Lucas et al., 1988; Jedrecy et al., 1990).

Figure 2-34 summarises some of the results of a study by Ryan et al. (1987) in which the structure of 23 nm layer of InGaAs (nominal composition $In_{0.53}Ga_{0.47}As$) deposited on InP was examined. The sample (a) was prepared by molecular beam epitaxy (MBE) and was studied using a glancing angle of incidence ($\approx 2.5°$) corresponding to a penetration depth of ≈ 240 nm. The triple axis (i.e. monochromator, sample and analyser) diffraction configuration enables a very high resolution in the two dimensions of the scattering plane so that the 2-d intensity distribution (b) of scattered X-rays around a Bragg reflection can be seen. It can be seen that the main Bragg peak itself is split due a slight lattice mismatch between the Fe doped InP substrate and its MBE grown InP buffer layer which induces a slight (0.03%) tetragonal distortion in the epitaxial overlayer. The slight skewing of the line of satellite peaks from the crystallographic axis indicates that the crystal face has been cut at an angle of $\approx 2°$ InP(100) in the [110] direction.

Examination of the environment around the InP 440 Bragg reflection of the sample reveals a series of weak satellite peaks arising from the capping layer of InP. The amplitude and periodicity of these satellites are modulated by a similar set of weaker and broader satellites (c) with roughly three times the period of the capping layer fringes originating from the InGaAs quantum-well layer. Fitting this experimental X-ray pattern produces a structural model for the interface which is in good agreement with that predicted from MBE growth experiments viz. InP cap thickness 68 (± 1) nm, InGaAs layer thickness 23 (± 1) nm, lattice parameter mismatch ($\Delta a/a$) $2.73(1) \times 10^{-3}$, rms surface roughness 2 (± 0.2) nm and rms interlayer interface roughness 1.2 (± 0.2) nm.

2.4.5 Ge/Si (001) Strained-Layer Semiconductor Interfaces

The properties of hetero-epitaxial thin films prepared by MBE for electronic device applications depend on a number of factors including layer perfection. Crystal strain provided by deliberately mismatching the lattices of the substrate and epitaxial film can be used to modify and optimise the electronic or optical properties of the material (e.g., O'Reilly, 1989). The degree of strain that can be accommodated depends inherently upon the layer thickness, with mismatch dislocations being formed for thicker overlayers. Studies have been made of systems such as lattice-matched $NiSi_2/Si(111)$ (Robinson et al., 1988) and Ge/Si(001) (Williams et al., 1989; Macdonald et al., 1990; Macdonald, 1990).

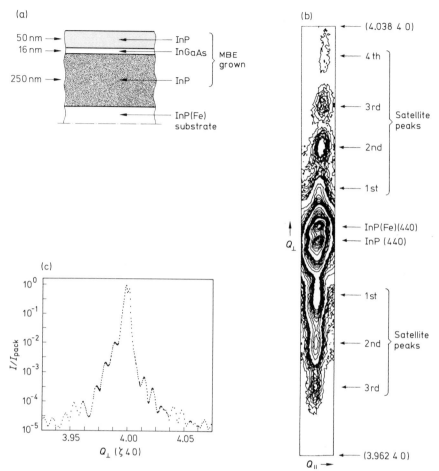

Figure 2-34. Surface diffraction investigation of a single quantum well structure after Ryan et al. (1987) (a) Sample details comprising InP capping layer, InGaAs quantum well and InP buffer layer, epitaxially grown by MBE on Fe doped InP. The layer thicknesses were estimated from the growth conditions. (b) An iso-intensity contour plot of the observed X-ray scattering in the region of the InP 440 Bragg reflection. The data are presented on a quasi-logarithmic scale over four decades of intensity to avoid over-contouring the main peak. $Q_\perp$, $Q_\parallel$ are the scattering vectors normal, parallel to the surface. (c) The intensity distribution along the [100] axis showing the modulation of the satellite Bragg intensities.

Figure 2-35 shows some of the results of a surface diffraction study of the effects of coverage of Ge films on Si(001) substrates (Williams et al., 1989; Macdonald et al., 1990; Macdonald, 1990). The data show a series of radial scans through the 200 Bragg peak as a function of the numbered monolayers (ML) of Ge deposited in situ from a Knudsen cell using a dedicated MBE system (Vlieg et al., 1987). Note that 1 ML ≈ 0.14 nm. It can be seen that the diffraction peak profile remains unchanged for a coverage less than 3 ML which indicates that the Ge epitaxy is coherent to the Si substrate. The lack of peak broadening illustrates the high crystalline quality of the overlayer. At layer thicknesses of ≈ 4 ML the development of a weak secondary feature at a lower scattering vector demonstrates the onset of strain relaxation. This

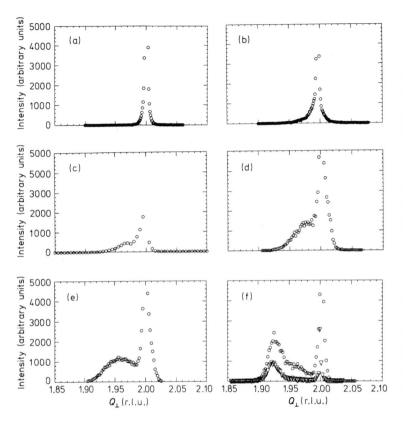

Figure 2-35. Surface X-ray diffraction scans as a function of coverage of MBE grown epitaxial Ge on Si (001) (after Macdonald, 1990). The intensity is plotted on an arbitrary scale with the Bragg peak intensity being $\approx 10^5$ for 11 ML, circles denote data for angles of incidence and exit of 0.09° and 0.13° respectively whilst triangles represent the same scan for 0.07° and 0.05° respectively leading to a reduced effective penetration depth. (a) Clean Si, (b) 3.9 ML of Ge, (c) 4.7 ML of Ge, (d) 5.5 ML of Ge, (e) 7.1 ML of Ge, (f) 11.0 ML of Ge. "r.l.u." stands for "reciprocal lattice unit".

increases in intensity and in separation from the first peak upon further deposition for coverages up to ≈ 11 ML, at which point the peak closely matches that expected for bulk Ge. Such a gradual shift indicates a gradual relaxation of strain after exceeding the 3–4 ML critical thickness. The lack of definition in the diffraction pattern between these two features is evidence for a strain profile within the epilayer. By reducing the effective penetration depth from 10 to 4 nm (Fig. 2-35f – the scan profile is denoted by triangles) the atomic layers close to the interface are still strained while the uppermost layers are almost fully relaxed.

2.4.6 In Situ Structure of Au (100) Under Electrochemical Potential Control

The structure of the electrode interface under conditions of potential control remains one of the fundamental goals for electrochemists. In an electrochemical environment, the electric field or charge at the surface can be controlled by changing the potential across the polarised "double layer" with fields as high as 10^7 V/cm being easily accessible. In electrochemical experiments the corresponding change in surface charge can easily exceed 0.1 electron per surface atom and can provide the driving force for the kind of surface reconstruction which is well known to exist on well-characterised metal surfaces under ultra-high vacuum conditions.

Surface diffraction through its capability to probe in situ through the electrolyte environment is the only technique which can examine surface structure at the electrochemical interface. Recent developments have seen the use of surface diffraction to probe a number of electrochemical systems (Melroy et al., 1988; Samant et al., 1988; Ocko et al., 1990).

Ocko et al. (1990) have investigated the surface structure of an Au (100) electrode surface in an aqueous solution of perchloric acid, as a function of the applied electrode potential. Cyclic voltammetry of the Au (100) surface in $HClO_4$ solution exhibits an anodic current peak at ≈ 0.7 V versus a Ag/AgCl electrode. From a number of other measurements (Hamelin, 1982; Kolb and Schneider, 1985, Friedrich et al., 1989; Zei et al., 1989) it has been inferred that a change in potential can induce a relaxation of the expected (e.g., Palmberg and Rhodin, 1967) hexagonal surface reconstruction. Figure 2-36 shows some of the data from the study of Ocko et al. (1990). Reflectivity data (Fig. 2-36a) confirms that the hexagonal reconstruction, as observed in ultra high vacuum, does indeed take place in solution at negative electrode potentials but, as implied from the cyclic voltametry, this reconstruction disappears at positive potentials to yield a lower density surface layer. Returning to a positive potential results in the recovery of the hexagonal reconstruction.

Additional and complementary information is obtained from surface diffraction measurements (Fig. 2-36b) which show formation of surface domains which have a coherence length of ≈ 30 nm. These are misoriented $\approx 0.8°$ with respect to the underlying Au (100) lattice.

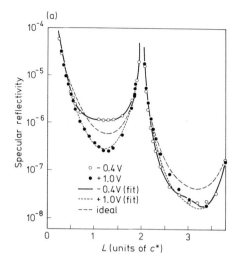

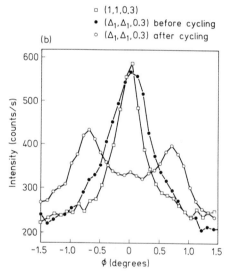

Figure 2-36. In situ surface examination of Au (100) under conditions of potential control (after Ocko et al., 1990). (a) Absolute reflectivity data at an electrode potential of -0.4 V (open circles) and 1.0 V (solid circles). The long-dashed line is for ideal termination with no rms displacement amplitude. The solid and short-dashed lines are fits as described in the text. (b) A comparison between glancing-angle X-ray diffraction rocking curves at an electrode potential of -0.4 V before (closed circles) and after cycling. The potential induces the formation of surface domains rotated with respect to the uncycled lattice of $\approx \pm 0.8°$.

2.4.7 Crystal/Solution Interface Structure of ADP (100) During Growth

Despite the seminal work of Burton, Cabrera and Frank (1951) on the structure of the crystal growth interface there is a critical absence of detailed experimental data to verify many of the currently proposed models. This, for the most part, has been due to the lack of a suitable in situ structural probe of the crystal structure of the growing crystalline interface under a layer of solution. Whilst microscopic observations can reveal screw dislocation sources of surface steps and surface roughening the influence of surface chemistry and interface structure on these processes remains totally unknown.

Experiments by Cunningham et al. (1990a and 1990b) have demonstrated that surface X-ray diffraction can be successfully used for the in situ assessment of growth and dissolution processes. This investigation involved an in situ examination of the (100) surfaces of ammonium dihydrogen phosphate ($NH_4H_2PO_4$ or ADP). X-ray rocking curves of the crystal were recorded under a variety of surface conditions; in air, under a covering of n-hexane and under a saturated aqueous solution of ADP. In air, the rocking curves measured for the ADP crystals were found to be extremely sharp with a half width of slightly less than 4 seconds of arc. Allowing for the slight angular dispersion in the triple-axis mode when the diffraction planes of the monochromator and sample crystals are not exactly matched this compares favourably with a theoretically predicted value of 2.5 seconds of arc. When covered with an insoluble solvent, in this case n-hexane, there was no noticeable variation in the diffraction profile, but on replacing the n-hexane with a saturated solution of ADP the in situ reflection curve broadened asymmetrically to a width of about 14 seconds of arc.

On dissolution of the crystal induced by a step increase in the growth cell temperature, the reflection width decreased again showing the process to be reversible. Rapid recording of the rocking curve as dissolution proceeded showed this process to occur within several minutes and to reach equilibrium within one hour. Figure 2-37 shows the data taken during the subsequent regrowth of the crystal surface initiated through cooling of the cell. The results mirror observations of the dissolution experiment.

These experiments show evidence for an interfacial layer, of substantial thickness, when the crystal surface is in equilibrium with its saturated solution. An interfacial layer has also been observed by other groups using light scattering (Steininger and Bilgram, 1990). It might be expected

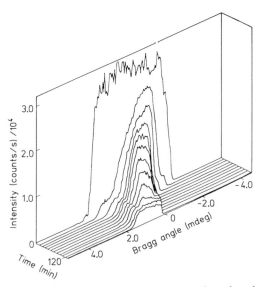

Figure 2-37. 200 rocking curves (wavelength ≈ 0.062 nm) recorded in situ from a nearly perfect single crystal of ADP under ≈ 6 mm of saturated aqueous solution showing the reduction of mosaic spread during crystal growth (after Cunningham et al., 1990a).

that such an interface would consist of weakly bonded layers. Additionally these layers would be highly labile and at the outset of crystallisation or dissolution thermal fluctuations would destroy the interfacial ordering and hence the diffraction conditions for this region. This supposition is supported by the study of the dynamic change in reflection curve shape and width. The transformation is comparatively rapid. During both growth and dissolution experiments the rocking curves reduce to half their initial width in under three minutes. After this initial reduction in width there follows a longer period as the interface returns to an equilibrium state – possibly involving an ordered interfacial structure.

2.4.8 Thin Polyphenylene Films Prepared from Different Precursors

Polyphenylene films formed on indium/tin oxide (ITO) electrode layers have been found to be effective substrates for super birefringent-effect liquid crystal displays (Scheffer and Nehring, 1984). These displays have been found to offer a significant advance in performance, such as improved contrast ratio and viewing angle, than that provided by the twisted nematic displays used in current technology. The polyphenylene inner substrate pre-orientates the liquid crystal molecules up to 25° (Nerhing et al., 1987; Holmes et al., 1989). This pre-tilting enables twist angles between the liquid crystal layers of up to 270° to be achieved without causing two dimensional instabilities in the liquid crystal matrix. An important benchmark in assessing the likely polymer film performance is the film density which reflects the packing density of the polymer molecules. A high density implies a tighter molecular packing, a smaller pre-tilt angle and hence less optimum device performance.

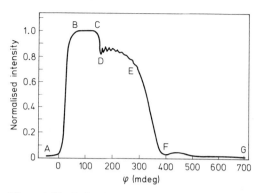

Figure 2-38. Reflectivity curve for polyphenylene film (carbonate precursor) at 0.1286 nm (after Roberts et al., 1990).

Figure 2-38 shows some of the results (Roberts et al., 1990) of the structural examination of polyphenylene thin films on ITO coated glass substrates deposited from either methylacetate and methylcarbonate precursors (Ballard et al., 1988). The reflectivity profile exhibits a number of distinct regions, i.e., total reflection from the polyphenylene film surface (AB), Kiessig fringes from the polyphenylene film (BC) and Kiessig fringes from the ITO substrate. The fringe positions were fitted using the method of Koenig and Carron (1967) and the film thickness and densities calculated (Table 2-6). The film density for the films prepared from the methylacetate precursor is much lower than that from the methylcarbonate precursor which implies that the latter has a much lower molecular packing density. This analysis is also in good agreement with the optical characterisation of the pre-tilt angles and demonstrates the utility of this non-destructive technique.

2.4.9 Photodissolution of Silver on Chalcogenide Glasses

Cowley and Lucas (1990) used X-ray surface reflectivity to characterise the diffu-

Table 2-6. Critical angles, thicknesses ([a]), electron densities and film densities for polyphenylene thin films prepared from carbonate and acetate precursors (after Roberts et al., 1990). Thickness data produced by elipsometry measurements ([b]) are also presented.

Precursor	Critical angles		Film thickness		Electron density (Ne)	Density	Pre-tilt angle
	(mdeg) 0.1286 nm	(mdeg) 0.1349 nm	([a]) (nm)	([b]) (nm)	(m^{-3})	($kg\ m^{-3}$)	(°)
carbonate	147.5	150.7	166.9	170.0	3.99×10^{-30}	1260 (+/−2)	16
acetate	140.1	143.1	153.0	157.0	3.60×10^{-30}	1130 (+/−2)	20

sion of silver into chalcogenide glasses under optical stimulation which is of topical interest for its potential as a lithographic system. Figure 2-39 shows a series of measurements (Lucas, 1989) made as a function of the photo-stimulation of a sample comprising 200 Å of silver deposited onto a 320 nm thick layer of $As_{30}S_{70}$ on float glass. Analysis of the reflectivity spectra using Parratt's method (Parrat, 1954) indicate that before illumination (a) the surface layer comprises ≈ 2.3 nm of oxide and 18.4 nm of silver on the chalcogenide glass.

After 100 s of illumination (b) the reflectivity oscillations are reduced in amplitude but maintain essentially the same angular positions, which implies that whilst the silver layer thickness has not significantly changed during initial illumination the density of this layer has. This reflects the initial diffusion of silver into the chalcogenide glass through grain boundaries.

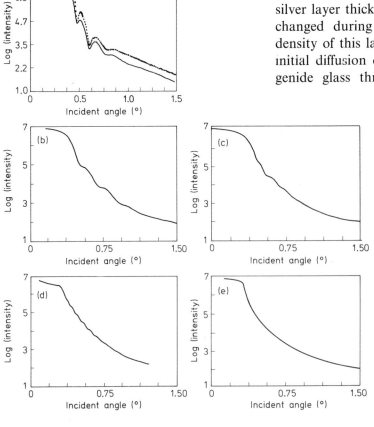

Figure 2-39. X-ray reflectivity studies of the photo-stimulated diffusion of silver into $As_{30}S_{70}$ glass (after Cowley and Lucas, 1990). (a) Experimental reflectivity curve (●●●●) before illumination overlaied with computer fit (———) using Parratt's equation; (b–e) reflectivity curves after illumination with a 200 W mercury lamp for 100, 160, 320 and 600 s respectively.

With increasing illumination times (c–e) the change in the fringe positions clearly shows the reduction in silver film thickness associated with the inter-diffusion process and the resultant formation of a silver/glass reacted layer which is ≈ 4 times thicker than the initial silver film.

On the basis of this work Cowley and Lucas (1990) suggested that the photostimulated interdiffusion takes place initially at grain boundaries or voids. The average thickness of the silver layer remains constant but its average electron density is reduced to give a very inhomogeneous reaction product layer. As the reaction proceeds from 100 to 300 s, silver islands are formed, while the product layer develops into a reasonably homogeneous layer. Even at the end, silver or silver-oxide islands remain. Upon further illumination the reaction product layer moves into the chalcogenide region destroying any sharp interface between these layers.

2.4.10 Freely Supported Films of Octylcyanobiphenyl

The use of reflectivity techniques need not be restricted to investigations of surfaces. Gierlokta et al. (1990) have demonstrated an alternative way of characterising the structure of the liquid/air interfaces using freely suspended films. Using this approach films with thicknesses varying from a few to many thousands of molecular layers can be assessed. Figure 2-40 shows the kind of data that can be obtained from systems such as thin films of octylcyanobiphenyl.

The data shows two characteristic features in these reflectivity profiles; firstly the smectic interlayer separation gives rise to a Bragg-like central peak which is broadened due to the finite number of layers. Secondly, away from the Bragg peak well resolved Kiessig fringes due to interference of the X-rays reflected from the two liquid/air interfaces can be seen and can be used to determine the total thickness of the layer. As this method allows independent determination of both intermolecular layer separation and the total film thickness the approach is very sensitive to the out-of-plane structure of the surface layers which in Gierlotka et al.'s (1990) study indicates an extended top layer at both of the interfaces with a thickness ≈ 1.36 times the spacing of the interior layers.

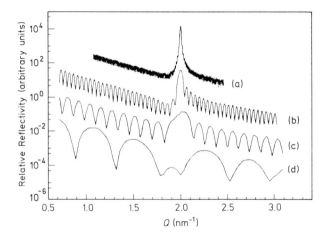

Figure 2-40. Reflectivity data for free standing films of octylcyanobiphenyl for 160 (a), 43 (b), 16 (c) and 4 (d) molecular layers in thickness; (a), (b) and (c) are shifted for reasons of clarity (after Gierlokta et al., 1990).

2.5 X-Ray Standing Wave Spectroscopy

2.5.1 Introduction

X-ray standing waves (XSW) are generated through dynamical interference between the incident and diffracted X-rays during Bragg diffraction from nearly perfect single crystals (Batterman, 1964). This is shown schematically in Fig. 2-41 which also illustrates that by adjusting the crystal over its range of diffraction causes the X-ray standing waves to lie between the atomic layers (inset A) of the structure or to be coincident with them (inset B). Experimentally (c), by recording the reflected and fluorescent radiation emitted from the crystal *at the same time*, the position of the incorporated atoms, surface films etc. with respect to the crystal lattice can be determined (Cowen et al., 1980; Bedzyk et al., 1984; Materlik and Zegenhagen, 1984; Materlik et al., 1984; Hertel et al., 1985; Funke and Materlik, 1985). In this respect XSW and EXAFS (see Sec. 2.2) are complementary techniques as they provide information concerning both the long and short range correlations for specific atoms within crystalline solids.

2.5.2 Surface Coordination of Adsorbed Br on Si (111)

Funke and Materlik (1985) have used XSW to investigate the atomic coordination of adsorbed Br atoms on Si(111). The result of the standing wave measurement is shown in Fig. 2-42 in which the Bragg reflectivity is shown superimposed with total bromine $K\alpha$ fluorescence yield. The best fit of the experimental data is to a model which assumes a single dominant adsorption site with an atomic coverage of 35 ($\pm$ 2)% with the Br atoms occupying a position $\approx$ 0.201 ($\pm$ 0.003) nm from the Si(111) diffraction

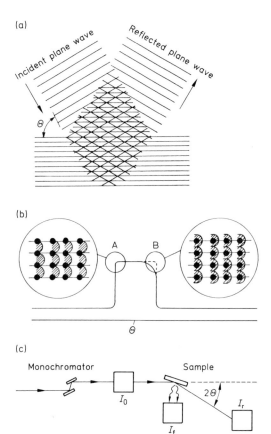

Figure 2-41. X-ray standing waves: (a) interference of incident and Bragg diffracted X-rays to give rise to a standing wave field; (b) changes to the angle of incidence move the standing waves from lying between the atomic layers (inset A) of the structure to be coincident with them (inset B); (c) experimental set up for standing wave measurements. I_0, I_r and I_f are the incident, reflected and fluorescent intensities respectively, Θ is the Bragg angle (after Cunningham et al., 1991b).

planes. This result shows that bromine is not adsorbed at the position on-top of the outermost surface Si atoms since assuming a Si-Br covalent bond length of 0.218 nm would give a coherent position of 0.257 nm with respect to the Si(111) surface.

Figure 2-43 shows a (112) atomic projection of the Si lattice indicating the position of the incorporation site. The 0.201 nm position determined from the XSW data

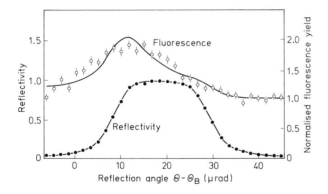

Figure 2-42. Angular dependence of the Si(111) reflectivity yield overlaied with fluorescence yield. In this case 45 L (2.10^{-7} mbar for 300 s) Br were deposited through the gas inlet valve. Measuring time was 30 minutes.

is indicated by point-dashed lines. The dashed circle is centred at this distance above an open threefold coordination site and illustrates the close agreement between the measured adsorption site and a model in which bromine is adsorbed ionically as Br$^-$ with an ionic radius of 0.196 nm. Such a model is also in agreement with the measured coherent coverage of 0.35 ML, which due to steric hindrance or electronic repulsion factors, allows at most, one third of a monolayer to be adsorbed as long as adjacent sites are not occupied.

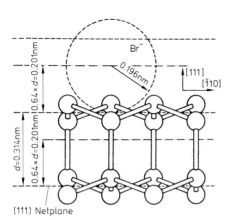

Figure 2-43. Side view of the unrelaxed Si(111)-(1×1) surface (after Funke and Materlik, 1985). The threefold open site of the adsorbed Br$^-$ above the outermost Si(111) diffraction plane is indicated by a dashed circle.

2.5.3 Structural Environment Around Cu^{2+} Ionic Habit Modifiers in Ammonium Sulfate

In industrial crystallisers, control of particle size and shape is often accomplished with the aid of habit modifiers. The selection and definition of the additives, which in ionic systems take the form of trace metallic ions, is currently restricted by the lack of quantitative structural data concerning the bonding of the ionic additive to the host system. Whilst fluorescence EXAFS can be used to assess the concentration and the local atomic environment around the habit modifying ions (Barrett et al., 1989c; Armstrong et al., 1990; Cunningham et al., 1991; Cunningham, 1991a), X-ray standing waves need to be used to define the relationship between the ion and its underlying crystal lattice.

The addition of trace amounts of Cu^{2+} ions to crystallising solutions of ammonium sulfate results in the habit modification of the product from a prismatic morphology elongated along the c-axis to one elongated along the pseudo-hexagonal a-axis. X-ray fluorescence analysis of the habit faces confirmed that the adsorbed ions were not sectorially distributed but were uniformly segregated throughout the crystals whilst EXAFS and ab initio quantum chemistry calculations revealed that

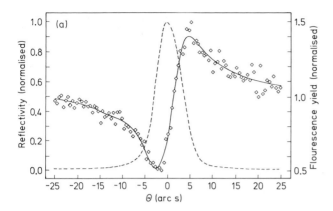

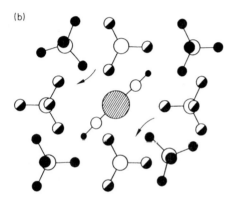

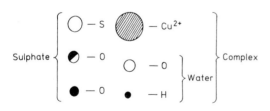

Figure 2-44. XSW data from examination of Cu^{2+} in ammonium sulfate (after Armstrong et al., 1991). (a) Normalised 002 X-ray standing wave spectra. The data show the overlay of the X-ray rocking curve (-----) and the angular dependance of the Cu K fluorescence intensity (◊◊◊◊). The latter is displaced to the right with respect to the former which is indicative of a substitutional incorporation of Cu^{2+} for SO^{2-}. (b) Proposed structural model (the NH_4^+ ions have been removed for clarity) for the incorporation of Cu as a $Cu(H_2O)_4^{2+}$ complex which removes the SO_4^{2-} and four NH_4^+ ions to maintain a charge neutrality.

Cu^{2+} is closely correlated with six oxygen atoms in a distorted octahedral configuration. The radial distribution function similar to that around Cu^{2+} in copper sulfate pentahydrate in that it appears strongly coordinated to four water molecules in a square planar arrangement with weaker interactions with two sulfate ions.

XSW spectra (Fig. 2-44) recorded from the doped ammonium sulfate showed (a) that the copper ion occupies a lattice site close to the plane of highest electron density. This is consistent with an incorporation site close to the SO_4^{2-} ion and indicates that adsorption results in a substantial structural rearrangement and an associated redistribution of ionic charge. The exact mechanism of incorporation is associated with the similarity in charge distributions which allows the water ions to successfully mimic the ammonium ions during adsorption. This substitution of one SO_4^{2-} by Cu^{2+} requires the removal of four NH_4^+ ions to maintain strict charge neutrality. The resulting structure (b) forms a distorted octahedral environment around the copper ion with weak binding to the two sulphate ions above and below a copper tetrahydrate square planar complex which lies close to the (101) crystal plane.

2.6 Characterisation of Lattice Defects Using X-Ray Topography

2.6.1 Introduction

The group of techniques which go under the general title of X-ray diffraction topography (or more usually just X-ray topography) are used for imaging and characterising dislocations and other extended defects in relatively large crystals with a defect content typically less than 10^4 to 10^5 dislocations cm^{-2}. These techniques are applicable to a diverse range of crystalline materials satisfying the above criteria, but have proved particularly successful for analysis in electronic materials where a high level of perfection is an essential requirement for optimal performance. Unlike comparable defect imaging methods used in electron microscopy, these techniques do not require magnification and are therefore not restricted to a narrow field of view, do not require ultra thin samples and can in principle be used with the sample crystal in a variety of environments. The complementary nature of X-ray topography and electron microscopy when used to investigate processes occuring over a wide dimensional range is intimated below.

X-ray topographic techniques take advantage of the remarkable sensitivity of X-ray diffraction intensity to spatial and orientation distortion of the crystal lattice. Depending on the perfection of the crystal sample and its thickness, images can be produced as a consequence of kinematical or dynamical scattering processes leading to direct or dynamical images respectively. The theoretical basis for image formation is to be found in comprehensive review articles such as that of Batterman and Cole (1964) or in the book by Tanner (1976). The techniques, developed 30 to 40 years ago, employed the broad band monochromatic radiation characteristic of standard laboratory sources. These are still used extensively today. Although some of the methodology when using synchrotron radiation is essentially the same, there are significant differences in the characteristics of synchrotron radiation which allow a variety of incomparable materials science experiments to be carried out.

Synchrotron radiation techniques fall into two broad categories, white radiation topography and double crystal or narrow band monochromatic topography. Each is capable of providing its own unique brand of information.

2.6.2 White Radiation Synchrotron Topography

2.6.2.1 Introduction

The principal advantages of white radiation synchrotron topography are illustrated by the following features which emphasise the important differences between synchrotron and conventional characteristic X-ray sources, specifically: spectral range, high photon flux, beam divergence and time structure.

2.6.2.2 Instrumentation

In many respects the experimental arrangement employed for white radiation synchrotron topography bears a closer resemblence to that for the Laue method than that used for conventional characteristic radiation topography. Figure 2-45 shows a schematic diagram of the white radiation topographic camera (Bowen et al., 1982) on station 7.6 at the Daresbury synchrotron source. The four circle instrument is large enough to enable the mounting of large and heavy apparatus to achieve specific sample environments. Images are recorded on X-ray film or plates and at

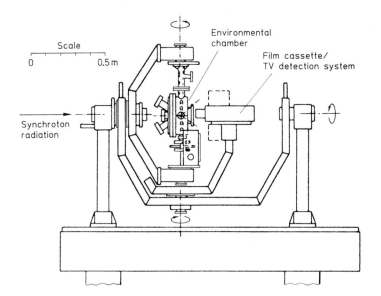

Figure 2-45. White radiation topographic camera on station 7.6 at Daresbury Laboratory (after Bowen et al., 1982).

lower resolution on a TV imaging system. Unlike a conventional Lang Camera, the large dimensions of the synchrotron beam eliminate the need for scanning mechanisms for most samples. Furthermore, the diffracted beam slit necessary to shield the film from the primary beam in topographs recorded on laboratory sources, is not generally required for synchrotron topography. The large specimen to film distances which can be used in the latter case adequately separate most images from the transmitted beam. To capture simultaneously the maximum number of images from different reflections, the film may be placed normal to the primary beam. As discussed below, however, optimum resolution is achieved by setting the film normal to the diffracted beam from one selected reflection.

2.6.2.3 Advantages of the Broad Spectral Range

As a consequence of the continuous nature of the synchrotron spectrum, white radiation topographs almost inevitably comprise superimposed images from several harmonic reflections. The integrated diffracted intensity (I) due to a given harmonic, and therefore its relative contribution to the topographic image, can be calculated from (Tuomi et al., 1974)

$$I = [|F_{hkl}| P(\lambda) \lambda^3 \exp(-\mu t)]/\sin^2 \Theta_B \quad (2\text{-}8)$$

where $|F_{hkl}|$ is the modulus of the structure factor, $P(\lambda)$ is the photon flux, λ is the wavelength, Θ_B is the Bragg angle, μ is the linear absorption coefficient and t is the sample thickness. The presence of harmonics within the topographic image has two major consequences. Firstly, since the intrinsic reflection range decreases with decreasing wavelength, the strain sensitivity is greater for high order reflections. The presence of high order harmonics in the image therefore results in a loss of resolution. The second major effect of harmonic contamination is observed in highly perfect crystals exhibiting Pendellösung fringes. In such cases complex fringe patterns can occur due to interference between wavefields of different harmonics and overlap of fringes associated with the various reflections (Tuomi et al., 1978). In this context it should be noted that the synchrotron radiation is highly polarised in the plane of the electron orbit. Consequently the periodic fading on Pendellösung fringes which is observed in conventional topographs (Hart and Lang, 1965), due to incoherent superposition of the σ and π polarisation components, does not occur in synchrotron radiation topographs when the plane of polarisation is either parallel or normal to the plane of incidence (Hart and Lang, 1965).

An important advantage over characteristic radiation topographs is the fact that the continuous spectrum employed in white radiation topography images warped and misoriented crystals in their entirety. In the conventional Lang technique, for example, only those regions of such samples which lie within the angular range defined by the beam divergence can be imaged. Increasing the beam divergence in characteristic source experiments allows more of the crystal to be imaged but at the cost of inferior resolution.

The contrast of defects in continuous radiation synchrotron topographs is substantially more complex than that observed in conventional Lang topography, comprising both diffraction (extinction) and orientation contrast components. As a consequence of orientation contrast effects the nature of the image is strongly dependent upon the sample to film distance since the diffracted beams arising from misorientated regions within the crystal are spatially separated. As demonstrated by Hart (1975) for lithium fluoride, this image separation can provide an extremely sensitive measure of subgrain misorientations, with tilts of less than 25 seconds of arc readily detectable.

The potential for wavelength tuning, possible because of the broad synchrotron spectrum, greatly enhances the flexibility of topography. It enables, for example, optimisation of diffraction geometries and conditions (Petroff and Sauvage 1978; Halfpenny and Sherwood, 1990) and depth profiling in reflection topography (Dudley, 1990). The need for the flexibility which wavelength tuning provides is well demonstrated in the case of synchrotron section topography of large uncut crystals (Halfpenny and Sherwood, 1990). In such studies two competing factors determine the optimum diffraction conditions. Firstly, to minimise the geometrical distortion of the section topograph, the Bragg angle should be as close to 45° as possible, either through the use of high order reflections or long wavelengths. The former are generally weak with narrow intrinsic reflection ranges and so yield inferior spatial resolution. The second controlling factor is absorption which for a given sample thickness will limit the wavelength which can be employed. In general the wavelength achieving the optimum compromise between these two factors will not correspond to a characteristic line. For many samples it is only through the wavelength tunability of synchrotron topography that such conditions can be realised. Figure 2-46 shows a white radiation synchrotron section topograph of a crystal of the organic non-

Figure 2-46. Synchrotron white radiation section topograph of a 2 cm thick crystal of (−)-2-(α-methylbenzylamino)-5-nitropyridine (MBA-NP) (after Halfpenny and Sherwood, 1990).

linear optical material (–)-2-(α-methylbenzylamino)-5-nitropyridine (MBA-NP). The low linear absorption coefficients of organic materials such as this enable relatively thick samples to be topographed (up to 2 cm in thickness in this case). The image in Fig. 2-46 is formed principally from the 040 reflection at 0.08 nm and the 060 at 0.055 nm at a Bragg angle of 15°, the relative contributions from these harmonics being approximately 79% and 21% respectively. At 0.08 nm the linear absorption coefficient of MBA-NP is about 1.7 cm^{-1} giving a product μt of around 3. The image is compressed parallel to the diffraction vector by a factor of $\sin 2\Theta$ (i.e., 50%). Longer wavelengths, though decreasing image distortion substantially degrade defect resolution through increased dynamical contrast contributions. The conditions employed represent an optimum compromise in the present case, yielding well defined images of dislocations, inclusions, growth sector boundaries and delineating clearly the seed crystal from which the sample was grown.

The observation of misfit dislocations in (GaAlAsP)/GaAs epitaxial layers (Petroff and Sauvage, 1978) provides another important illustration of the use of wavelength tuning in white radiation synchrotron topography. In this case the wavelength was tuned to just above the K absorption edge of gallium. A substantial reduction in absorption was thus achieved, similar to that for silver $K\alpha$ radiation. The longer wavelength employed, however, resulted in considerably narrower image widths and therefore enhanced resolution.

2.6.2.4 Influence of Beam Flux, Dimensions and Divergence

A crystal placed in the synchrotron beam yields a pattern of diffraction spots, as in the conventional Laue method. Because of the extended beam in the synchrotron case, however, each spot is in fact a topograph of the crystal. In this way the time required for defect characterisation by invisibility criteria in different reflections can be greatly reduced. The X-ray flux from a synchrotron source is some 3 to 4 orders of magnitude greater than a conventional sealed tube X-ray source, producing dramatic reductions in exposure times for topography. In contrast to Lang topography using a laboratory source, the large dimensions of the incident synchrotron beam allow large areas to be imaged without the need for sample scanning, thereby further enhancing the speed advantage of synchrotron topography.

It can be shown that the geometrical resolution R_g is given by

$$R_g = S x/L \qquad (2-9)$$

where S is the source size, x the sample to film distance and L the sample to source distance. With a source size of 0.15 mm × 1.55 mm and a sample to source distance of 80 metres (topography station 7.6 at the SRS, Daresbury U.K.) it is apparent that a geometrical resolution comparable with that of laboratory Lang topography can be achieved with relatively large sample to film distances. The geometrical resolution defined above relates to a diffracted beam normal to the X-ray film. For a film normal to the incident beam, the geometrical resolution then depends primarily upon the Bragg angle of the reflection in question. Those with large Bragg angles are incident on the film at shallow angles and therefore suffer substantial distortion and loss of resolution.

The high photon flux, together with the fact that no sample scanning mechanism is required and the possibility of high resolution despite large specimen to film distances, together make white radiation synchrotron topography ideal for dynamic and in situ studies of a wide range of physical and chemical phenomena, including recrystallisation (Gastaldi and Jourdan, 1978 and 1979; Jourdan and Gastaldi, 1979), plastic deformation (Miltat and Bowen, 1979; Ahamad et al., 1987; Ahamad and Whitworth, 1988), motion of magnetic domains (Tanner et al., 1976; Safa and Tanner, 1978; Sery et al., 1978; Clark et al., 1979; Chikaura and Tanner, 1979; Stephenson et al., 1979; Toumi et al., 1979; Clark et al., 1983) and solid-state reactivity (Dudley et al., 1983 and 1986).

Studies of the recrystallisation of aluminium (Gastaldi and Jourdan, 1978 and 1979; Jourdan and Gastaldi, 1979) provide a clear demonstration of the power of white radiation synchrotron topography. The principal alternatives to X-ray methods for such studies are TEM and optical microscopy. While the high resolution of TEM enables the observation of the nucleation stage of the recrystallisation, the relatively small area imaged (about 10 µm across) precludes observation of the growth of nuclei into large grains. Furthermore, the growth behaviour of grains in the thin samples required for TEM is likely to differ substantially from that in bulk samples. Optical microscopy suffers from the disadvantage of being a purely surface technique. Grain growth rates can be as high as a millimetre per minute so for X-ray studies the rapid exposure times achievable with white radiation synchrotron topography are essential for dynamic observations. The continuous spectrum of the synchrotron source is also vital since the orientation of the growing grains cannot be predicted. Utilising these advantages, white radiation topographic studies of grain growth in aluminium have provided information on the formation and evolution of growth defects and have revealed a distinction between the structure of moving and stationary grain boundaries. Although the resolution of X-ray topography is insufficient to observe nucleation, the transmission from nucleation to grain growth has been examined.

Short exposure times and the ability to record several reflections simultaneously are also important advantages in studies of plastic deformation and dislocation motion. The work of Whitworth and co-workers (Ahamad et al., 1987; Ahamad and Whitworth, 1988) on dislocation motion in ice illustrates well the capabilities of synchrotron topography in such studies. Rapid exposures (around 20 seconds) enabled dynamic observations of dislocation motion. Dislocation loops gliding on the basal plane form hexagonal loops comprising screw and 60° segments, indicating the existence of a Peierls barrier in these orientations. Straight dislocations in both orientations were found to move at approximately equal velocities which varied linearly with applied stress. The velocity of glide of edge dislocations on non-basal planes, however, was an order of magnitude higher. Of particular note is the fact that substantial recovery was observed after periods as short as 18 minutes, thereby illustrating the need for rapid exposures and the dynamic, in situ capabilities of white radiation synchrotron topography.

The additional space available around the sample, without loss of resolution, in the case of white radiation synchrotron topography offers considerable flexibility in the control of sample environment, whether it be temperature (furnaces or cryostats), pressure, magnetic or electric fields or the introduction of corrosive or other reactive ambients. Figure 2-47 shows an environmental chamber used for in situ X-ray topography under well defined environmen-

Figure 2-47. Environmental chamber for synchrotron radiation topography (after Gillespie et al., 1989).

tal conditions on the topography station 7.6 at the Daresbury Synchrotron Radiation Source (Gillespie et al., 1989). The water cooled cylindrical chamber is constructed from stainless steel UHV components. Two ports fitted with beryllium windows form the X-ray entrance and exit windows of the chamber. In addition a further 12 radial and 4 angled ports provide access for the furnace, sample holder, inputs for gas supplies and the pumping system as well as electrical connections. The furnace, constructed from tubular stainless steel with thermocoax (nichrome in iconel) windings, fits directly over the sample holder and has two 2 cm apertures to allow passage of the X-ray beam. The chamber/furnace combination can operate up to 1270 K at a pressure of around 10^{-7} torr. At higher pressures the maximum operating temperature is limited to about 770 K to avoid damage to the beryllium windows. The best achievable vacuum is 10^{-7} torr.

This chamber has been employed in a range of X-ray topographic studies of solid state chemical reactivity. Figure 2-48 shows the changes in structural perfection of a crystal of nickel sulfate hexahydrate during dehydration at reduced pressure. Figure 2-48a shows a white radiation synchrotron topograph of the crystal at room temperature and atmospheric pressure. After 30 minutes at a pressure of 0.1 Pa (also at room temperature) regions of surface strain have developed, associated with numerous dehydration sites, as shown in Fig. 2-48b. It is apparent that the strain centres and therefore the decomposition sites are not associated with the grown-in dislocations visible in the topograph, in contradiction, in this case, to the earlier suggestions that dislocations play a key role in solid-state decomposition. Further studies performed using this chamber include observations of the initial stages of the thermal decomposition of calcite at temperatures up to 748 K; thermal dehydration of nickel sulfate hexahydrate at atmospheric pressure and an examination of the alpha/beta phase transformation in quartz at low pressure. Despite considerable potential, however, the application of white radiation synchrotron topography to studies of the role of defects and structural perfection in solid state chemical reactivity has been rather limited.

A significant problem which can seriously limit the application of synchrotron white radiation topography is that of radiation-induced damage. Exposure of many materials to an unfiltered white synchrotron beam can result in the formation of colour centres, sample curvature, genera-

2.6 Characterisation of Lattice Defects Using X-Ray Topography

Figure 2-48. White radiation topograph of nickel sulfate hexahydrate recorded using the environmental chamber (a) at atmospheric pressure and (b) after pumping at 0.1 Pa for 30 minutes (after Gillespie et al., 1989).

also accounts for the curvature observed in many radiation damaged samples. Since the X-ray flux reaching the exit surface of a crystal is diminished by absorption, the induced damage also decreases through the sample thickness. Differing levels of decomposition and point defect generation through the sample therefore result in a non-uniform stress which relaxes by generating sample curvature and in some instances dislocations. The use of, for example, an aluminium filter can substantially reduce the level of radiation damage by attenuating the strongly absorbed long wavelength components of the beam which are largely responsible for the radiation damage. A major drawback of this, however, is that the relative contribution of high order harmonics (resulting from the short wavelengths) to the topographic image is increased, thereby impairing resolution.

2.6.2.5 Experiments Involving the Time Structure

A unique aspect of synchrotron radiation which has been greatly underexploited is its time structure. The radiation generated by a synchrotron storage ring occurs in the form of a regular sequence of pulses as the electron bunches orbit around the ring. The frequency of these pulses is determined by the number of electron bunches and the dimensions of the ring. Stable operating frequencies in the range from a few MHz to GHz can be achieved. This well defined time structure has been successfully exploited in stroboscopic topography of periodic phenomena including the observation of surface acoustic waves in lithium niobate and quartz (Whatmore et al., 1982; Goddard et al., 1983). Since X-ray topography also permits the simultaneous observation of crystal defects, interactions

tion of slip dislocations and ultimately, complete radiolytic decomposition, leading to void formation. The extent of radiation damage depends upon the sensitivity of the material and its X-ray absorption coefficient. It is interesting to note that many organic materials, despite their relative chemical instability, exhibit surprisingly greater resistance to radiation induced damage than inorganic materials. This is simply a result of the lower absorption coefficients of the former. Absorption between these and surface acoustic waves can also be examined.

2.6.3 Double Crystal Topography

2.6.3.1 Introduction

In a white radiation experiment, the intrinsic angular width and wavelength dispersion of a reflection defined by the rocking curve is generally less than the divergence of the incident beam and its wavelength spread. The resulting overlap of diffracted beams from adjacent parts of the crystal results in an integrated intensity which reduces strain sensitivity and prevents quantitative strain mapping. Slight misorientation or dilation of the lattice can occur between growth sectors of a crystal as a consequence of the different growth mechanism, or within a single growth sector because of the fluctuations of temperature or supersaturation during growth. This is usually associated with impurity incorporation. Although X-ray topography is inherently capable of detecting strain to below 1 part in 10^5 these differences will not generally be imaged in a white beam experiment. This is unfortunate since impurities at low concentrations may be of significance to the performance of a device into which such crystalline material is incorporated. Therefore, although white beam topography is probably the preferred technique for defect analysis and for studying for example the temporal evolution of defects, it is not sensitive enough for certain applications.

In order to improve the strain sensitivity of X-ray topography, the double crystal arrangement is used. Bonse (1958) has provided the theoretical background for the technique and the criteria for optimal use has been reiterated in a review on the use of synchrotron radiation in topography by Miltat and Sauvage-Simkin (1984). The detailed diffractometry associated with rocking curve analysis which can provide information about the mosaic spread, curvature, defect content and thickness (particularly with multilayer structures) of a specimen and the topographic image obtained with the double crystal configuration are intimately linked.

2.6.3.2 Instrumentation

In general use, the system is set up in the $(+-)$ parallel configuration as shown in Figs. 2-49 a and 2-49 b. When the two crystals are of the same material and the same reflections are used, the doubly diffracted beam is parallel to the incident beam and there is no wavelength dispersion. As the second (specimen) crystal is also illuminated by what is essentially a narrow band monochromatic plane wave coming from the first crystal, maximum strain sensitivity is achieved. The technique has been much

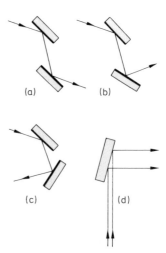

Figure 2-49. (a) Parallel non-dispersive double crystal geometry in $(+-)$ configuration when identical crystals and reflections are used; (b) non-parallel dispersive double crystal geometry in $(+-)$ configuration with non identical crystals; (c) double crystal geometry in the $(++)$ configuration; (d) use of an assymetric reflection for beam expansion.

used for studies of silicon because of the high degree of crystal perfection which can be attained with this material, a necessary requirement for obtaining rocking curve widths close to the intrinsic value. Although ideally the same material should be used for both crystals, this is not possible for general use, bearing in mind the necessity for a high degree of perfection in the monochromatic crystal. This means that there will be a mismatch between the lattice plane spacings for the two reflections and there will be some wavelength dispersion introduced, thereby reducing strain sensitivity. However, provided reflections are chosen in which the difference in lattice plane spacings is small, then the reduction in strain sensitivity is kept to a minimum and the technique still proves highly satisfactory. One bonus is that since higher order reflections may not coincide for the first and second crystal, harmonic content may be reduced or even eliminated. Because of the high levels of perfection available, germanium and sometimes quartz in addition to silicon are suitable as first crystal materials.

The use of synchrotron radiation should permit in situ double crystal experiments to be carried out using environmental stages. Up to the present time, few experiments have been reported, perhaps due to the general perception that imaging crystals which are being distorted due to tensile or compressive loading, thermal anisotropy and chemical processes is difficult with monochromatic radiation. However, George and Michot (1982) report using a high temperature tensile stage operating under a controlled atmosphere to study dislocation velocities in silicon and the movement of dislocations from a crack tip under load. High strain sensitivity was not a particular issue, so the wavelength dispersive (+ +) crystal configuration appears to have been used. Also, Michot et al. (1984) have employed a hot stage to follow the evolution of fluid inclusions in synthetic quartz up to decrepitation.

Double crystal diffraction from two silicon crystals can be used as a tuneable non dispersive source of plane wave radiation for topographic studies of a specimen crystal positioned on a third axis. Kuriyama and his group at National Institute for Science and Technology, U.S.A., have developed this technique for their dynamical diffraction imaging (Kuriyama et al., 1989). In addition the use of assymetrically cut crystals can be used to expand (or decrease) the cross section of the incident beam. With synchrotron radiation, the advantages of video recording and processing techniques for real time experiments cannot be overstated. Unfortunately the resolution of even the best videcon detector is limited to about 10 μm. By using two orthogonal, asymetrically cut crystals to diffract the imaging beam from the sample, Kuriyama et al. have been able to produce images magnified up to 150 × thus overcoming the resolution problem. As these authors state, the magnifier can be used as a zoom lens since a pair of asymetrically cut crystals provides a continuously changing magnification as the energy of the incident beam is changed.

2.6.3.3 Applications

Strain mapping using synchrotron double crystal techniques, with a major emphasis on differences between growth sectors, has been reported for quartz (Fig. 2-50, Zarka, 1984), strontium nitrate (Miltat and Sauvage-Simkin, 1984) and diamond (Lang et al., 1987). Such differences can be very small. By rotating the sample crystal about the Bragg reflection, topographs can be recorded at various positions on the rock-

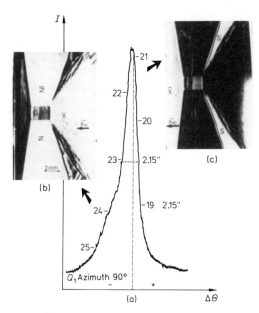

Figure 2-50. Double crystal topographs of a Y-cut of a quartz crystal (1000 reflection) showing contrast changes in images recorded at two different positions on the rocking curve (after Zarka, 1984). Such changes allow variations in the lattice parameter $\approx 10^{-5} - 10^{-6}$ and lattice tilts of $\approx 0.2''$ to be determined.

ing curve. Using this technique, the variations of contrast were used to estimate changes of lattice parameter to less than 1 part in 10^5. If both lattice dilation and tilting are present, the effects of each can be separated by rotating the sample crystal about the surface normal. At the other extreme, melt grown compound semiconductor crystals such as GaAs contain misorientations which are larger than the rocking curve width. Strain associated with these crystals has been assessed by mapping the contours of misorientation recorded at a series of fixed angular settings (Kitano et al., 1985).

Double crystal diffractometry and topography have also been applied successfully to epitaxial layers. An analysis of rocking curves gives the lattice mismatch, lattice tilt between epilayer and substrate, layer thickness and compositional variation with depth, substrate and layer perfection, and curvature. As with the white radiation technique, topography can be used to study misfit dislocations. In addition moiré patterns observed in topographs of GaAlAs/GaAs layer structures have been used to determine lattice mismatch to 1 part in 10^6 (Chu and Tanner 1987).

2.7 Conclusions and Future Perspective

The foregoing examples demonstrate well the breadth of novel experiments which can be carried out using synchrotron X-radiation and the improved working which can be achieved compared with traditional radiation. There is obviously much scope for broadening these applications to a wider range of experiments.

For the future, however, current developments in synchrotron sources would lead to the possibility of an even greater range of opportunities. The principal potential advances are two-fold.

For the machines themselves the aim is to build sources of even greater intensity and energy and with tunable radiation available principally from insertion devices. This will allow considerably reduced exposure and recording times; sub seconds rather than minutes/hours, and the easier use of the time structure of the beam for stroboscopic type experiments. Quasi-forbidden reflections will become accessible. Weaker scattering processes can be studied more satisfactorily.

The increased energy range will give a greater penetrability for in situ dynamic studies of a variety of diffraction processes. It will also allow diffraction studies of higher atomic number materials and provide access to a wider range of experimentation using anomalous dispersion X-ray diffraction.

In parallel, developments in electronic recording will result in greater ease of image and pattern detection thus contributing to improved and more accurate operation.

There is no doubt that synchrotrons will continue to develop and occupy an essential place in the armoury of the materials scientists.

2.8 Acknowledgements

The authors gratefully acknowledge the support of the U.K. Science and Engineering Research Council for a number of the projects described in this article. More particularly, however, we thank the Director and staff of the SERC's Daresbury Laboratory for the provision of much of the wide range of facilities described and of which we have had use in carrying out these experiments.

2.9 References

Ahamad, S., Otomo, M., Whitworth, R. W. (1987), *J. Phys. (Paris) Coll. 48*, 175.
Ahamad, S., Whitworth, R. W. (1988), *Phil. Mag. A 57*, 749.
Akhtar, M. J., Catlow, C. R. A., Clark, S. M., Temmerman, W. M. (1988), *J. Phys. C: Solid State Phys. 21*, L917.
Aleshko-Ozhevskij, O. P., (1982), *Sov. Phys. Cryst. 27*, 673.
Aleshko-Ozhevskij, O. P. (1983), *Ferroelectrics 48*, 157.
Amoros, J. L., Buerger, M. J., Amoros, M. C. (1975), *The Laue Method*. New York: Academic Press.
Andrews, S. J., Hails, J. E., Harding, M. M., Cruickshank, D. W. J. (1987), *Acta Cryst. A 43*, 70.
Andrews, S. J., Papiz, M. Z., McMeeking, R., Blake, A. J., Lowe, B. M., Frankling, K. R., Helliwell, J. R., Harding, M. M. (1988), *Acta Cryst. B 44*, 73.
Anwar, J., Barnes, P., Clark, S. M., Dooryhee, E., Hausermann, D., Tarling, S. E. (1990), *J. Mat. Sci. Lett. 9*, 436.

Armstrong, D. R., Cunningham, D. A. H., Roberts, K. J., Sherwood, J. N. (1991), *Proc. 6th Int. Conf. on X-ray Absorption Fine Structure (XAFS VI)*. Ellis Horwood, Chichester, pp. 435–437.
Ballard, D. G. H., Courtis, A., Shirley, I. M., Taylor, S. C. (1988), *Macromolecules 21*, 294.
Barnes, P., Clark, S. M., Hauserman, D., Fentiman, C. H., Rashid, S., Muhamad, N. N., Henderson, E. (1992), *Phase Transitions*, in press.
Barrett, N. T., Gibson, P. N., Greaves, G. N., Mackle, P., Roberts, K. J., Sacchi, M. (1989a), *J. Phys. D: Appl. Phys. 22*, 542.
Barrett, N. T., Gibson, P. N., Greaves, G. N., Roberts, K. J., Sacchi, M. (1989b), *Physica B 158*, 690.
Barrett, N. T., Lamble, G. M., Roberts, K. J., Sherwood, J. N., Greaves, G. N., Davey, R. J., Oldman, R. J., Jones, D. (1989c), *J. Crystal Growth 94*, 689.
Barrett, N. T., Greaves, G. N., Pizzini, S., Roberts, K. J. (1990), *Surf. Sci. 227*, 337.
Batterman, B. W. (1964), *Phys. Rev. 133*, 759.
Batterman B. W., Cole, H. (1964), *Rev. Mod. Phys. 36*, 681.
Beaven, P. A., Frisius, F., Kampmann, R., Wagner, R. (1986), *Proc. of the Second International Symposium on Environmental Degradation of Materials in Nuclear Power Systems – Water Reactors;* Monterey, Sept. 1985, (ANS).
Bednorz, J. G., Mueller, K. A. (1986), *Z. Phys. B: Cond. Matter 64*, 189.
Bedzyk, M. J., Materlik, G., Kovalchuk, M. V. (1984), *Phys. Rev. B 30*, 2453.
Bhat, H. L., Sheen, D. B., Sherwood, J. N. (1982), *The Application of Synchrotron Radiation to Problems in Materials Science:* Proc. Daresbury Study Weekend DL/SCI/R19, pp. 90–95.
Bhat, H. L., Herley, P. J., Sheen, D. B., Sherwood, J. N. (1984), in: *Applications of X-ray Topographic Methods to Materials Science:* Weissmann, S., Balibar, F., Petroff, J. (Eds.). New York: Plenum Press, pp. 401–411.
Bhat, H. L., Littlejohn, A., McAllister, J. M. R., Shaw, J., Sheen, D. B., Sherwood, J. N. (1985), *Materials Science Monographs 286*, 707.
Bhat, H. L., Clark, S. M., El Korashy, A., Roberts, K. J. (1990a), *J. Appl. Cryst. 23*, 545.
Bhat, H. L., Roberts, K. J., Sacchi, M. (1990b), *J. Phys.: Condens. Matter 2*, 8557.
Bianconi, A. (1988), in: *X-ray Absorption, Principles, Applications, Techniques of EXAFS, SEXAFS and XANES:* Köningsberger, D. C., Prins, R. (Eds.). New York: Wiley, pp. 573–662.
Binsted, N., Greaves, G. N., Henderson, C. M. B. (1986), *J. de Physique C8*, 837.
Bonse, U. (1958), *Zeit. Phys. 153*, 278.
Bordas, J., Glazer, A. M., Hauser, H. (1975), *Phil. Mag., 32*, 471.
Bosio, L., Cortes, R., Defrain, A., Froment, M. J. (1984), *J. Electroan. Chem. 180*, 265.
Bowen, D. K., Clark, G. F., Davies, S. T., Nicholson, J. R. S., Roberts, K. J., Sherwood, J. N., Tanner, B. K. (1982), *Nucl. Instrum. Methods 192*, 277.

Bragg, W. L. (1949), *The Crystalline State,* Volume 1, *General Survey.* Bell and Sons, p. 27.

Bragg, W. L. (1975), in: *The Development of X-ray Analysis,* p. 137.

Burton, W. K., Cabrera, N., Frank, F. C. (1951), *Philos. Trans. Soc. London 243,* 299.

Buswell, J. T., Brock, J. M. (1987), *CEGB Report TPRD/B/0922/R87.*

Buswell, J. T., Little, E. A., Jones, R. B., Sinclair, R. N. (1986), *Proceedings of the Second International Symposium on Environmental Degradation of Materials in Nuclear Power Systems – Water Reactors;* Monterey, Sept. 1985, (ANS).

Buswell, J. T., English, C. A., Hetherington, M. H., Phythian, W. J., Smith, G. D. W., Worrall, G. M. (1988), *Proc. of the 14th International Symposium on Effects of Radiation on Materials, Andover, Massachusetts,* Vol. 2. American Society for Testing Metals.

Cernik, R. J., Clark, S. M., Pattison, P. (1989), to be published in: *Advances in X-ray Analysis.*

Cernik, R. J., Cheetham, A. K., Prout, C. K., Watkin, D. J., Wilkinson, A. P., Willis, B. T. M. (1990a), to be published in J. Appl. Cryst.

Cernik, R. J., Murray, P. K., Pattison, P. (1990b), *J. Appl. Cryst. 23,* 292.

Chikaura, Y., Tanner, B. K. (1979), *Jap. J. Appl. Phys. 18,* 1389.

Chu, X., Tanner, B. K. (1987), *Mater. Lett. 5,* 153.

Chu, C. W., Hor, P. H., Gao, L., Huang, Z. J. (1987), *Science 235,* 567.

Clark, S. M. (1989), *Nucl. Instrum. Meth. A 276,* 381.

Clark, S. M., Miller, M. C. (1990), to be published in: *Rev. Sci. Inst.*

Clark, G. F., Tanner, B. K., Sery, R. S., Savage, H. T. (1979), *J. de Physique 40 C5,* 183.

Clark, G. F., Goddard, P. A., Nicholson, J. R. S., Tanner, B. K., Wanklyn, B. M. (1983), *Phil. Mag. B 47,* 307.

Clearfield, A. (1964), *Inorganic Chemistry 3,* 146.

Clucas, J. A., Harding, M. M., Maginn, S. J. (1988), *J. Chem. Soc. Chem. Comm.,* 185.

Cocco, G., Enzo, S., Barrett, N. T., Roberts, K. J. (1989), *J. Less Comm. Metals 154,* 177.

Cocco, G., Enzo, S., Barrett, N. T., Roberts, K. J. (1990), in: *Proc. of the 2nd European Conference on Progress in Synchrotron Radiation Research.* Società Italiana di Fisica, Conference Proceedings, Vol. 25, pp. 693–696.

Cocco, G., Enzo, S., Barrett, N. T., Roberts, K. J. (1992), *Phys. Rev. B,* in press.

Cowen, P. L., Golovenchko, J. A., Robbins, M. F. (1980), *Phys. Rev. Lett. 44,* 1680.

Cowley, R. A., Lucas, C. A. (1990), *Faraday Discuss. Chem. Soc. 89,* 181.

Cox, D. E., Hasting, J. B., Thomlinson, W., Prewitt, C. T. (1983), *Nucl. Instrum. Methods. 208,* 573.

Cruickshank, D. W. J., Helliwell, J. R., Moffat, K. (1987), *Acta Cryst. A 43,* 656.

Cunningham, D. A. H. (1991), *Ph.D. Thesis, University of Strathclyde,* Glasgow.

Cunningham, D. A. H., Davey, R. J., Roberts, K. J., Sherwood, J. N., Shripathi, T. (1990a), *J. Cryst. Growth 99,* 1065.

Cunningham, D. A. H., Davey, R. J., Doyle, S. E., Gerson, A. R., Hausermann, D., Herron, M., Robinson, J., Roberts, K. J., Sherwood J. N., Shripathi, T., Walsh, F. C. (1990b), in: *Proc. 2nd Workshop on Synchrotron Light: Applications and Related Techniques:* Craievitch, A. (Ed.). Singapore: World Scientific Press, pp. 230–244.

Cunningham, D. A. H., Davey, R. J., Gerson, A. R., Ristić, R., Roberts, K. J., Sherwood, J. N., Shripathi, T. (1991a), in: *Particle Design via Crystallisation,* Amer. Inst. Chem. Eng. Symposium Series, 84, Vol. 87: Ramanarayanan, R., Kern, W., Larsen, M., Sikdar, S. (Eds.); pp. 104–113.

Cunningham, D. A. H., Gerson, A. R., Roberts, K. J., Sherwood, J. N., Wojciochowski, K. (1991b), in: *Advances in Industrial Crystallisation:* Garside, J., Davey, R. J., Jones, A. G. (Eds.). London: Butterworths, pp. 105–130.

Docherty, R., El Korashy, A., Jennissen, H.-D., Klapper, H. Roberts, K. J., Scheffen-Lauenroth, T. (1988), *J. Appl. Cryst. 21,* 406.

Doyle, S. E., Gerson, A. R., Roberts, K. J., Sherwood, J. N., Wroblewski, T. (1991), *J. Crystal Growth 112,* 302–307.

Dubois, J. M. (1988), *J. Less Comm. Metals 145,* 309.

Dudley, M. (1990), *J. X-ray Sci. and Technol. 2,* 195.

Dudley, M., Sherwood, J. N., Ando, D. J., Bloor, D. (1983), *Mol. Cryst. Liq. Cryst. 93,* 223.

Dudley, M., Sherwood, J. N., Ando, D. J., Bloor, D. (1985), in: *Polydiacetylenes:* Bloor, D., Chance, R. R. (Eds.). Dordrecht: Martin Nijhoff, pp. 87–92.

Dudley, M., Sherwood, J. N., Bloor, D. (1986), *Proc. Am. Chem. Soc. Div. Polym. Mater. Sci. Eng. 54,* 426.

Dudley, M., Baruchel, J., Sherwood, J. N. (1990), *J. Appl. Crystallog. 23,* 186.

Dudley, M., Sherwood, J. N., Bloor, D. (1991), *Proc. Roy. Soc. A 434,* 243.

Durham, P. J. (1988), in: *X-ray Absorption. Principles, Applications, Techniques of EXAFS, SEXAFS and XANES:* Köningsberger, D. C., Prins. R. (Eds.). New York: J. Wiley and Sons, pp. 53–84.

Edwards, B., Garner, C. D., Roberts, K. J. (1990), *Proc. 2nd European Conference on Progress in X-ray Synchrotron Radiation Research: Società Italiana di Fisica,* Conference Proceedings, Vol. 24, pp. 415–418.

El Korashy, A. (1988), *Ph.D. Thesis, University of Assiut,* Egypt.

El Korashy, A., Roberts, K. J., Scheffen-Lauenroth, T., Dam, B. (1987), *J. Appl. Cryst. 20,* 512.

Frahm, R. (1989a), *Rev. Sci. Instrum. 60 (7),* 2515.

Frahm, R. (1989b), *Physica B 158,* 342.

Friedrich, W., Knipping, P., von Laue, M. (1912), *Ber. Bayer. Acad. Wiss.,* 303.

Friedrich, A., Pettinger, B., Kolb, D. M., Lupke, G., Steinhoff, R., Marowsky, G. (1989), *Chem. Phys. Lett. 163*, 123.
Funke, P., Materlik, G. (1985), *Solid State Comm. 54*, 921.
Gastaldi, J., Jourdan, C. (1978), *Phys. Stat. Sol. (a) 49*, 529.
Gastaldi, J., Jourdan, C. (1979), *Phys. Stat. Sol. (a) 52*, 139.
George, A., Michot, G. (1982), *J. Appl. Cryst. 15*, 412.
Gerson, A. R. (1990), Ph.D. Thesis, University of Strathclyde.
Gerson, A. R., Roberts, K. J., Sherwood, J. N. (1991a), *Acta Cryst B 47*, 280.
Gerson, A. R., Roberts, K. J., Sherwood, J. N. (1991b), *Powder Technology 65*, 243.
Gesi, K. (1978), *J. Phys. Soc. Japan 45*, 1431.
Gesi, K., Iizumi, M. (1979), *J. Phys. Soc. Japan 46*, 697.
Gierlokta, S., Lambooy, P., de Jeu, W. H., (1990), *Europhys. Lett. 12*, 341.
Gillespie, K., Litlejohn, A., Roberts, K. J., Sheen, D. B., Sherwood, J. N. (1989), *Rev. Sci. Instrum. 60*, 2498.
Goddard, P. A., Clark, G. F., Tanner, B. K., Whatmore, R. W. (1983), *Nucl. Instrum., Methods 208*, 705.
Greaves, G. N., Davis, E. A. (1974), *Phil. Mag. 29*, 1201.
Greaves, G. N., Barrett, N. T., Antonini, M., Thornley, R., Willis, B. T., Steele, A. (1989), *J. Amer. Chem. Soc. 111*, 4313.
Grunes, L. A. (1983), *Phys. Rev. B 27*, 2111.
Gurman, S. J. (1982), *J. Mat Sci. 14*, 1541.
Gurman, S. J. (1988), *J. Phys. C 21*, 3699.
Hajdu, J., Machin, P. A., Campbell, J. W., Greenough, T. J., Clifton, J. J., Zurek, S., Glover, S., Johnson, L. N., Elder, M. (1987), *Nature 329*, 178.
Hajdu, J., Acharya, K. R., Stuart, D., Barford, D., Johnson, L. N. (1988), *Trends Biochem. Sci. 13*, 104.
Halfpenny, P. J., Sherwood, J. N. (1990), *Phil. Mag. Lett. 62*, 1.
Halfpenny, P. J., Ristić, R. I., Sherwood, J. N. (1991), in preparation.
Hamelin, A. (1982), *J. Electroanal. Chem. 142*, 299.
Hansen, M. (1958), *Constitution of Binary Alloys*. New York: McGraw-Hill.
Harding, M. M., Magin, S. J., Campbell, J. W., Clifton, I., Machin, P. A. (1988), *Acta Cryst. B 44*, 142.
Hart, M. (1975), *J. Appl. Cryst. 8*, 436.
Hart, M., Lang, A. R. (1965), *Acta Cryst. 19*, 73.
Hart, M., Parrish, W. (1986), *Materials Science Forum 9*, 39.
Hart, M., Parrish, W. (1989), *J. Mater. Res.*, in press.
Hart, M., Cernik, R. J., Parrish, W., Toraya, H. (1990), *J. Appl. Cryst. 23*, 286.
Hastings, J. B., Thomlinson, W., Cox, D. E. (1984), *J. Appl. Cryst. Growth 17*, 85.

Hausermann, D., Salvador, G., Sherman, W. F. (1989), in: *High Pressure Science and Technology: Proc. XIth AIRAFT Int. Conf., Kiev:* Novikov, N. V. (Ed.); pp. 186–194.
Hausermann, D., Daghooghi, M. R., Sherman, W. F. (1990), *High Pressure Research 4*, 414.
Hayes, T. M., Boyce, J. B. (1982), *Solid State Phys. 37*, 173.
Heald, S. (1988), in: *X-ray Absorption: Principles, Applications, Techniques of EXAFS, SEXAFS and XANES:* Köningsberger, D. C., Prins, R. (Eds.). New York: Wiley, pp. 87–118.
Helliwell, J. R. (1984), *Reports on Prog. in Physics 47*, 1403.
Helliwell, J. R. (1991), *Macromolecular Crystallography with Synchrotron Radiation*. Cambridge: Harvard University Press.
Helliwell, J. R., Habash, J., Cruickshank, D. W. J., Harding, M. M., Greenough, T. J., Campbell, J. W., Clifton, I. J., Elder, M., Machin, P., Papiz, M., Zurek, S. (1989), *J. Appl. Cryst. 22*, 483.
Hertel, N., Materlik, G., Zegenhagen, Z. (1985), *J. Z. Phys. B 58 (3)*, 199.
Holmes, P. A., Nevin, A., Nehring, J., Amstutz, H. (1989), *European Patent Application No. 0299664*.
Houchin, M. R. (1987), *Patent wo87/07885*, U.S.A.
Iglesia, E., Boudart, M. (1984), *Journal of Catalysis 88*, 325.
Isherwood, B. J. (1977), *GEC Journal of Science and Technology 43*, 111.
Itoh, K., Kataoka, T., Matsunaga, H., Nakamura, E. (1980), *J. Phys. Soc. Japan. 48*, 1039.
Jacobi, H. (1972), *Z. Krist. 135*, 467.
James, R. J. (1958), *The Crystalline State*, Vol. 2, *The Optical Principles of the Diffraction of X-rays*. G. Bell and Sons.
Jedrecy, N., Sauvage-Simkin, M., Pinchaux, R., Geiser, N., Massies, J., Etgens, V. H. (1990), *Società Italiana di Fisica Conference Proceedings 25*, 481.
Johnson, W. L. (1986), *Prog. Mat. Sci. 30*, 81.
Jourdan, C., Gastaldi, J. (1979), *Scripta Met. 13*, 55.
Kiessig, H. (1931), *Ann. der Physik 10*, 769.
Kitano, T., Matsui, J., Ishikawa, T. (1985), *Jap. J. Appl. Phys. 24*, L948.
Koch, E. E. (Ed.) (1983), *Handbook on Synchrotron Radiation*, Vol. 1. Amsterdam: North Holland.
Koenig, J. H., Carron, J. G. (1967), *Mat. Res. Bull. 2*, 509.
Kolb, D. M., Schneider, J. (1985), *Surf. Sci. 162*, 764.
Köningsberger, D. C. (1988), *X-ray Absorption: Principles, Applications, Techniques of EXAFS, SEXAFS and XANES:* Köningsberger, D. C., Prins, R. (Eds.). New York: Wiley.
Kucharczyk, D., Paciorek, W., Kalincinska-Karut, J. (1982), *Phase Transitions 2*, 277.
Kunz, C. (Ed.) (1979), *Synchrotron Radiation: Techniques and Applications, Topics in Current Physics*, Vol. 10. Berlin: Springer.
Kuriyama, M., Steiner, B. W., Dobbyn, R. C. (1989), *Ann. Rev. Mater. Sci. 19*, 183.

Lang, A. R., Kowalski, G., Makepiece, A. P. W., Moore, M., Clackson, S., Yacoot, A. (1987), *Synchrotron Radiation, Appendix to the Daresbury Annual Report 1986/1987*, SERC Daresbury Laboratory.
Lee, P. A., Citrin, P. H., Eisenberger, P., Kincaid, B. M. (1981), *Rev. Mod. Phys. 53*, 769.
Lim, G., Parrish, W., Oritz, C., Bellotto, M., Hart, M. (1987), *J. Mater. Res. 2*, 471.
Longo, J. M., Raccah, P. M. (1973), *J. Solid State Chem. 6*, 526.
Lucas, C. A. (1989), Ph.D. Thesis, Edinburgh University.
Lucas, G. E., Odette, G. R. (1986), *Proceedings of the Second International Symposium on Environmental Degradation of Materials in Nuclear Power Systems – Water Reactors*, Monterey, Sept. 1985, (ANS).
Lucas, C. A., Hatton, P. D., Bates, S., Ryan, T. W., Miles, S., Tanner, B. K. (1988), *J. Appl. Phys. 63 (6)*, 1936.
Luth, H., Nyburg, S. C., Robinson, P. M., Scott, H. G. (1974), *Mol. Cryst. Liq. Cryst. 27*, 337.
Macdonald, J. E. (1990), *Faraday Discuss. Chem. Soc. 89*, 191.
Macdonald, J. E., Williams, A. A., van Filfhout, R., van der Veen, J. F., Finney, M. S., Johnson, A. D., Norris, C. (1990), in: *Proceedings of NATO ARW, Kinetics of Ordering at Surfaces*. New York: Plenum Press, in press.
Mamott, G. T., Barnes, P., Tarling, S. E., Jones, S. L., Norman, C. J. (1991), *J. Mater. Sci.*, in press.
Martens, G., Rabe, P. (1980), *Phys. Stat. Sol. (a) 58*, 415.
Materlik, G., Zegenhagen, Z. (1984), *J. Phys. Lett. 104A*, 47.
Materlik, G., Frahm, A., Bedzyk, M. J. (1984), *Phys. Rev. Lett. 52*, 441.
McGinn, S. T. (1990), Ph.D. Thesis, Liverpool University.
McSkimin, H. J., Bond, W. L. (1957), *Phys. Rev. 105*, 116.
Melroy, O. R., Toney, M. F., Borges, G. L., Samant, M. G., Blum, L., Kortright, J. B., Ross, P. N. (1988), *Phys. Rev. B38*, 10962.
Michot, G., Weil, B., George, A. (1984), *J. Cryst. Growth 69*, 627.
Mikhail, I., Peters, K. (1979), *Acta Cryst. B35*, 1200.
Milia, F., Kind, R., Slak, J. (1983), *Phys. Rev. B27*, 6662–6668.
Miltat, J., Bowen, D. K. (1979), *J. de Physique 40*, 389.
Miltat, J., Sauvage-Simkin, M. (1984), in: *Applications of X-ray Topographic Methods to Materials Science*: Weissman, S., Balibar, F., Petroff, J. F. (Eds.). New York: Plenum Press, pp. 185–209.
Mitrovic, M. M., Ristić, R. I., Ciric, I. (1990), *Appl. Phys. A 51*, 374.
Murray, A. D., Crockroft, A. J., Fitch, A. N. (1990), *MPROF – a Program for multiplatten Rietveld refinement of crystal structures from X-ray powder data*. To be published.

Nehring, J., Amstutz, H., Holmes, P. A., Nevin, A. (1987), *Appl. Phys. Lett. 51*, 1283.
Nyburg, S. C., Potworowski, J. A. (1973), *Acta. Cryst. B29*, 347.
Ocko, B. M., Wong, J., Davenport, A., Isaacs, H. (1990), *Phys. Rev. Lett. 65*, 1466.
Oldman, R. J. (1986), *J. de Physique C8*, 321.
O'Reilly, E. P. (1989), *Semicond. Sci. and Tech. 4*, 121.
Palmberg, W. W., Rhodin, T. N. (1967), *Phys. Rev. 161*, 586.
Parratt, L. G. (1954), *Phys. Rev. 95*, 359.
Parrish, W., Hart, M., Huang, T. C. (1986), *J. Appl. Cryst. Growth 19*, 92.
Petroff, J. F., Sauvage, M. (1978), *J. Cryst. Growth 43*, 628.
Phythian, W. J., English, C. A. (1991), *UKAEA Harwell Report AERE-R-13632*, to be published.
Pizzini, S., Roberts, K. J., Greaves, G. N., Harris, N., Moore, P., Pantos, E., Oldman, R. J. (1989), *Rev. Sci. Instrum. 60*, 2525.
Pizzini, S., Roberts, K. J., Phythian, W. J., English, C. A. (1990a), *Phil. Mag. Lett. 61*, 223.
Pizzini, S., Roberts, K. J., Phythian, W. J., English, C. A. (1990b), *Proc. of the 6th International Conference on X-Ray Absorption Fine Structure (XAFS VI)*. Ellis Horwood, Chichester, pp. 530–532.
Pizzini, S., Roberts, K. J., Greaves, G. N., Barrett, N. T., Dring, I., Oldman, R. J. (1990c), *Faraday Discuss. Chem. Soc. 89*, 51.
Pizzini, S. (1990d), Ph.D. Thesis, University of Strathclyde.
Polk, D. E., Bordeaux, D. S. (1973), *Phys. Rev. Lett. 31*, 92.
Quilichini, M., Mathieu, J. P., Le Postollec, M., Toupry, N. (1982), *J. de Physique 43*, 787.
Reynolds, C. D., Stowell, B., Joshi, K. K., Harding, M. M., Maginn, S. J., Dodson, G. G. (1988), *Acta Cryst. B44*, 512.
Rietveld, H. M. (1969), *J. Appl. Cryst. 2*, 65.
Ristić, R. I., Sherwood, J. N., Wojciechowski, K. (1988), *J. Crystal Growth 91*, 163.
Ristić, R. I., Sherwood, J. N., Shripathi, T. (1991), in: *Advances in Industrial Crystallisation*: Garside, J., Davey, R. J. (Eds.). London: Butterworths, pp. 77–91.
Roberts, K. J., Sherwood, J. N., Bowen, D. K., Davies, S. T. (1982), *Materials Letters 2*, 300.
Roberts, K. J., Sherwood, J. N., Bowen, D. K., Davies, S. T. (1983), *Materials Letters 1*, 104.
Roberts, K. J., Sherwood, J. N., Shripathi, T., Oldman, R. J., Holmes, P. A., Nevin, A. (1990), *J. Physics D: Applied Physics 23*, 255.
Robinson, I. K., Tung, R. T., Feidenhans'l, R. (1988), *Phys. Rev. B38*, 3632.
Ruoff, A. L. (1991), *Materials Science and Technology*, Vol. 5: Cahn, R. W., Haasen, P., Kramer, E. J. (Eds.). Weinheim: VCH, pp. 473–498.
Ryan, T. W., Hatton, P. D., Bates, S., Watt, M., Sotomayor-Torres, C., Claxton, P. A., Roberts, J. S. (1987), *Semicond. Sci. Technol.*, 241.

Sacchi, M., Antonini, G. M., Barrett, N. T., Greaves, G. N., Thornley, F. R. (1989), *Proc. 2nd Int. Workshop on Non-Crystalline Solids, San Sebastian (Spain)*, July 1989.

Safa, M., Tanner, B. K. (1978), *Phil. Mag. B 37*, 739.

Samant, M. G., Toney, M. F., Borges, G. L., Blum, L., Melroy, O. R. (1988), *J. Phys. Chem. 92*, 220.

Samwer, K. (1988), *Phys. Rep. 161*, 1.

Sayers, D. E., Stern, E. A., Lytle, F. M. (1971), *Phys. Rev. Lett. 27*, 1204.

Scheffer, T. J., Nehring, J. (1984), *Appl. Phys. Lett. 45*, 1021.

Schultz, L. (1987), *Mater. Sci. Engin. 93*, 213.

Sery, R. S., Savage, H. T., Tanner, B. K., Clark, G. F. (1978), *J. Appl. Phys. 49*, 2010.

Shirane, G., Endoh, Y., Birgeneau, R. J., Kastner, M. A., Hidaka, Y., Oda, M., Suzuki, M., Muarakami, T. (1987), *Phys. Rev. Lett. 59*, 1613.

Steininger, R., Bilgram, J. H. (1990), *J. Crystal Growth 99*, 98.

Stephenson, J. D., Kelha, V., Tilli, M., Tuomi, T. (1979), *Phys. Stat. Sol. (a) 51*, 93.

Stern, E. A. (1974), *Phys. Rev. B 10*, 3027.

Stocks, G. M., Temmerman, W. M., Szotek, Z., Sterne, P. A. (1988), *Supercond. Sci. Technol. 1*, 57.

Stuhrmann, H. B. (Ed.) (1982), *Uses of Synchrotron Radiation in Biology*. London: Academic Press.

Tanner, B. K. (1976), *X-ray Diffraction Topography*. Oxford: Pergamon Press.

Tanner, B. K., Safa, M., Midgley, D., Bordas, J. (1976), *J. Magn. Mat. 1*, 337.

Teo, B. K., Chen, H. S., Wang, R., D'Antonio, M. R. (1983), *J. Non-Crystalline Solids 58*, 249.

Tuomi, T., Naukkarnin, K., Rabe, P. (1974), *Phys. Status Solidi A 25*, 93.

Tuomi, T., Tilli, M., Kelha, V., Stephenson, J. D. (1978), *Phys. Stat. Sol. (a) 50*, 427.

Tuomi, T., Stephenson, J. D., Tilli, M., Kelha, V. (1979), *Phys. Stat. Sol. (a) 53*, 571.

Vecher, A. A., Dalidovich, S. V., Gusev, E. A. (1985), *Thermochimica Acta 89*, 383.

Vlieg, E., van't Ent, A., de Jong, A. P., Neerings, H., van der Veen, J. F. (1987), *Nucl. Instrum. and Methods A 262*, 522.

Wagner, C. J., Boldrick, M. S., Keller, L., (1988), *Advances in X-ray Analysis 31*, 129.

Wainfan, N., Parratt, L. G. (1960), *J. Appl. Phys. 31*, 1331.

Wainfan, N., Nancy, J., Scott, N. J., Parratt, L. G. (1959), *J. Appl. Phys. 30*, 1604.

Waiti, G. C., Lundu, M. L., Ghosh, S. K., Banerjee, B. K. (1974), *Thermal Analysis*, Vol. 2, *Proc. 4th Intern. Conf. Therm. Anal.*, Budapest 1974.

Weeber, A. W., Bakker, H. (1988), *Physica B 153*, 93.

Whatmore, R. W., Goddard, P. A., Tanner, B. K., Clark, G. F. (1982), *Nature 299*, 44.

Williams, A. A., Macdonald, J. E., van Silfhout, R. G., van der Veen, J. F., Johnson, A. D., Norris, C. (1989), *J. Phys. Condens. Matt. 1*, 273.

Winick, H., Doniach, S. (1980), *Synchrotron Radiation Research*. New York: Plenum Press.

Wood, I. G., Thompson, P., Matthewman, J. C. (1983), *Acta Cryst. B 39*, 543.

Zarka, A. (1983), *J. Appl. Cryst. 16*, 354.

Zarka, A. (1984), in: *Applications of X-ray Topographic Methods to Materials Science:* Weissmann, S., Balibar, F., Petroff, J. F. (Eds.). New York: Plenum Press, pp. 487–500.

Zei, M. S., Lehmpfuhl, G., Kolb, D. M. (1989), *Surf. Sci. 221*, 23.

General Reading

Amoros, J. L., Buerger, M. J., Amoros, M. C. (1975), *The Laue Method*. New York: Academic Press.

Bowen, D. K. (Ed.) (1983), *The Application of Synchrotron Radiation to Problems in Materials Science*, SERC Daresbury Laboratory Report DL/SCI/R19.

Koch, E. E. (Ed.) (1983), *Handbook on Synchrotron Radiation*, Vol. 1. Amsterdam: North Holland.

Köningsberger, D. C., Prins, R. (Eds.) (1988), *X-ray Absorption, Principles, Applications, Techniques, of EXAFS, SEXAFS and XANES*. New York: Wiley.

Kunz, C. (Ed.) (1979), *Synchrotron Radiation: Techniques and Applications, Topics in Current Physics*, Vol. 10. Berlin: Springer.

Lee, P. A., Citrin, P. H., Eisenberger, P., Kincaid, B. M. (1981), *Rev. Modern Phys. 53*, 769.

Stuhrmann, H. B. (Ed.) (1982), Uses of Synchrotron Radiation in Biology. London: Academic Press.

Tanner, B. K. (1976), *X-ray Diffraction Topography*. Oxford: Pergamon Press.

Tanner, B. K. (1977), *Progress in Crystal Growth and Characterisation 1*, 23.

Tanner, B. K., Bowen, D. K. (Eds.) (1979), *Characterisation of Crystal Growth Defects by X-ray Methods*. New York: Plenum Press.

Weissmann, S., Balibar, F., Petroff, J.-F. (Eds.) (1984), *Application of X-ray Topographic Methods to Materials Science*. New York: Plenum Press.

Winick, H., Doniach, S. (Eds.) (1980), *Synchrotron Radiation Research*. New York, London: Plenum Press.

3 X-Ray Fluorescence Analysis

Ron Jenkins

International Centre for Diffraction Data, Newtown Square, PA, U.S.A.

3.1	**Introduction**	172
3.2	**Historical Development of X-ray Spectrometry**	173
3.3	**X-ray Wavelengths**	175
3.3.1	Continuous Radiation	175
3.3.2	Characteristic Radiation	175
3.3.3	Selection Rules	176
3.3.4	Nomenclature for X-ray Wavelengths	178
3.4	**Properties of X-radiation**	179
3.4.1	Absorption of X-rays	179
3.4.2	Coherent and Incoherent Scattering	179
3.4.3	Interference and Diffraction	180
3.5	**Instrumentation for X-ray Fluorescence**	181
3.5.1	Sources	181
3.5.2	Detectors	182
3.5.3	Types of Spectrometer Used for X-ray Fluorescence	185
3.5.3.1	Wavelength Dispersive Systems	188
3.5.3.2	Energy Dispersive Systems	189
3.5.3.3	Total Reflection Spectrometers (TRXRF)	190
3.5.3.4	Synchrotron Source X-ray Fluorescence (SSXRF)	191
3.5.3.5	Proton Excited X-ray Fluorescence (PIXE)	193
3.6	**Qualitative Analysis with the X-ray Spectrometer**	193
3.7	**Accuracy of X-ray Fluorescence**	194
3.7.1	Counting Statistical Errors	194
3.7.2	Matrix Effects	196
3.7.3	Specimen Preparation for X-ray Fluorescence	197
3.8	**Quantitative Analysis**	198
3.8.1	Measurement of Pure Intensities	201
3.8.2	Use of Internal Standards	202
3.8.3	Type Standardization	202
3.8.4	Influence Correction Methods	203
3.8.5	Fundamental Methods	204
3.9	**Trace Analysis**	205
3.9.1	Analysis of Low Concentrations	205
3.9.2	Analysis of Small Amounts of Sample	207
3.10	**References**	208

3.1 Introduction

X-rays are a short wavelength form of electromagnetic radiation discovered by Wilhelm Röntgen just before the turn of the century (1898). X-ray photons are produced following the ejection of an inner orbital electron from an excited atom, and subsequent transition of atomic orbital electrons from states of high to low energy. A beam of X-rays passing through matter is subject to three processes, absorption, scatter, and fluorescence. The absorption of X-rays varies as the third power of the atomic number of the absorber. Thus, when a polychromatic beam of X-rays is passed through a heterogeneous material, areas of high average atomic number will attenuate the beam to a greater extent than areas of lower atomic number. Thus the beam of radiation emerging from the absorber has an intensity distribution across the irradiation area of the specimen, which is related to the average atomic number distribution across the same area. It is upon this principle that all methods of X-ray radiography are based. Study of materials by use of the X-ray absorption process is the oldest of all of the X-ray methods in use and Röntgen himself included a radiograph of his wife's hand in his first published X-ray paper. Today, there are many different forms of X-ray absorptiometry in use, including industrial radiography, diagnostic medical and dental radiography, and security screening.

X-rays are scattered mainly by the loosely bound outer electrons of an atom. Scattering of X-rays may be coherent (same wavelength) or incoherent (longer wavelength). Coherently scattered photons may undergo subsequent mutual interference, leading in turn to the generation of diffraction maxima. The angles at which the diffraction maxima occur can be related to the spacings between planes of atoms in the crystal lattice and hence, X-ray diffraction patterns can be used to study the structure of solid materials. Following the discovery of the diffraction of X-rays in 1912 by Friedrich, Knipping and Von Laue the use of this method for materials analysis has become very important both in industry and research, to the extent that, today, it is one of the most useful techniques available for the study of structure dependent properties of materials. Fluorescence occurs when the primary X-ray photons are energetic enough to create electron vacancies in the specimen, leading in turn to the generation of secondary (fluorescence) radiation produced from the specimen. This secondary radiation is characteristic of the elements making up the specimen. The technique used to isolate and measure individual characteristic wavelengths following excitation by primary X-radiation is called X-ray fluorescence spectrometry.

X-ray spectrometric techniques provided important information for the theoretical physicist in the first half of this century and since the early 1950s they have found an increasing use in the field of materials characterization. While most of the early work in X-ray spectrometry was carried out using electron excitation (Von Hevesey, 1932), today, use of electron-excited X-radiation is restricted mainly to X-ray spectrometric attachments to electron microscopes. Most modern stand-alone X-ray spectrometers use X-ray excitation sources rather than electron excitation. X-ray fluorescence spectrometry typically uses a polychromatic beam of short wavelength X-radiation to excite longer wavelength characteristic lines from the sample to be analyzed. Modern X-ray spectrometers use either the diffracting power of a single crystal to isolate narrow wavelength bands, or a proportional detector to isolate narrow energy bands, from the polychromatic radiation (including characteristic radiation) excited in the sample. The first of these methods is called *wavelength dispersive spectrometry* and the second, *energy dispersive spectrometry*. Because the relationship between emission wavelength and atomic number is known, isolation of individual characteristic lines allows the unique identification of an element to be made and elemental concentrations can be estimated from characteristic line intensities. Thus this technique is a means of materials characterization in terms of chemical composition.

3.2 Historical Development of X-ray Spectrometry

X-ray fluorescence spectrometry provides the means of the identification of an element by measurement of its characteristic X-ray emission wavelength or energy. The method allows the quantization of a given element by first measuring the emitted characteristic line intensity and then relating this intensity to elemental concentration. While the roots of the method go back to the early part of this century, it is only during the last thirty years or so that the technique has gained major significance as a routine means of elemental analysis (Birks, 1976). The first use of the X-ray spectrometric method dates back to the classic work of Henry Moseley (1912). In Moseley's original X-ray spectrometer, the source of primary radiation was a cold cathode tube in which the source of electrons was residual air in the tube itself, the specimen for analysis forming the target of the tube. Radiation produced from the specimen then passed through a thin gold window, onto an analyzing crystal whence it was diffracted to the detector. One of the major problems in the use of electrons for the excitation of characteristic X-radiation is that the process of conversion of electron energy into X-rays is relatively inefficient; about 99% of the electron energy is converted to heat energy. This means in turn that it may be difficult to analyze specimens which are volatile or tend to melt. Nevertheless, the technique seemed to hold some promise as an analytical tool and one of the first published papers on the use of X-ray spectroscopy for real chemical analysis appeared as long ago as 1922, when Hadding described the use of the technique for the analysis of minerals.

In 1925, a practical solution to the problems associated with electron excitation was suggested by Coster and Nishina, who used primary X-ray photons for the excitation of secondary characteristic X-ray spectra. The use of X-rays, rather than electrons, to excite characteristic X-radiation avoids the problem of the heating of the specimen. It is possible to produce the primary X-ray photons inside a sealed X-ray tube under high vacuum and efficient cooling conditions, which means the specimen itself need not be subject to heat dissipation problems or the high vacuum requirements of the electron beam system. Use of X-rays rather than electrons represented the beginnings of the technique of X-ray fluorescence as we know it today. The fluorescence method was first employed on a practical basis in 1928 by Glocker and Schreiber. Unfortunately, data obtained at that time were rather poor because X-ray excitation is even less efficient than electron excitation. Also the detectors and crystals available at that time were rather primitive and thus the fluorescence technique did not seem to hold too much promise. In any event, widespread use of the technique had to wait until the mid 1940s when X-ray fluorescence was rediscovered by Friedman and Birks (1976). The basis of their spectrometer was a diffractometer that had been originally designed for the orientation of quartz oscillator plates. A Geiger counter was used as a means of measuring the intensities of the diffracted characteristic lines and quite reasonable sensitivity was obtained for a very large part of the atomic number range.

The first commercial X-ray spectrometer became available in the early 1950s and although these earlier spectrometers operated only with an air path, they were able to provide qualitative and quantitative information on all elements above atomic number 22 (titanium). Later versions allowed use of helium or vacuum paths that extended the lower atomic number cut-off. Most modern commercially available X-ray spectrometers have a range from about 0.4 to 20 Å (40 to 0.6 keV) and this range will allow measurement of the K series from fluorine ($Z = 9$) to lutetium ($Z = 71$), and the L series from manganese ($Z = 25$) to uranium ($Z = 92$). Other line series arise from the M and N levels but these have little use in analytical X-ray spectrometry. In practice, the number of vacancies in electron levels resulting in the production of characteristic X-ray photons is less than the total number of vacancies created in the excitation process, because the atom can also regain its initial state by reorganization of atomic electrons without the emission of X-ray photons (the Auger process). This is an important factor in determining the absolute number of counts that an element will give under a certain set of experimental conditions. It is mainly for this reason that the sensitivity of the X-ray spectrometric technique is rather poor for the very low atomic number elements, since fluorescent yields for these low atomic numbers are very small.

Today, nearly all commercially available X-ray spectrometers use the fluorescence excitation method and employ a sealed X-ray tube as the primary excitation source. Some of the simpler systems may use a radio-isotope source, because of considerations of cost and/or portability. While electron excitation is generally not used in stand-alone X-ray spectrometers, it is the basis of X-ray spectrometry carried out on electron column instruments. The ability to focus the primary electron beam allows analysis of extremely small areas down to a micron or so in diameter, or even less in specialized instruments. This, in combination with imaging and electron diffraction, offers an extremely powerful method for the examination of small specimens, inclusions, grain boundary phenomena, etc. The instruments used for this type of work may be in the form of a specially designed electron microprobe analyzer (e.g., Birks, 1963) or simply an energy or wavelength dispersive attachment to a scanning electron microscope.

Over the past thirty years or so, the X ray fluorescence method has become one of the most valuable methods for the qualitative and quantitative analysis of materials. Many methods of instrumental elemental analysis are available today and among the factors that will generally be taken into consideration in the selection of one of these methods are accuracy, range of application, speed, cost, sensitivity and reliability. While it is certainly true that no one technique can ever be expected to offer all of the features that a given analyst might desire, the X-ray method has good overall performance characteristics. In particular, the speed, accuracy and versatility of X-ray fluorescence are the most important features among the many that have made it the method of choice in over 20 000 laboratories all over the world. Both the simultaneous wavelength dispersive spectrometer and the energy dispersive spectrometers lend themselves admirably to the qualitative and quantitative analysis of solid materials and solutions.

3.3 X-ray Wavelengths

3.3.1 Continuous Radiation

When a high energy electron beam is incident upon a specimen, one of the products of the interaction is an emission of a broad wavelength band of radiation called continuum, also referred to as white radiation or bremsstrahlung. This white radiation is produced as the impinging high energy electrons are decelerated by the atomic electrons of the elements making up the specimen. The intensity/wavelength distribution of this radiation is typified by a minimum wavelength λ_{min}, which is proportional to the maximum accelerating potential V of the electrons, i.e. 12.4/V (keV). The intensity distribution of the continuum reaches a maximum at a wavelength 1.5 to 2 times greater than λ_{min}. Increasing the accelerating potential causes the intensity distribution of the continuum to shift towards shorter wavelengths. Most X-ray tubes employ a heated tungsten filament as a source of electrons, and a layer of pure metal such as chromium, rhodium or tungsten, as the anode. The broad band of white radiation produced by this type of tube is ideal for the excitation of the characteristic lines from a wide range of atomic numbers. In general, the higher the atomic number of the anode material, the more intense the beam of radiation produced by the tube. Conversely, however, because the higher atomic number anode elements generally require thicker exit windows (to provide adequate dissipation of heat from the electrons scattered by the anode) in the tube, the longer wavelength output from such a tube is rather poor and so these high atomic number anode tubes are less satisfactory for excitation of longer wavelengths from low atomic number samples.

3.3.2 Characteristic Radiation

In addition to electron interactions leading to the production of white radiation, there are also electron interactions which produce characteristic radiation. If a high energy particle, such as an electron, strikes a bound atomic electron, and the energy of the particle is greater than the binding energy of the atomic electron, it is possible that the atomic electron will be ejected from its atomic position, departing from the atom with a kinetic energy $(E-\psi)$ equivalent to the difference between the energy E of the initial particle and the binding energy ψ of the atomic electron. Where the exciting particles are X-ray photons, the ejected electron is called a photoelectron and the interaction between primary X-ray photons and atomic electrons is called the photoelectric effect. As long as the vacancy in the shell exists, the atom is in an unstable state and there are two processes by which the atom can revert back to its original state. The first of these involves a rearrangement that does not result in the emission of X-ray photons, but in the emission of other photoelectrons from the atom. The effect is known as the Auger effect (Auger, 1925), and the emitted photoelectrons are called Auger electrons.

The second process by which the excited atom can regain stability is by transference of an electron from one of the outer orbitals to fill the vacancy. The energy difference between the initial and final states of the transferred electron may be given off in the form of an X-ray photon. Since all emitted X-ray photons have energies proportional to the differences in the energy states of atomic electrons, the lines from a given element will be characteristic of that element. The relationship between the wavelength of a characteristic X-ray photon and the atomic number Z of the excited element was first established by Moseley. Moseley's law is written:

$$\frac{1}{\lambda} = K \times (Z - \sigma)^2 \quad (3\text{-}1)$$

in which K is a constant that takes on different values for each spectral series. σ is the shielding constant that has a value of just less than unity. The wavelength λ of the X-ray photon is inversely related to the energy E of the photon according to the relationship:

$$\lambda(\text{Å}) = \frac{12.4}{E(\text{keV})} \quad (3\text{-}2)$$

Since there are two competing effects by which an atom may return to its initial state, and since only one of these processes will give rise to the production of a characteristic X-ray photon, the intensity of an emitted characteristic X-ray beam will be dependent upon the relative effectiveness of the two processes within a given atom. As an example, the number of quanta of K series radiations emitted per ionized atom is a fixed ratio for a given atomic number, this ratio being called the fluorescent yield. The fluorescent yield varies as the fourth power of atomic number and approaches unity for higher atomic numbers. Fluorescent yield values are several orders of magnitude less for the very low atomic numbers. In practice this means that if, for example, one were to compare the intensities obtained from pure barium ($Z = 56$) and pure aluminium ($Z = 13$), all other things being equal, pure barium would give about 50 times more counts than would pure aluminium. The L fluorescent yield for a given atomic number is always less by about a factor of three than the corresponding K fluorescent yield.

3.3.3 Selection Rules

An excited atom can revert to its original ground state by transferring an electron from an outer atomic level to fill the vacancy in the inner shell. An X-ray photon is emitted from the atom as part of this de-excitation step, the emitted photon having an energy equal to the energy difference between the initial and final states of the transferred electron. Each unique atom has a number of available electrons which can take part in the transfer and since millions of atoms are typically involved in the excitation of a given specimen, all possible de-excitation routes are taken. These de-excitation routes can be defined by a simple set of selection rules that account for the majority of the observed wavelengths. Each electron in an atom can be defined by four quantum numbers. The first of these quantum numbers is the principal quantum number n, that can take on all integral values. When n is equal to 1, the level is referred to as the K level; when n is 2, the L level, and so on. l is the angular quantum number and this can take all values from $(n-1)$ to zero. m is the magnetic quantum number that can take values of $+l$, zero and $-l$. s is the spin quantum number with a value of $+\frac{1}{2}$. The total momentum "J" of an electron is given by the vector sum of $(l+s)$. Since no two electrons within a given atom can have the same set of quantum numbers, a series of levels or shells can be constructed. Table 3-1 gives the atomic structures of the first three principal shells. The first shell, the K shell, has a maximum of two electrons and these are both in the 1s level (orbital). Note that the "s" refers, in this context, to the orbital shape (s = sharp) taken by the electrons and not the spin quantum number! Since the value of J must be positive in this instance the only allowed value is $\frac{1}{2}$. In the second shell, the L shell, there are eight electrons, two in the 2s level and six in the 2p levels. In this instance J has a value of $\frac{1}{2}$ for the 1s level and $\frac{3}{2}$ or $\frac{1}{2}$ for the 2p level, thus giving a total of three possible L transition levels.

Table 3-1. Atomic structures of the first three principal shells.

Shell (electrons)	n	m	l	s	Orbitals	J
K(2)	1	0	0	$\pm\frac{1}{2}$	1s	$\frac{1}{2}$
L(8)	2	0	0	$\pm\frac{1}{2}$	2s	$\frac{1}{2}$
	2	1	1	$\pm\frac{1}{2}$	2p	$\frac{1}{2},\frac{3}{2}$
	2	0	1	$\pm\frac{1}{2}$	2p	$\frac{1}{2},\frac{3}{2}$
	2	−1	1	$\pm\frac{1}{2}$	2p	$\frac{1}{2},\frac{3}{2}$
M(18)	3	0	0	$\pm\frac{1}{2}$	3s	$\frac{1}{2}$
	3	1	1	$\pm\frac{1}{2}$	3p	$\frac{1}{2},\frac{3}{2}$
	3	0	1	$\pm\frac{1}{2}$	3p	$\frac{1}{2},\frac{3}{2}$
	3	−1	1	$\pm\frac{1}{2}$	3p	$\frac{1}{2},\frac{3}{2}$
	3	2	2	$\pm\frac{1}{2}$	3d	$\frac{3}{2},\frac{5}{2}$
	3	−1	2	$\pm\frac{1}{2}$	3d	$\frac{3}{2},\frac{5}{2}$
	3	0	2	$\pm\frac{1}{2}$	3d	$\frac{3}{2},\frac{5}{2}$
	3	−1	2	$\pm\frac{1}{2}$	3d	$\frac{3}{2},\frac{5}{2}$
	3	−2	2	$\pm\frac{1}{2}$	3d	$\frac{3}{2},\frac{5}{2}$

These levels are referred to as L_I, L_{II} and L_{III} respectively. In the M level, there are a maximum of eighteen electrons, two in the 3s level, eight in the 3p level and ten in the 3d level. Again with the values of $\frac{3}{2}$ or $\frac{1}{2}$ for J in the 3p level; and $\frac{5}{2}$ and $\frac{3}{2}$ in the 3d level, a total of five M transition levels are possible. Similar rules can be used to build up additional levels, N, O etc.

The selection rules for the production of normal (diagram) lines require that the principal quantum number must change by at least one, the angular quantum number must change by only one, and the J quantum number must change by zero or one. Transition groups may now be constructed, based on the appropriate number of transition levels. Application of the selection rules indicates that in, for example, the K series, only $L_I \rightarrow K$ and $L_{III} \rightarrow K$ transitions are allowed for a change in the principal quantum number of one. There are equivalent pairs of transitions for $n=2$, $n=3$, $n=4$, etc. Figure 3-1 shows the lines that are observed in the K series. Three groups of lines are indicated. The normal lines are shown on the left hand side of the figure and these consist of three pairs of lines from the L_I/L_{III}, M_{II}/M_{III} and N_{II}/N_{III} sub-shells respectively. While most of the observed fluorescent lines are normal, certain lines may also occur in X-ray spectra that do not at first sight fit the basic selection rules. These lines are called forbidden lines and are shown in the center portion of the figure. Forbidden lines typically arise from outer orbital levels where there is no sharp energy distinction between orbitals. As an example, in the transition elements, where the 3d level is only partially filled and is energetically similar to the 3p levels, a weak forbidden transition (the β_5) is observed. A third type of line may also occur − called satellite lines, which arise from dual ionizations. Following the ejection of the initial electron in the photoelectric process, a short, but finite, period of time elapses before the vacancy is filled. This time period is called the lifetime of the excited state. For the lower atomic number elements, this lifetime increases to such an extent that there is a significant probability that a second electron can be ejected from the atom before the first vacancy is filled. The loss of the second electron modifies the energies of the electrons in the surrounding sub-shells, and other pairs of X-ray emission lines are produced, corresponding to the $K_{\alpha1}/K_{\alpha2}$. In the K series the most common of these satellite lines are the $K_{\alpha3}/K_{\alpha4}$ and the $K_{\alpha5}/K_{\alpha6}$ doublets. These lines are shown at the right hand side of the figure. Although, because they are relatively weak, neither forbidden transitions nor satellite lines have great analytical significance, they may cause some confusion in qualitative interpretation of spectra and may even be misinterpreted as coming from trace elements.

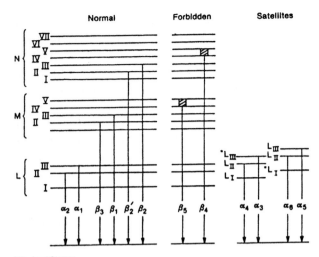

Figure 3-1. Observed lines in the K series.

* Ionized States

3.3.4 Nomenclature for X-ray Wavelengths

The classically accepted nomenclature system for the observed lines is that proposed by Siegbahn in the 1920s. Figure 3-2 shows plots of the reciprocal of the square root of the wavelength as a function of atomic number, for the K, L and M series. As indicated by Moseley's law (Eq. 3-1) such plots should be linear. A scale directly in wavelength is also shown, to indicate the range of wavelengths over which a given series occurs. In practice, the number of lines observed from a given element will depend upon the atomic number of the element, the excitation conditions and the wavelength range of the spectrometer employed. Generally, commercial spectrometers cover the range 0.3 to 20 Å (newer instruments may

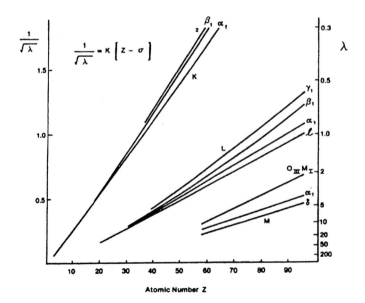

Figure 3-2. Moseley diagrams for the K, L and M series.

allow measurements in excess of 100 Å) and three X-ray series are covered by this range, the K series, the L series and the M series, corresponding to transitions to K, L and M levels respectively Each series consists of a number of groups of lines. The strongest group of lines in the series is denoted α, the next strongest β, and the third γ. There are a much larger number of lines in the higher series and for a detailed list of all of the reported wavelengths the reader is referred to the work of Bearden (1964). In X-ray spectrometry most of the analytical work is carried out using either the K or the L series wavelengths. The M series may, however, also be useful, especially in the measurement of higher atomic numbers.

3.4 Properties of X-radiation

3.4.1 Absorption of X-rays

When a beam of X-ray photons of intensity $I_0(\lambda)$ falls onto a specimen a fraction of the beam will pass through the absorber, this fraction being given by the expression:

$$I(\lambda) = I_0(\lambda) \times \exp(\mu \varrho x) \qquad (3\text{-}3)$$

where μ is the mass attenuation coefficient of absorber for the wavelength and ϱ the density of the specimen. x is the distance travelled by the photons through the specimen. It will be seen from Eq. (3-3) that a number $(I_0 - I)$ of photons has been lost in the absorption process. Although a significant fraction of this loss may be due to scatter, by far the greater loss is due to the photoelectric effect. Photoelectric absorption occurs at each of the energy levels of the atom, thus the total photoelectric absorption is determined by the sum of each individual absorption within a specific shell. Where the absorber is made up of a number of different elements, as is usually the case, the total absorption is made up of the sum of the products of the individual elemental mass attenuation coefficients and the weight fractions of the respective elements. This product is referred to as the total matrix absorption. The value of the mass attenuation referred to in Eq. (3-3) is a function of both the photoelectric absorption and the scatter. However, the photoelectric absorption influence is usually large in comparison with the scatter and to all intents and purposes the mass attenuation coefficient is equivalent to the photoelectric absorption. A plot of the mass attenuation coefficient as a function of wavelength contains a number of discontinuities, called absorption edges, at wavelengths corresponding to the binding energies of the electrons in the various subshells. Between absorption edges, as the wavelength of the incident X-ray photons become longer, the absorption increases. This particular effect is very important in quantitative X-ray spectrometry because the intensity of a beam of characteristic photons leaving a specimen is dependent upon the relative absorption effects of the different atoms making up the specimen. This effect is called a matrix effect and is one of the reasons why a curve of characteristic line intensity as a function of element concentration may not be a straight line.

3.4.2 Coherent and Incoherent Scattering

Scattering occurs when an X-ray photon interacts with the electrons of the target element. Where this interaction is elastic, i.e., no energy is lost in the collision process, the scattering is referred to as coherent (Rayleigh) scattering. Since no energy change is involved, the coherently scattered radiation will retain exactly the same wavelength as that of the incident beam. It can also happen that the scattered photon gives up a small part of its energy during the collision, especially where the electron with which the photon collides is only loosely bound. In this instance, the scatter is referred to as incoherent (Compton scattering). Compton scattering is best presented in terms of the corpuscular nature of the X-ray photon. In this instance, an X-ray photon collides with a loosely bound outer atomic electron. The electron recoils under the impact, removing a small portion of the energy of the primary photon, which is then deflected with the corresponding loss of energy, or increase of wavelength.

3.4.3 Interference and Diffraction

X-ray diffraction is a combination of two phenomena – coherent scatter and interference. At any point where two or more waves cross one another, they are said to interfere. Interference does not imply the impedance of one wave train by another, but rather describes the effect of superposition of one wave upon another. The principal of superposition is that the resulting displacement, at any point and at any instant, may be formed by adding the instantaneous displacements that would be produced at the same point by independent wave trains, if each were present alone. Under certain geometric conditions, wavelengths that are exactly in phase may add to one another, and those that are exactly out of phase, may cancel each other out. Under such conditions coherently scattered photons may constructively interfere with each other giving diffraction maxima. As illustrated in Fig. 3-3(a), a crystal lattice consists of a regular arrangement of atoms, with layers of high atomic density running throughout the crystal structure. Planes of high atomic density means, in turn, planes of high electron density. Since scattering occurs between impinging X-ray photons and the loosely bound outer orbital atomic electrons, when a monochromatic beam of radiation falls onto the high atomic density layers scattering will occur. In order to satisfy the requirement for constructive interference, it is necessary that the scattered waves originating from the individual atoms i.e. the scattering points, be in phase with one another. The

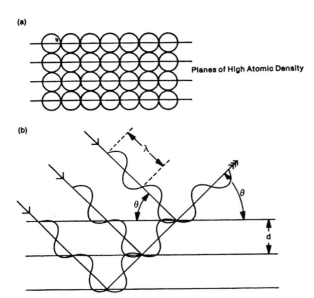

Figure 3-3. The crystal lattice and the origin of X-ray diffraction. The structure of an ideal crystal is shown in (a). The geometric condition for Bragg diffraction is shown in the lower portion of the figure (b).

geometric conditions for this condition to occur are illustrated in Fig. 3-3(b). Here, a series of parallel rays strike a set of crystal planes at an angle θ and are scattered as previously described. Reinforcement will occur when the difference in the path lengths of the two interfering waves is equal to a whole number of wavelengths. This path length difference is equal to $2d \times \sin\theta$, where d is the interplanar spacing; hence the overall condition for reinforcement is that:

$$n\lambda = 2d \times \sin\theta \qquad (3\text{-}4)$$

where n is an integer. Equation (3-4) is a statement of Bragg's law. Bragg's law is important in wavelength dispersive spectrometry since by using a crystal of fixed $2d$, each unique wavelength will be diffracted at a unique diffraction angle (Bragg angle). Thus, by measuring the diffraction angle θ, knowledge of the d-spacing of the analyzing crystal allows the determination of the wavelength. Since there is a simple relationship between wavelength and atomic number, as given by Moseley's law, one can establish the atomic number(s) of the element(s) from which the wavelengths were emitted.

3.5 Instrumentation for X-ray Fluorescence

3.5.1 Sources

Several different types of source have been employed for the excitation of characteristic X-radiation, including those based on electrons, X-rays, γ-rays, protons and synchrotron radiation. Sometimes a bremsstrahlung X-ray source is used to generate specific X-radiation from an intermediate pure element sample called a secondary fluorescer. By far the most common source today is the X-ray photon source. This source is used in primary mode in the wavelength and primary energy dispersive systems, and in secondary fluorescer mode in secondary target energy dispersive spectrometers. A γ-source is typically a radio-isotope that is used either directly, or in a mode equivalent to the secondary fluorescer mode in energy dispersive spectrometry. Most conventional wavelength dispersive X-ray spectrometers use a high power (2 to 4 kW) X-ray bremsstrahlung source. Energy dispersive spectrometers use either a high power or low power (0.5 to 1.0 kW) primary source, depending on whether the spectrometer is used in the secondary or primary mode. In all cases, the primary source unit consists of a very stable high voltage generator, capable of providing a potential of typically 40 to 100 kV. The current from the generator is fed to the filament of the X-ray tube that is typically a coil of tungsten wire. The applied current causes the filament to glow, emitting electrons in all directions. A portion of this electron cloud is accelerated to the anode of the X-ray tube, which is typically a water cooled block of copper with the required anode material plated or cemented to its surface. The impinging electrons produce X-radiation, a significant portion of which passes through a thin beryllium window to the specimen.

In order to excite a given characteristic line the source must be run at a voltage V_o well in excess of the critical excitation potential V_c of the element in question. The relationship between the measured intensity of the characteristic line I, the tube current i and the operating and critical excitation potentials is as follows:

$$I = K \times i \times (V_o - V_c)^{1.6} \qquad (3\text{-}5)$$

The product of i and V_o represents the maximum output of the source in kilowatts. The optimum value for V_o/V_c is 3 to 5. This optimum value occurs because at very high operating potentials, the electrons striking the target in the X-ray tube penetrate so deeply into the target that self absorption of target radiation becomes significant. Since it is the intention to eventually equate the value of I for a given wavelength or energy to the concentration of the corresponding analyte element, it is vital that, over the short term (1 to 2 hours), both the tube current and voltage be stabilized to better than a tenth of percent. The current from the generator is fed to the tungsten filament of the X-ray tube. A sealed X-ray tube has an anode of Cr, Rh, W, Ag, Au or Mo and delivers an intense source of continuous and characteristic radiation, which then impinges onto the analyzed specimen, where characteristic radiation is generated. In general, most of the excitation of the longer wavelength characteristic lines comes from the longer wavelength characteristic lines from the tube, and most of the short wavelength excitation, from the continuous radiation from the tube.

3.5.2 Detectors

An X-ray detector is a transducer for converting X-ray photon energy into voltage pulses. Detectors work through a process of photo-ionization in which interaction between the entering X-ray photon and the active detector material produces a number of electrons. The current produced by these electrons is converted to a voltage pulse by a capacitor and resistor, such that one digital voltage pulse is produced for each entering X-ray photon. In addition to being sensitive to the appropriate photon energies, i.e. being applicable to a given range of wavelengths or energies, there are two other important properties that an ideal detector should possess. These properties are proportionality and linearity. Each X-ray photon entering the detector produces a voltage pulse and where the size of the voltage pulse is proportional to the photon energy, the detector is said to be proportional. Proportionality is needed where the technique of pulse height selection is to be used. Pulse height selection is a means of electronically rejecting pulses of voltage levels other than those corresponding to the characteristic line being measured. This technique is a very powerful tool in reducing background levels and the influence of overlapping lines from elements other than the analyte (Jenkins and de Vries, 1970). X-ray photons enter the detector at a certain rate and where the output pulses are produced at this same rate the detector is said to be linear. Linearity is important where the various count rates produced by the detector are to be used as measures of the photon intensity for each measured line. The properties of proportionality and linearity are, to a certain extent, controllable by the electronics associated with the actual detector. To this extent, while it is common practice to refer to the characteristics of detectors, the properties of the associated pulse processing chain should always be included.

A gas-flow proportional counter consists of a cylindrical tube about 2 cm in diameter, carrying a thin (25 to 50 µm) wire along its radial axis. The tube is filled with a mixture of inert gas and quench gas – typically 90% argon/10% methane (P-10). The cylindrical tube is grounded and about +1800 volts is applied to the central wire. The anode wire is connected to a resistor shunted by a capacitor forming the pulse producing circuit. An X-ray photon entering the detector produces a number of ion pairs, each comprising one electron and one argon positive ion. The average energy e_i required to produce one ion pair in argon is equal to 26.4 eV. Thus the number of ion pairs n produced by a photon of energy W will equal

(E/e_i). Following ionization, the charges separate with the electrons moving towards the (anode) wire and the argon ions to the grounded cylinder. As the electrons approach the high field region close to the anode wire they are accelerated sufficiently to produce further ionization of argon atoms. Thus a much larger number N of electrons will actually reach the anode wire. This effect is called gas gain, or gas multiplication, and its magnitude is given by M, where M is equal to N/n. For gas flow proportional counters used in X-ray spectrometry, M typically has a value of around 10^5. Provided that the gas gain is constant, the size of the voltage pulse V produced is directly proportional to the energy E of the incident X-ray photon.

In practice not all photons arising from photon energy E will be exactly equal to V. There is a random process associated with the production of the voltage pulses and the resolution of a counter is related to the variance in the average number of ion pairs produced per incident X-ray photon. The resolution is generally expressed in terms of the full width at half maximum of the pulse amplitude distribution. The theoretical resolution R_t, of a flow counter can be derived from:

$$R_t(\%) = \frac{38.3}{\sqrt{E}} \quad (3\text{-}6)$$

While the gas flow proportional counter is ideal for the measurement of longer wavelengths it is rather insensitive to wavelengths shorter than about 1.5 Å. For this shorter wavelength region it is common to use the scintillation counter. The scintillation counter consists of two parts, the phosphor (scintillator) and the photo-multiplier. The phosphor is typically a large single crystal of sodium iodide that has been doped with thallium. When X-ray photons fall onto the phosphor, blue light photons are produced, where the number of blue light photons is related to the energy of the incident X-ray photon. These blue light photons produce electrons by interaction with a photo-surface in the photo-multiplier, and the number of electrons is linearly increased by a series of secondary surface dynodes, electrically charged with respect to each other, in the photomultiplier. The current produced by the photo-multiplier is then coverted to a voltage pulse, as in the case of the gas flow proportional counter. Since the number of electrons is proportional to the energy of the incident X-ray photon, the scintillation counter is a proportional counter. Because of inefficiencies in the X-ray/blue-light/electron conversion processes, the average energy to produce a single event in the scintillation counter is more than a magnitude greater than the equivalent process in the flow counter. For this reason, the resolution of the scintillation counter is about four times worse than that of the flow counter.

The Si(Li) detector consists of a small cylinder (about 1 cm diameter and 3 mm thick) of p-type silicon that has been compensated by lithium to increase its electrical resistivity. A Schottky barrier contact on the front of the silicon disk produces a p-i-n type diode. In order to inhibit the mobility of the lithium ions and to reduce electronic noise, the diode and its preamplifier are cooled to the temperature of liquid nitrogen. While some progress has been made using miniature refrigerator compressors for this cooling process, the common practice remains to mount the detector assembly on a cold finger that is in turn kept cool with a reservoir of liquid nitrogen. By applying a reverse bias of around 1000 volts, most of the remaining charge carriers in the silicon are removed. Incident X-ray photons interact to produce a specific number of electron hole pairs. The average energy to produce one ion pair is equal to about 3.8 eV for cooled silicon. The charge produced is swept from the diode by the bias voltage to a charge sensitive preamplifier. A charge loop integrates the charge on a capacitor to produce an output pulse as in the case of the flow proportional counter, although in this case the gain is equal to unity since the Si(Li) detector does not have an equivalent property to gas gain.

The resolution R of the Si(Li) detector is given in eV by:

$$R = (\sigma_{noise}^2 + [2.35\,(e_i \times F \times E)]^2)^{1/2} \quad (3\text{-}7)$$

F is a variable called the Fano factor, which for the Si(Li) detector has a value of about 0.12 (Walter, 1971). The noise contribution is about 100 eV. Using a value of 3.8 for e_i, the calculated resolution for Mn K_α radiation ($E = 5.895$ eV) is about 160 eV.

Table 3-2 summarizes some of the characteristics of the three detectors commonly employed in X-ray spectrometry and also compares the resolution of these detectors with that of the crystal spectrometer. As was previously shown, the number of ion pairs produced is directly proportional to the energy of the X-ray photon and inversely proportional to the average energy to produce one ion pair. Going from the Si(Li) detector to the flow counter to the scintillation counter, the average energy to produce an ion pair increases by roughly an order of magnitude each time. Since the resolution of the detector is related to the square root of the number of electrons per photon, the resolution of the three detectors varies by about $\sqrt{10}$ or about three times, as indicated in the data for Cu K_α shown in the table. Another important practical parameter is the actual number of electrons produced for one photon of Cu K_α that eventually arrive at the anode. The size of the voltage pulse produced by the cathode follower of the detector is proportional to the current produced, i.e. to the number of electrons reaching the collector. In a detector that has internal gain, this number of electrons will be the product of the initial number of electrons and the gain of the detector. In the case of the gas flow proportional counter, this is the gas gain, and in the scintillation counter, the photo-multiplier gain. It will be remembered that the Si(Li) detector has no internal gain. The Si(Li) detector has a high number of initial electrons/photon (therefore, a high resolution), but, relative to the other two detectors, has a small number of final

Table 3-2. Characteristics of common X-ray detectors.

Detector type	Useful range (Å)	Average energy per ion pair (eV)	Cu K_α				
			Initial electrons	Gain	Final electrons	(eV)	(Res.) (%)
Si(Li)	0.5–8	3.8	2116	1	2×10^3	160	2.0
Flow proportional	1.5–50	26.4	305	6×10^4	2×10^7	1086	13.5
Scintillation	0.2–2	350	23	10^6	2×10^7	3638	45.3
Spectrometer*							
LiF(200)						31	0.39
LiF(220) 1st order						22	0.27
LiF(220) 2nd order						12	0.15

* Based on a 10 mm long primary collimator of 150 µm spacing and a 5 cm long secondary collimator with 120 µm spacing.

electrons. This means that normal external amplification of voltage pulses from the Si(Li) detector cannot be used because this would also amplify noise from the detector. For this reason a cooled charge-sensitive preamplifier is used for Si(Li) spectrometers rather than a simple linear electronic amplifier. Finally, the table shows the actual resolution of the detectors used on their own, in comparison with a crystal spectrometer. While the resolution of the Si(Li) detector for Cu K_α is worse than the crystal spectrometers as the energy of the analyte photon increases, the resolution difference between detectors alone and crystal spectrometers decreases. Note too that the resolution of both gas flow and scintillation counters is almost never sufficient for use without a crystal spectrometer.

3.5.3 Types of Spectrometer Used for X-ray Fluorescence

The basic function of the spectrometer is to separate the polychromatic beam of radiation coming from the specimen in order that the intensities of each individual characteristic line can be measured. A spectrometer should provide sufficient resolution of lines to allow such data to be taken, at the same time providing a sufficiently large response above background to make the measurements statistically significant, especially at low analyte concentration levels. It is also necessary that the spectrometer allow measurements over the wavelength range to be covered. Thus, in the selection of a set of spectrometer operating variables, four factors are important: resolution, response, background level and range. Owing to many factors, optimum selection of some of these characteristics may be mutually exclusive; as an example, attempts to improve resolution invariably cause lowering of absolute peak intensities.

X-ray spectrometers fall roughly into two major categories – Wavelength Dispersive Instruments and Energy Dispersive Instruments. The wavelength dispersive system was introduced commercially in the early 1950s and over the past 25 years or so has developed into a widely accepted analytical tool. The analytical chemist has available to him a wide range of instruments for the qualitative and quantitative analysis of multi-element samples and in the choice of technique he will generally consider such factors as sensitivity, speed, accuracy, cost, range of applicability, and so on. Within the two major categories of X-ray spectrometers specified, there is of course a wide diversity of instruments available but most of these will fit the four basic types:

1. Simultaneous wavelength dispersive
2. Sequential wavelength dispersive
3. Bremsstrahlung source energy dispersive
4. Secondary target energy dispersive

Although these different types of instrument may bear little physical resemblance to one another, for the purpose of comparison they differ only in the type of source used for excitation, the number of elements which they are able to measure at one time, the speed at which they collect data and finally, their price range. All of the instruments listed above are, in principle at least, capable of measuring all elements in the periodic classification from $Z=9(F)$ and upwards. All can be fitted with multi-sample handling facilities and all can be automated by use of mini-computers. All are capable of precision of the order of a few tenths of one percent and all have sensitivities down to the low p.p.m. level. As far as the analyst is concerned, they differ only in their speed, cost and number of elements measurable at the same time.

Single channel wavelength dispersive spectrometers are typically employed for both routine and non-routine analysis of a wide range of products, including ferrous and non-ferrous alloys, oils, slags and sinters, ores and minerals, thin films, and so on. These systems are very flexible but relative to the multi-channel spectrometers are somewhat slow. The multi-channel wavelength dispersive instruments are used solely for routine, high throughput, analyses where the great need is for fast accurate analysis, but where flexibility is of no importance. Energy dispersive spectrometers have the great advantage of being able to display information on all elements at the same time. They lack somewhat in resolution over the wavelength dispersive systems but also find great application in quality control, trouble shooting problems, and so on. They have been particularly effective in the fields of scrap alloy sorting, in forensic science and in the provision of elemental data to supplement X-ray powder diffraction data.

Wavelength dispersive spectrometers employ diffraction by a single crystal to separate characteristic wavelengths emitted by the sample. Energy dispersive spectrometers use the proportional characteristics of a photon detector, typically lithium drifted silicon, to separate the characteristic photons in terms of their energies. Since there is a simple relationship between wavelength and energy, these techniques each provide the same basic type of information, and the characteristics of the two methods differ mainly in their relative sensitivities and the way in which data is collected and presented. Generally speaking the wavelength dispersive system is roughly one to two orders of magnitude more sensitive than the energy dispersive system. Against this, however, the energy dispersive spectrometer measures all elements within its range at essentially the same time, whereas the wavelength dispersive system identifies only those elements for which it is programmed. To this extent, the energy dispersive system is more useful in recognizing unexpected elements. Both wavelength and energy dispersive spectrometers typically employ a primary X-ray photon source operating at 0.5 to 3 kW. A disadvantage with this type of source is that the specimen scatters the white radiation from the source leading to significant background levels that tend to become one of the major limitations in the determination of low concentration levels. Typical analysis times vary from about 10 seconds to three minutes per element. The minimum sample size required is of the order of a few milligrams, although typical sample sizes are probably around several grams. Good accuracy is obtainable and in favorable cases, standard deviations of the order of a few tenths of one percent are possible. This is because the matrix effects in X-ray spectrometry are well understood and relatively easy to overcome. The sensitivity is fair and determinations down to the low parts per million level are possible for most elements.

While the various types of specialized X-ray spectrometers are not generally available to the average user, they do have important roles to play in special areas of application. Included within this category are total reflection spectrometers (TRXRF), synchrotron source spectrometers (SSXRF), and proton induced X-ray emission spectrometers (PIXE). Two things that each of these three special systems have in common are a very high sensitivity and ability to work with extremely low concentrations and/or small specimens. The TRXRF system makes use of the fact that at very low glancing angles, primary X-ray photons are almost completely absorbed within thin specimens and the high background that would generally occur due to scatter from the sample support is absent. The recent development of high intensity synchrotron radiation beams has led to interest in their application for X-ray fluorescence analysis. The high intensity available in synchrotron beams allows use of very narrow bandpath monochromators between source and specimen, giving, in turn, a high degree of selective excitation. This selectivity overcomes one of the major disadvantages of the classical EDS approach and allows excellent detection limits to be obtained. The proton induced X-ray emission system differs from conventional energy dispersive spectroscopy in that a proton source is used in place of the photon source. The proton source is typically a Van de Graf generator or a cyclotron, giving protons in the energy range of about 2 to 3 MeV. In addition to being an intense source relative to the conventional photon source, the proton excitation system generates relatively much lower backgrounds. In addition, the cross section for characteristic X-ray production is quite large and good excitation efficiency is possible.

One of the problems with any X-ray spectrometer system is that the absolute sensitivity (i.e. the measured c/s per % of analyte element) decreases significantly as the lower atomic number region is approached. The three main reasons for this are: decrease in the fluorescence yield with decrease in atomic number; the absolute number of useful long wavelength X-ray photons from a bremsstrahlung source decreases with increase of wavelength; and thirdly, absorption effects generally become more severe with increase of the wavelength of the analyte line. The first two of these problems are inherent to the X-ray excitation process and to constraints in the basic design of conventional X-ray tubes. The third, however, is a factor that depends very much on the instrument design, and, in particular, upon the characteristics of the detector. The detector that is used in long wavelength spectrometers is typically a gas flow proportional counter, in which an extremely thin, high-transmission window is employed. The detector typically employed in energy dispersive systems is the Si(Li) diode which has an electrical contact layer on the front surface typically a 0.02 µm thick layer of gold, followed by a 0.1 µm thick dead layer of silicon. The absorption problems caused by these two layers become most significant for low energy X-ray photons, which have a high probability of being absorbed in the dead layer. Probably the biggest source of absorption loss in the Si(Li) detector is that due to the thin beryllium window that is part of the liquid nitrogen cryostat. A combination of these facts causes a loss in the sensitivity of a typical energy dispersive system of almost an order of magnitude, for the K lines of sulfur ($Z=16$) to sodium ($Z=11$). The equivalent number with a gas flow counter would be about a factor of two. Until very recently, the lowest atomic number usefully detectable with a typical energy dispersive spectrometer has been magnesium ($Z=12$). New developments in ultra-thin windows for the Si(Li) detector now allows measurements down to oxygen ($Z=8$). In comparison, the lower atomic number limit for the conventional wavelength dispersive system is fluorine ($Z=9$), and by use of special crystals this can be extended down to beryllium ($Z=4$).

A compromise that must always be made in the design and setup of any spectrometer is that between intensity and resolution, resolution being defined as the ability of the spectrometer to separate lines. In the flat-crystal wavelength dispersive system this resolution is dependent upon the angular dispersion of the analyzing crystal and the divergence allowed by the collimators (Jenkins, 1974). In the energy dispersive system the resolution is dependent only upon the detector and detector amplifier. In absolute terms, the resolution of the wavelength dispersive system typically lies in the range 10 to 100 eV, compared to a value of 150 to 200 eV for the energy dispersive system. An advantage of the wavelength dispersive spectrometer in this context is that the resolution/intensity measuring selection is much more controllable. In the case of secondary fluorescer systems some local modifications to the measured spectrum can be made by use of filters either in the primary beam or in the secondary fluoresced beam and this does offer some flexibility.

3.5.3.1 Wavelength Dispersive Systems

A wavelength dispersive spectrometer may be a single channel instrument in which a single crystal and a single detector are used for the measurement of a series of wavelengths sequentially; or a multichannel spectrometer in which many crystal/detector sets are used to measure elements simultaneously. Of these two basic types, the sequential systems are the most common. A typical sequential spectrometer system consists of the X-ray tube, a specimen holder support, a primary collimator, an analyzing crystal and a tandem detector. The gas flow counter is ideal for the measurement of the longer wavelengths and the scintillation counter is best for the short wavelengths. A portion of the characteristic fluorescence radiation from the specimen is passed via a collimator or slit onto the surface of an analyzing crystal, where individual wavelengths are diffracted to the detector in accordance with the Bragg law. A goniometer is used to maintain the required θ to 2θ relationship between crystal and detector. Typically, six or so different analyzing crystals and two different collimators are provided in this type of spectrometer, giving the operator a wide range of choice of dispersion conditions. In general, the smaller the 2d-spacing of the crystal, the better the separation of the lines but the smaller the wavelength range that can be covered. Since the maximum achievable angle θ on a wavelength dispersive spectrometer is generally around 73°, the maximum wavelength that can be diffracted by a crystal of spacing 2d is equal to about 1.9d. The separating power of a crystal spectrometer is dependent upon the divergence allowed by the collimators (that to a first approximation determines the width of the diffracted lines), and the angular dispersion of the crystal. Since mechanical limitations prevent wide selectability of line shape just by selection of collimator divergence, in practice, the resolution of the spectrometer is typically determined by the angular dispersion of the analyzing crystal, albeit with some influence of the breadth of the diffracted line profile. The angular dispersion $d\theta/d\lambda$ of a crystal of spacing 2d is given by:

$$\frac{d\theta}{d\lambda} = \frac{n}{2 \times d \times \cos\theta} \qquad (3\text{-}8)$$

It will be seen from Eq. (3-8) that the angular dispersion will be high when the 2d-spacing is small. This is unfortunate as far as the range of the spectrometer is concerned, because a small value of 2d means in turn a small range of wavelengths coverable. Thus, as with the resolution and peak intensities, the obtaining of high dispersion can only be obtained at the expense of cutting down the wavelength range covered by a particular crystal. In order to circumvent this problem, it is likely that several analyzing crystals will be employed in the coverage of a number of analyte elements. Many different analyzing crystals are available, each having its own special characteristics, but three or four crystals will generally suffice for most applications. Table 3-3 gives a short list of the more common-

Table 3-3. Analyzing crystals used in wavelength dispersive X-ray spectrometry.

Crystal	Range			
	Planes	2d (Å)	K lines	L Lines
Lithium fluoride	(220)	2.848	>Ti(22)	>La(57)
Lithium fluoride	(200)	4.028	>K(19)	>Cd(48)
Penta erythritol	(002)	8.742	Al(13)–K(19)	–
Thallium acid phthalate	(001)	26.4	F(9)–Na(11)	–
LSM's	–	50–120	Be(4)–F(9)	–

ly used crystals. While the maximum wavelength covered by traditional spectrometer designs is about 20 Å, recent developments now allow the extension of the wavelength range significantly beyond this value.

Classically, large single crystals have been used as diffracting structures in the wavelength dispersive spectrometer. The three dimensional lattice of atoms is oriented and fabricated such that Bragg planes form the interatomic 2d-spacing for the wavelength in question. The selection of crystals for the longer (>8 Å) wavelength region is difficult, however, mainly because there are not too many crystals available for work in this region. The most commonly employed crystal is probably TAP, thallium acid phthalate (2d=26.3 Å), and this allows measurement of the K lines of elements including magnesium, sodium, fluorine and oxygen. Several alternatives to single crystals as diffracting structures have been sought over the years including the use of complex organic materials with large 2d-spacings, gratings, and specular reflectors, metal disulphides, organic intercalation complexes such as graphite, molybdenum disulphide, mica and clays. For a comprehensive list of analyzing crystals see Bertin (1975).

In addition to crystals and multilayer films, a third alternative has recently become available and this is the layered synthetic micro-structure, LSM. LSM's are constructed by applying successive layers of atoms or molecules on a suitably smooth substrate. In this manner, both the 2d-spacing and composition of each layer are selected for optimum diffraction characteristics, thus, to a certain extent, they can be designed and fabricated to give optimum performance for special applications (Barbee, 1985). Typical, experimental data shows factors of about four to six times improvement in peak intensities compared to TAP, for the range of elements measured (Nicolosi et al., 1985).

3.5.3.2 Energy Dispersive Systems

The energy dispersive spectrometer consists of the excitation source and the spectrometer/detection system. The spectrometer/detector is typically a Si(Li) detector which is a proportional detector of high intrinsic resolution. A multi-channel analyzer is used to collect, integrate and display the resolved pulses. While similar properties are sought from the energy dispersive system as with the wavelength dispersive system, the means of selecting these optimum conditions are very different. Since the resolution of the energy dispersive system is equated directly to the resolution of the detector, this feature is of paramount importance. The output from an energy dispersive spectrometer is generally displayed on a cathode ray tube and the operator is able to dynamically display the contents of the various channels as an energy spectrum and provision is generally made to allow zooming in on portions of the spectrum of special interest, to overlay spectra, to subtract background, and so on. Even though the Si(Li) detector is the most common detector used in energy dispersive X-ray spectrometry, it is certainly not the only one. As an example, the higher absorbing power of germanium makes it an alternative for the measurement of high energy spectra. Both cadmium telluride, CdTe, and mercuric iodide, HgI_2, show some promise as detectors capable of operating satisfactorily at room temperature. As an example, a HgI_2 based energy dispersive spectrometer, in which both detector and FET were cooled using a Peltier cooler, has been used in a scanning electron microscope and a resolution of 225 eV obtained for Mn K_α (5.9 keV) and 195 eV for Mg K_α (1.25 keV) (Iwanczyk et al., 1986).

All of the earlier energy dispersive spectrometers were operated in what is called the primary mode and consisted simply of the excitation source, typically a closely coupled low-power, end-window X-ray tube, and the detection system. In principle, this primary excitation system offered the possibility of a relatively inexpensive instrument, with two significant advantages over the wavelength dispersive system. Firstly the ability to collect and display the total emission spectrum from the sample at the same time, giving great speed in the acquisition and display of data. Secondly, offering mechanical simplicity since there is almost no need at all for moving parts. In practice, however, there is a limit as to the maximum count rate that the spectrometer can handle and this led, in the mid-1970s, to the development of the secondary mode of operation. In the secondary mode, a carefully selected pure element standard is interposed between primary source and specimen, along with absorption filters where appropriate, such that a selectable energy range of secondary photons is incident upon the sample. This allows selective excitation of certain portions of the energy range, thus increasing the ratio of useful to unwanted photons entering the detector. While this configuration does not completely eliminate the count rate and resolution limitations of the primary system, it certainly does reduce them.

3.5.3.3 Total Reflection Spectrometers (TRXRF)

One of the major problems that inhibits the obtaining of good detection limits in small samples is the high background due to scatter from the sample substrate support material. The suggestion to overcome this problem by using total reflection of the primary beam was made long ago, but the absence of suitable instrumentation prevented real progress being made until the late 1970s (Knoth and Schwenke, 1978). Mainly owing to the work of Schwenke and his coworkers, good sample preparation and presentation procedures are now available, making TRXRF a valuable technique for trace analysis (Michaelis et al., 1984). As illustrated in Fig. 3-4, the TRXRF method is essentially an energy dispersive technique in which the Si(Li) detector is placed close to (about 5 mm), and directly above, the sample. Primary radiation enters the sample at a glancing angle of a few seconds of arc. The sample itself is typically presented as a thin film on the optical flat surface of a quartz plate. In the instrument described by Michaelis a series of reflectors is employed to aid in the reduction of background. Here, a beam of radiation from a sealed X-ray tube passes through a fixed aperture onto a pair

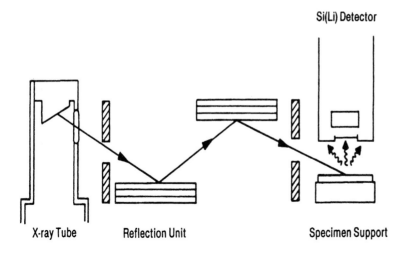

Figure 3-4. The Total Reflection X-ray spectrometer.

of reflectors that are placed very close to each other. Scattered radiation passes through the first aperture to impinge on the sample at a very low glancing angle. Because the primary radiation enters the sample at an angle barely less than the critical angle for total reflection, this radiation barely penetrates the substrate media; thus scatter and fluorescence from the substrate are minimal. Because the background is so low, picogram amounts can be measured or concentrations in the range of a few tenths of a p.p.b. can be obtained without recourse to preconcentration (Aiginger and Wobranschek, 1985).

One area in which the TRXRF technique has found great application is in the analysis of natural waters (West and Nurrenberg, 1988). The concentration levels of, for example, transition metals in rain, river and sea waters, are normally too low to allow estimation by standard X-ray fluorescence techniques, unless preconcentration is employed. Using TRXRF concentration levels down to less than 10 µg/L are achievable. While the TRXRF method is most applicable to homogeneous liquid samples in which the sample is evaporated onto the optical flat, success has also been achieved in the application of the method to solids including particulates, sediments, air dusts and minerals. In these instances the sample is first digested in concentrated nitric acid and then diluted to a calibrated volume with ultra-pure water, after the addition of an internal standard. Where undissolved material is still present, this may be dispersed using an ultrasonic bath before the specimen is taken. In addition to the advantage of ease of specimen preparation and the ability to handle milligram quantities of material, the TRXRF method is also relatively simple to apply quantitatively. Because the specimen is only a few microns thick, one generally does not observe the rather complicated matrix effects usually encountered with thick samples. Thus the only standard required is that to establish the sensitivity of the spectrometer, in terms of c/s per %, for the element(s) in question. This standard is generally added to the sample to be analyzed during the specimen preparation procedure.

3.5.3.4 Synchrotron Source X-ray Fluorescence (SSXRF)

The availability of intense, linearly polarized synchroton radiation beams (Sparks,

1980) has prompted workers in the field of X-ray fluorescence (e.g., Gilfrich, 1983) to explore what the source has to offer over more conventional excitation media. In the synchrotron, electrons with kinetic energies of the order of several billion electron volts (typically 3 GeV at this time), orbit in a high vacuum tube between the poles of strong (about 10^4 Gauss) magnets. A vertical field accelerates the electrons horizontally, causing the emission of synchrotron radiation. Thus synchrotron source radiation can be considered as magnetic bremsstrahlung in contrast to normal electronic bremsstrahlung produced when electrons are decelerated by the electrons of an atom. In the case of both fluorescence and diffraction it has been found that because the primary source of radiation is so intense, it is possible to use a high degree of monochromatization between source and specimen, giving a source that is wavelength (and, therefore, energy) tunable, as well as being highly monochromatic. There are several different excitation modes that can be used using SSXRF including: direct excitation with continuum, excitation with absorber modified continuum, excitation with source crystal monochromatized continuum, excitation with absorber modified continuum, excitation with source crystal monochromatized continuum, excitation with source radiation scattered by a mirror, and reflection and transmission modes.

The intensity of the synchrotron beam is probably 4 to 5 orders of magnitude greater than the conventional bremsstrahlung source sealed X-ray tubes. This, in combination with its energy tunability and polarization in the plane of the synchroton ring, allows very rapid bulk analyses to be obtained on small areas. Because the synchrotron beam has such a high intensity and small divergence, it is possible to use it as a microprobe of high special resolution (about 10 µm). Absolute limits of detection around 10^{-14} µg have been reported using such an arrangement (Petersen et al., 1986). Synchrotron source X-ray fluorescence has also been used in combination with TRXRF. Very high signal/background ratios have been obtained employing this arrangement for the analysis of small quantities of aqueous solutions dried on the reflector, with detection limits of <1 p.p.b. or 1 p.g. Additional advantages accrue because synchrotron radiation is highly polarized and background due to scatter can be greatly reduced by placing the detector at 90° to the path of the incident beam and in the plane of polarization. A disadvantage of the SSXRF technique is that the source intensity decreases with time, but this can be overcome by bracketing analytical measurements between source standards and/or by continuously monitoring the primary beam. In addition to this, problems can also arise from the occurrence of diffraction peaks from highly ordered specimens. Giauque et al. have described experiments (Giauque, 1986), using the Stanford Synchrotron Radiation Laboratory, to establish what minimum detectable limits (MDL's) could be obtained under optimum excitation conditions. Using thin film standards on stretched tetrafluoro-polyethylene mounts, it was found that for counting times of the order of a few hundred seconds, MDL's of the order of 20 p.p.b. could be obtained on a range of elements including Ca, Ti, V, Mn, Fe, Cu, Zn, Rb, Ge, Sn and Pb. The optimum excitation energy was found to be about one and a quarter times the absorption edge energy of the element in question. By working with a fixed excitation energy of 18 keV, the average MDL was found to be about 100 p.p.b.

3.5.3.5 Proton Excited X-ray Fluorescence (PIXE)

While the use of protons as a potential source for the excitation of characteristic X-rays has been recognized since the early 1960s, it is only since the 1980s that the technique has come into its own (Garten, 1984). The PIXE method uses a beam of fast ions (protons) of primary energies in the range 1 to 4 MeV. In addition to the ion accelerator, the system contains an energy defining magnetic deflection field, a magnetic or electrostatic lens along the excitation beam pipe, a high vacuum target chamber for the specimen(s) and an energy dispersive detector/analyzer. The great advantage of the PIXE method over other sources is that it generates only a small amount of background and is thus applicable to very low concentration levels and the analysis of very small samples. As an example, the use of conventional X-ray fluorescence and PIXE have been compared with special reference to applications in art and archeology. Together, they seem to offer the museum scientist and archaeologist excellent tools for non-destructive testing (Malmqvist, 1986). Comparison of the PIXE method has been made with many other spectroscopic techniques for the analysis of twenty-two elements in ancient pottery (Bird et al., 1986). The high sensitivity of PIXE has also been of great use in the field of forensic science and in medicine (Cesareo, 1982). Applications include in vitro analysis of trace elements in human body fluids and normal pathological tissues, in vivo analysis of iodine in the thyroid, lead in the skeleton and cadmium in the kidney. Trace elements in blood serum of patients with liver cancer has also been studied by PIXE.

3.6 Qualitative Analysis with the X-ray Spectrometer

Both the simultaneous (scanning) wavelength dispersive spectrometer and the energy dispersive spectrometers lend themselves admirably to the qualitative analysis of materials, since there is a simple relationship between the wavelength or energy of a characteristic X-ray photon and the atomic number of the element from which the characteristic emission line occurs. Thus by measuring the wavelengths, or energies, of a given series of lines from an unknown material, the atomic numbers of the excited elements can be established. The inherent simplicity of characteristic X-ray spectra make the process of allocating atomic numbers to the emission lines relatively easy, and the chance of making a gross error is rather small. There are only 100 or so elements, and within the range of the conventional spectrometer each element gives, on an average, only half a dozen lines. If one compares the X-ray emission spectrum with the ultra-violet emission spectrum, since the X-ray spectrum arises from a limited number of inner orbital transitions, the number of X-ray lines is similarly rather few. Ultra-violet spectra, on the other hand, arise from transitions to empty levels, of which there may be many, leading to a significant number of lines in the UV emission spectrum. A further benefit of the X-ray emission spectrum for qualitative analysis is that because transitions do arise from inner orbitals, the effect of chemical combination, or valence state is almost negligible.

The output from a wavelength dispersive spectrometer may be either analog or digital. For qualitative work an analog output is traditionally used and the digital output from the detector amplifier is fed through a d/a converter, called a rate meter, to an x/t recorder which is synchronously coupled with the goniometer scan speed. The recorder thus records an intensity/time diagram, in terms of an intensity/(2θ) diagram. It is generally more convenient to employ digital counting for quantitative work and a timer/scale combination is provided which will allow pulses to be integrated over a period of several tens of seconds and then displayed as count or count rate. In more modern spectrometers a scaler/timer may also take the place of the rate meter, using step-scanning. In this process the contents of the scaler are displayed on the x-axis of the x/t recorder as a voltage level. The scaler/timer is then reset and started to count for a selected time interval. At the end of this time the timer sends a stop pulse to the scaler that now holds a number of counts equal to the product of the counting rate and the count time. The contents of the scaler are then displayed as before, the goniometer stepped to its next position and the whole cycle repeated. Generally, the process is completely controlled by a microprocessor or a personal computer.

Qualitative analysis has traditionally been performed by scanning the goniometer synchronously at a fixed angular speed. However, most scanning spectrometers are slow in this sequential angular/intensity data collection mode, not only because the data are taken sequentially, but also because in order to cover the full range of elements a series of scans must be made with different conditions. In addition to this, scanning at a fixed speed is somewhat inefficient because in the crystal dispersive system, atomic number varies as a function of one over the square root of the angle. In effect this means that in the low atomic number region much scanning time is wasted in scanning angular space that contains no characteristic line data. Although some unique designs have brought about some reduction in data acquisition times (e.g., Jenkins et al., 1985), this remains a major limitation of wavelength dispersive spectrometers. Data interpretation involves identifying each line in the measured spectrogram. However, since the relationship between atomic number and Bragg angle is rather complicated, it is common practice to use sets of tables for interpretation relating wavelength and atomic number, with diffraction angle for specific analyzing crystals (e.g., White and Johnson, 1970). Some automated wavelength dispersive spectrometers provide the user with software programs for the interpretation and labelling of peaks e.g., Garbauskas and Goehner, 1983.

3.7 Accuracy of X-ray Fluorescence

3.7.1 Counting Statistical Errors

The production of X-rays is a random process that can be described by a Gaussian distribution. Since the number of photons counted is nearly always large, typically thousands or hundreds of thousands, rather than a few hundred, the properties of the Gaussian distribution can be used to predict the probable error for a given count measurement. There will be a random error $\sigma(N)$ associated with a measured value of N, this being equal to $\sqrt{N}$. As an example, if 10^6 counts are taken, the 1σ a standard deviation will be $\sqrt{10^6}=10^3$, or 0.1%. The measured parameter in wavelength dispersive X-ray spectrometry is generally the counting rate R and, based on what has been already stated, the magnitude of the random counting error associated with a given datum can be expressed as:

$$\sigma(\%) = \frac{100}{\sqrt{N}} = \frac{100}{\sqrt{R \times t}} \qquad (3\text{-}9)$$

Care must be exercised in relating the counting error (or indeed any intensity related error) with an estimate of the error in terms of concentration. Provided that the sensitivity of the spectrometer in c/s per %, is linear, a count error can be directly related to a concentration error. However, where the sensitivity of the spectrometer changes over the range of measured response, a given fractional count error may be much greater when expressed in terms of concentration.

In many analytical situations the peak lies above a significant background and this adds a further complication to the counting statistical error. An additional factor that must also be considered is that whereas with the scanning wavelength dispersive spectrometer the peaks and background are measured sequentially, in the case of the energy dispersive and the multichannel wavelength dispersive spectrometers, a single counting time is selected for the complete experiment, thus all peaks and all backgrounds are counted for the same time. To estimate the net counting error in the case of a sequential wavelength dispersive spectrometer it is necessary to consider the counting error of the net response of peak counting rate R_p, and background counting rate R_b, since the analyte element is only responsible for $(R_p - R_b)$.

Equation (3-9) must then be expanded to include the background count rate term:

$$\sigma(R_p - R_b) = \frac{100}{\sqrt{t}} \times \frac{1}{\sqrt{R_p} - \sqrt{R_p}} \quad (3\text{-}10)$$

One of the conditions for Eq. (3-10) is that the total counting time t, must be correctly proportioned between time spent counting on the peak t_p, and time spent counting on the background t_b:

$$\frac{t_p}{t_b} = \sqrt{\frac{R_p}{R_b}} \quad (3\text{-}11)$$

Several points are worth noting with reference to Eq. (3-10). Firstly, where the count time is limited – which is usually the case in most analyses, the net counting error is a minimum when $\left(\frac{1}{\sqrt{R_p} - \sqrt{R_p}}\right)$ is maximum. This expression can, therefore, be used as a figure of merit for the setting up of instrumental variables (Jenkins and de Vries, 1970). Secondly, it will be noted that as R_b becomes small relative to R_p, Eq. (3-10) approximates to Eq. (3-9). In other words, as the background becomes less significant relative to the peak, its effect on the net counting error becomes smaller. The point at which the peak to background value exceeds 10:1 is generally taken as that where background can be ignored completely. A third point to be noted is that as R_b approaches R_p the counting error becomes infinite, and this will be a major factor in determining the lowest concentration limit that can be detected.

In the case of the energy dispersive spectrometer, the peak and background are recorded simultaneously and the question of division of time between peak and background does not arise. A choice does, however, have to be made as to what portion of the complete recorded spectrum should be used for the measurement of peak and background. The associated net counting error σ_{net} associated is given by:

$$\sigma_{net} = \sqrt{P + B\left(1 + \frac{n_p}{n_b}\right)} \quad (3\text{-}12)$$

Where each number of background channels is chosen to be one half of the total number of peak channels, Eq. (3-12) reduces to $\sigma_{net} = \sqrt{(P + 2B)}$, or expressed as a percentage of the peak as:

$$\sigma_{net} = \frac{100 \times \sqrt{P + 2B}}{P} \quad (3\text{-}13)$$

3.7.2 Matrix Effects

In the conversion of net line intensity to analyte concentration it is necessary to correct for any absorption and/or enhancement effects which occur. Absorption effects include both primary and secondary absorption, and enhancement effects include direct enhancement, involving the analyte element and one enhancing element, plus third element effects, that involve additional element(s) beyond the analyte and enhancer. Primary absorption occurs because all atoms of the specimen matrix will absorb photons from the primary source. Since there is a competition for these primary photons by the atoms making up the specimen, the intensity/wavelength distribution of these photons available for the excitation of a given analyte element may be modified by other matrix elements. Secondary absorption refers to the effect of the absorption of characteristic analyte radiation by the specimen matrix. As characteristic radiation passes out from the specimen in which it was generated, it will be absorbed by all matrix elements, by amounts relative to the mass absorption coefficients of these elements. The mass absorption coefficient is a parameter which defines the magnitude of the absorption of a certain element for a specific X-ray wavelength. The total absorption a of a specimen, is dependent on both primary and secondary absorption. The total absorption by element i for an analyte wavelength λ_i is given by the following relationship:

$$\alpha_i = \mu_i(\lambda) + A[\mu_i(\lambda_j)] \quad (3\text{-}14)$$

The factor A is a geometric constant equal to the ratio of the sines of the incident and take-off angles of the spectrometer. This factor is needed to correct for the fact that the incident and emergent rays from the sample have different path lengths. The term λ in the equation refers to the primary radiation. Since most conventional X-ray spectrometers use a bremsstrahlung source, in practice λ is a range of wavelengths, although in simple calculations it may be acceptable to use a single equivalent wavelength value (Stephenson, 1971), where the equivalent wavelength is defined as a single wavelength having the same excitation characteristics as the full continuum.

There are a number of routes by which the analyte element can be excited or enhanced and these are illustrated in Figure 3-5a which shows the direct excitation of an analyte element i by the primary continuum P_1. There may also be excitation of the analyte by characteristic lines from the source, designated in Figure 3-5b by P_2. Both the continuous and characteristic radiation from the source may be somewhat modified by Compton scatter and the excitation by this modified source radiation is indicated in Figure 3-5c by P_3. Enhancement effects, Figure 3-5d, occur when a non-analyte matrix element A emits a characteristic line that has an energy just in excess of the absorption edge of the analyte element. This means that the non-analyte element in question is able to excite the analyte, giving characteristic photons over and above those produced by the primary continuum. This gives an increased, or enhanced, signal from the analyte. The third element effect is also shown in the figure. Here, a third element B, is also excited by the source. Not only can B directly enhance i, but it can also enhance A, thus increasing the enhancing effect of A on i. This last effect is called the third element effect. Table 3-4 shows some of the data published for the chromium/iron/nickel system (Shiraiwa and Fujino, 1967) and illustrates the relative importance of the various excitation routes. Relating these data to Figure 3-5, i is the element chromium, iron is the enhancer A, and nickel is the third element B. The data given are for three differ-

3.7 Accuracy of X-ray Fluorescence

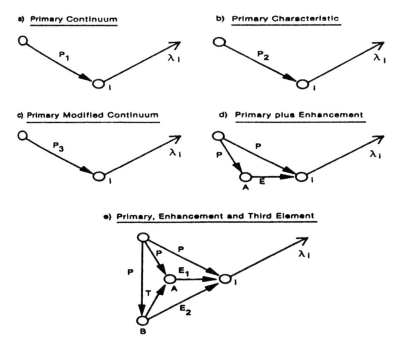

Figure 3-5. Various routes by which characteristic analyte radiation can be excited.

Table 3-4. Published data of the third element effect for the chromium/iron/nickel system.

Percentage of counts for Cr K_α, element proportions given for Ni:Fe:Cr

	Route	25:25:50	10:40:50	40:10:50
Primary	P	87.5	87.2	87.6
Secondary	E1	6.7	10.6	2.6
	E2	5.3	2.0	9.4
Third element	T	0.5	0.2	0.4

ent alloy compositions: 25/25/50; 10/40/50 and 40/10/50, nickel/iron/chromium in each case. Note that in each case, roughly 87% of the actual measured chromium K_α radiation comes from direct excitation by the source. In the case of direct enhancement by iron, each one percent of iron added increases the chromium radiation by about 0.26%. Direct excitation of chromium by nickel gives about 0.21% increase in chromium radiation per percent nickel added. It will also be seen that the third element effect is much less important and here the chromium intensity is increased by only about 0.02% per percent of nickel added. From this one would predict that a fourth element effect would be negligible.

3.7.3 Specimen Preparation for X-ray Fluorescence

Because X-ray spectrometry is essentially a comparative method of analysis, it is vital that all standards and unknowns be presented to the spectrometer in a reproducible and identical manner. Any method of specimen preparation must give specimens which are reproducible and which, for a certain calibration range, have similar physical properties including mass attenuation coefficient, density, particle size, and particle homogeneity. In addition the specimen preparation method must be rapid and cheap and must not introduce extra significant systematic errors, for example, the introduction of trace elements from contaminants in a diluent. Specimen preparation is an all important factor in the ultimate accuracy of any X-ray determination, and many papers have been published describing a multitude of methods and recipes for sample handling (e.g., Jenkins et al., 1981).

In general samples fit into three main categories:

1. Samples which can be handled directly following some simple pre-treatment such as pelletizing or surfacing. For example, homogeneous samples of powders, bulk metals or liquids.
2. Samples which require significant pre-treatment. For example, heterogeneous samples, samples requiring matrix dilution to overcome inter-element effects and samples exhibiting particle size effects.
3. Samples which require special handling treatment. For example, samples of limited size, samples requiring concentration or prior separation and radioactive samples.

The ideal specimen for X-ray fluorescence analysis is one in which the analyzed volume of specimen is representative of the total specimen, which is, itself, representative of the sample submitted for analysis. There are many forms of specimen suitable for X-ray fluorescence analysis, and the form of the sample as received will generally determine the method of pre-treatment. It is convenient to refer to the material received for analysis as the sample, and that which is actually analyzed in the spectrometer as the specimen. While the direct analysis of certain materials is certainly possible, more often than not some pre-treatment is required to convert the sample to the specimen. This step is referred to as specimen preparation. In general, the analyst would prefer to analyze the sample directly, because if it is taken as received, any problems arising from sample contamination that might occur during pre-treatment are avoided. In practice, however, there are three major constraints that may prevent this ideal circumstance from being achieved: sample size, sample size homogeneity and sample composition heterogeneity.

Problems of sample size are frequently severe in the case of bulk materials such as metals, large pieces of rock, etc. Problems of sample composition heterogeneity will generally occur under these circumstances as well, and in the analysis of powdered materials, heterogeneity must almost always be considered. The sample as received may be either homogeneous or heterogeneous; in the latter case, it may be necessary to render the sample homogeneous before an analysis can be made. Heterogeneous bulk solids are generally the most difficult kind of sample to handle, and it may be necessary to dissolve or chemically react the material in some way to give a homogeneous preparation. Heterogeneous powders are either ground to a fine particle size and then pelletized, or fused with a glass forming material such as borax. Solid material in liquids or gases must be filtered out and the filter analyzed as a solid. Where analyte concentrations in liquids or solutions are too high or too low, dilution or pre-concentration techniques may be employed to bring the analyte concentration within an acceptable range.

3.8 Quantitative Analysis

In the X-ray analytical laboratory the quantitative method of analysis employed will be typically predicated by a number of circumstances of which probably the four most common are: the complexity of the analytical problem; the time allowable; the computational facilities at the disposal of the analyst; and the number of standards available. It is convenient to break quantitative analytical methods down into two major categories: single element methods, and multiple element methods, as shown in Table 3-5. The simplest quantitative analysis situation to handle is the determination of a single element in a known matrix. A slightly more difficult case might be the determination of a single element where the matrix is unknown. As shown in the table, three basic methods are commonly employed in this situation; use of internal standards, use of standard addition, or use of a scattered line from the X-ray source. The most complex case is the analysis of all, or most, of the elements in a sample, about which little or nothing is known. In this case a full qualitative analysis would be required before any attempt is made to quantitate the matrix elements. Once the qualitative composition of the sample is known, again, one of three general techniques is typically applied; use of type standardization, use of an influence coefficient method, or use of a fundamental parameter technique. Both the influence coefficient and fundamental parameter technique require a computer for their application.

The great flexibility, sensitivity and range of the various types of X-ray fluorescence spectrometer make them ideal for quantitative analysis. In common with all analytical methods, quantitative X-ray fluorescence analysis is subject to a number of random and systematic errors that contribute to the final accuracy of the analytical result. Like all instrumental methods of analysis, the potentially high precision of X-ray spectrometry can only be translated into high accuracy if the various systematic errors in the analysis process are taken care of. The precision of a wavelength dispersive system, for the measurement of a single, well separated line, is typically of the order of 0.1%, and about 0.25% for the energy dispersive system. The major source of the random error is the X-ray source, with additional count time dependent errors arising from the statistics of the actual counting process. A good rule-of-thumb which can be used in X-ray fluorescence analysis to estimate the expected standard deviation σ at an analyte concentration level C, is given by:

$$\sigma = K \sqrt{C + 0.1} \tag{3-15}$$

where K varies between 0.005 and 0.05 (Jenkins, 1988). For example, at a concentration level $C=25\%$, the expected value of σ would be between about 0.025% and 0.25%. A K value of 0.005 would be considered very high quality analysis and a value of 0.05 rather poor quality. The value of K actually obtained under routine laboratory conditions depends upon many factors but with reasonably careful measurements a K value of around 0.02 to 0.03 can be obtained.

Table 3-6 lists the four main categories of random and systematic error encountered in X-ray fluorescence analysis. The first category includes the selection and preparation of the sample to be analyzed. Two stages are generally involved before the actual prepared specimen is presented to the spectrometer, these being sampling and speci-

Table 3-5. Quantitative procedures employed in X-ray fluorescence analysis.

Single element methods	internal standardization
	standard addition
	use of scattered source radiation
Multiple element methods	type standardization
	use of influence coefficients
	fundamental parameter techniques

Table 3-6. Sources of error in X-ray fluorescence analysis.

	Source	Random (%)	Systematic (%)
1	Sample preparation	0–1	0–5
	Sample inhomogeneity	–	0–50
2	Excitation source	0.05–0.2	0.05–0.5
	Spectrometer	0.05–0.1	0.05–0.1
3	Counting statistics	Time dependent	–
	Dead time	–	0–25
4	Primary absorption	–	0–50
	Secondary absorption	–	0–25
	Enhancement	–	0–15
	Third element	–	0–2

men preparation. The actual sampling is rarely under the control of the spectroscopist and it generally has to be assumed that the container containing the material for analysis does, in fact, contain a representative sample. It will be seen from the table that in addition to a relatively large random error, inadequate sample preparation and residual sample heterogeneity can lead to very large systematic errors. For accurate analysis these errors must be reduced by use of a suitable specimen preparation method. The second category includes errors arising from the X-ray source previously discussed. Source errors can be reduced to less than 0.1% by use of the ratio counting technique, provided that high frequency transients are absent. The third category involves the actual counting process and these errors can be both random and systematic. System errors due to detector dead time can be corrected either by use of electronic dead time correctors or by some mathematical approach. The fourth category includes all errors arising from inter-element effects. Each of the effects listed can give large systematic errors that must be controlled by the calibration and correction scheme.

The correlation between the characteristic line intensity of an analyte element and the concentration of that element is typically non-linear over wide ranges of concentration, due to inter-element effects between the analyte element and other elements making up the specimen matrix. However, the situation can be greatly simplified in the case of homogeneous specimens, where severe enhancement effects are absent, and here, the slope of a calibration curve is inversely proportional to the total absorption of the specimen for the analyte wavelength. In this instance the slope K of the calibration curve is taken as W/I where I is the line intensity and W the weight fraction of the analyte element. Thus, the following relationship holds:

$$W = \frac{I}{K} \qquad (3\text{-}16)$$

Single element techniques reduce the influence of the absorption term in Eq. (3-16), generally by referring the intensity of the analyte wavelength to a similar wavelength, arising either from an added standard, or from a scattered line from the X-ray tube. In certain cases, limiting the concentration range of the analyte may allow the assumption to be made that the absorption value does not significantly change over the concentration range and the calibration curve is essentially linear. This assumption is applied in the traditional type standardization technique. Type standardization was widely employed in the 1960s and 1970s, but now that computers are generally available, it is usually considered more desirable to work with general purpose calibration schemes, which are applicable to a variety of matrix types, over wide concentration ranges. Sherman (1955) showed that it was possible to express the intensity/concentration relationship in terms of independently determined fundamental parameters. Unfortunately, fundamental type methods require a fair degree of computation and suitable computational facilities were not generally available until the early 1970s. By the early 1960s, limited computational facilities were available at reasonable costs, and because of the clear need for some degree of fast mathematical correction of matrix effects, a number of so-called empirical correction techniques were developed which required far less computation and were, therefore, usable by the computers available at the time.

In principle, an empirical correction procedure can be described as the correction of an analyte element intensity for the influence of an interfering element(s) using the product of the intensity from the interfering element line and a constant factor, as the correction term (Beattie and Brissey, 1954). This constant factor is, today, generally referred to as an influence coefficient, since it is assumed to represent the influence of the interfering element on the analyte. Commonly employed influence coefficient methods may use either the intensity, or the concentration, of the interfering element as the correction term. These methods are referred to as intensity correction and concentration correction methods, respectively. Intensity correction models give a series of linear equations which do not require much computation, but they are generally not applicable to wide ranges of analyte concentration. Various versions of the intensity correction models found initial application in the analysis of non-ferrous metals where correction constants were applied as look-up tables. Later versions (Lucas-Tooth and Pyne, 1964) were supplied on commercially available computer controlled spectrometers and were used for a wider range of application. The Lachance-Traill model (1966) is a concentration model, which in effect requires the solving of a series of simultaneous equations, by regression analysis or matrix inversion techniques. This approach is more rigorous than the intensity models, and so they became popular in the early 1970s as suitable low cost mini-computers became available.

3.8.1 Measurement of Pure Intensities

The intensity of an analyte line is subject not only to the influence of other matrix elements, but also to random and systematic errors due to the spectrometer and counting procedure employed. Provided that a sufficient number of counts is taken, and provided that the spectrometer source is adequately calibrated, random errors from these sources are generally insignificant, relative to other errors. Systematic errors from these sources are, however, by no means insignificant, and effects such as counting dead time, background, and line overlap, can all contribute to the total experimental error in the measured intensity. A problem may arise in that incorrect conclusions about potential matrix effects may be drawn from data, which is subject to instrument systematic errors. As an example, under a certain set experimental conditions, a series of binary alloys may give a calibration curve of decreasing slope. The conclusion may be drawn that radiation from the analyte element was enhanced by the other matrix element, where in point of fact, the problem could have also been due to dead time loss in the counting circuitry because the count rate was too high. If a correction were applied for an enhancement effect, the procedure would break down if the count rates were changed, for example, by varying the source conditions. Problems of this type can be particularly troublesome in the application of influence correction methods. Unless the instrument dependent errors are completely separated from the matrix dependent terms, the instrument effects will tend to become associated with the influence correction terms. In practice, this may not be completely disastrous for a specific spectrometer calibrated for a particular application, since the method will probably work, provided that the experimental conditions do not change. However, one major consequence is that it will probably be impossible to transport a set of correction constants from one spectrometer to another. Mainly for this reason, it is common practice today to attempt to obtain intensities as free from systematic instrumental errors as possible. Such intensities are referred to as pure intensities.

3.8.2 Use of Internal Standards

One of the most useful techniques for the determination of a single analyte element in a known or unknown matrix is to use an internal standard. The technique is one of the oldest methods of quantitative analysis and is based on the addition of a known concentration of an element which gives a wavelength close to that of the analyte wavelength. The assumption is made that the effect of the matrix on the internal standard is essentially the same as the effect of the matrix on the analyte element. Internal standards are best suited to the measurements of analyte concentrations below about 10%. The reason for this limit arises because it is generally advisable to add the internal standard element at about the same concentration level as that of the analyte. When more than 10% of the internal standard is added, it may significantly change the specimen matrix and introduce errors into the determination. Care must also be taken to ensure that the particle sizes of specimen and internal standard are about the same, and that the two components are adequately mixed. Where an appropriate internal standard cannot be found it may be possible to use the analyte itself as an internal standard. This method is a special case of standard addition, and it is generally referred to as spiking.

3.8.3 Type Standardization

As has been previously stated, provided that the total specimen absorption does not vary significantly over a range of analyte concentrations, and provided that enhancement effects are absent and that the specimen is homogeneous, a linear relationship will be obtained between analyte concentration and measured characteristic line intensity. Where these provisos are met, type standardization techniques can be employed. It will also be clear from previous discussion that by limiting the range of analyte concentration to be covered in a given calibration procedure, the range in absorption can also be reduced. Type standardization is probably the oldest of the quantitative analytical methods employed, and the method is usually evaluated by taking data from a well characterized set of standards, and, by inspection, establishing whether a linear relationship is indeed observed. Where this is not the case, the analyte concentration range may be further restricted. The analyst of today is fortunate in that many hundreds of good reference standards are commercially available. While the type standardization method is not without its pitfalls, it is nevertheless extremely useful and is especially useful for quality control type applications where a finished product is being compared with a desired product.

Special reference standards may be made up for particular purposes, and these may serve the dual purpose of instrument calibration as well as establishing working curves for analysis. As an example, two thin glass film standard reference materials specially designed for calibration of X-ray spectrometers are available from the National Bureau of Standards in Washington, as Standard Reference Materials (SRM 1832 and 1833; Pella et al., 1986). They consist of a silica-base film deposited by focussed ion-beam coating onto a polycarbonate substrate. SRM 1832 contains aluminum, silicon, calcium, vanadium, manganese, cobalt and copper; and SRM 1833 contains silicon, potassium, titanium, iron, zinc and rhodium. The standards are especially useful for the analysis of particulate matter.

3.8.4 Influence Correction Methods

It is useful to divide influence coefficient correction procedures into three basic types: Fundamental, Derived and Regression. Fundamental models are those which require starting with concentrations, then calculating the intensities. Derived models are those which are based on some simplification of a fundamental method but which still allow concentrations to be calculated from intensities. Regression models are those which are semi-empirical in nature, and which allow the determination of influence coefficients by regression analysis of data sets obtained from standards. All regression models have essentially the same form and consist of a weight fraction term, W (or concentration C); an intensity (or intensity ratio) term, I; an instrument dependent term which essentially defines the sensitivity of the spectrometer for the analyte in question, and a correction term, which corrects the instrument sensitivity for the effect of the matrix. The general form is as follows:

$$\frac{W}{R} = constant + [model\ dependent\ term] \quad (3\text{-}17)$$

Where W is the weight fraction of the analyte and R is the ratio of the analyte intensity measured from a pure elemental standard to that of the analyte line intensity measured from the specimen, with each intensity corrected for instrumental effects.

The different models vary only in the form of this correction term. Figure 3-6 shows several of the more important of the commonly employed influence coefficient methods. All of these models are concentration correction models in which the product of the influence coefficient and the concentration of the interfering element are used to correct the slope of the analyte calibration curve. The Lachance-Traill model was the first of the concentration correction models to be published. Some years after the La-

Linear Model:
$$W_i/R_i = K_i$$

Lachance-Traill (1966):
$$W_i/R_i = K_i + \sum_j a_{ij} W_j$$

Claisse-Quintin (1967):
$$W_i/R_i = K_i + \sum_j a_{ij} W_j + \sum_j \gamma_{ij} W_j^2$$

Rasberry-Heinrich (1974):
$$W_i/R_i = K_i + \sum_j a_{ij} W_j + \sum_{k \neq j} \beta_{ik}(W_k/1 + W_i)$$

Lachance-Claisse (1980):
$$W_i/R_i = 1 + \sum_j a_{ij} W_j + \sum_j \sum_{k > j} a_{ijk} W_j W_k$$

Figure 3-6. Influence correction models.

chance-Traill paper appeared, Heinrich and his co-workers at the National Bureau of Standards, suggested an extension to the Lachance-Traill approach (Rasberry and Heinrich, 1974) in which absorbing and enhancing elements are separated as α and β terms. These authors suggested that the enhancing effect cannot be adequately described by the same hyperbolic function as the absorbing effect. A thorough study (Tertian, 1986) of Lachance-Traill coefficients based on theoretically calculated fluorescence intensities shows that all binary coefficients vary systematically with composition. Both the Claisse-Quintin (1967) and Lachance-Claisse models use higher order terms to correct for so-called crossed effects, which includes enhancement and third element effects. These models are generally more suited for very wide concentration range analysis.

In all of these methods one of three basic approaches is used to determine the values of the influence coefficients, following the initial measurement of intensities using a series of well characterized standards. The first approach is to use multiple regression analysis techniques to give the best fit for slope, background and influence coefficient terms. Alternatively, the same data set can be used to graphically determine individual influence coefficients. As an example, in the case of the Lachance-Traill equation, for a binary mixture a/b the expression for the determination of a would be:

$$\frac{W_a}{R_a} = 1 + (\alpha_{ab} \times W_b) \qquad (3\text{-}18)$$

By plotting data from a range of analyzed standards in terms of W_a/R_a as a function of W_b a straight line should be observed with a slope of α_{ab} and an intercept of unity. This approach is especially useful for visualizing the form of the influence coefficient correction (Lachance, 1985). Thirdly, the influence coefficient can be calculated using a fundamental type equation based on physical constants.

The major advantage to be gained by use of influence coefficient methods is that a wide range of concentration ranges can be covered using a relatively inexpensive computer for the calculations. A major disadvantage is that a large number of well analyzed standards may be required for the initial determination of the coefficients. However, where adequate precautions have been taken to ensure correct separation of instrument and matrix dependent terms, the correction constants are transportable from one spectrometer to another and, in principle, need only be determined once.

3.8.5 Fundamental Methods

Since the early work of Sherman there has been a growing interest in the provision of an intensity/concentration algorithm which would allow the calculation of the concentration values without recourse to the use of standards. Sherman's work was improved upon first by the Japanese team of Shiraiwa and Fujino (1967) and later, by the Americans, Criss and Birks (1968) with their program NRLXRF. The same group also solved the problem of describing the intensity distribution from the X-ray tube (Gilfrich and Birks, 1968). The problem for the average analyst in the late 1960s and early 1970s, however, remained that of finding sufficient computational power to apply these methods. In the early 1970s, de Jongh (1973) suggested an elegant solution in which he proposed the use of a large main-frame computer for the calculation of the influence coefficients, then use of a small minicomputer for their actual application using a concentration correction influence model. One of the problem areas remains that of adequately describing the intensity distribution form the X-ray tube. Gilfrich and Birks demonstrated an experimental approach to this problem by measuring the spectral distribution from the tube in an independent experiment. More recently, this work has been extended to calculate spectral distributions using data obtained from the electron microprobe (Pella et al., 1985). While software packages are available for fundamental type calculations using data obtained with the energy dispersive system, one major drawback remains in their application system which use a modified primary excitation spectrum. Most fundamental quantitative approaches in use today employ measured or calculated continuous radiation functions in the calculation of the primary absorption effect. Where sharp discontinuities or "breaks" in this primary spectrum occur, as in the case of the energy dispersive system, the calculation becomes very complicated.

3.9 Trace Analysis

3.9.1 Analysis of Low Concentrations

The X-ray fluorescence method is particularly applicable to the qualitative and quantitative analysis of low concentrations of elements in a wide range of samples, as well as allowing the analysis of elements at higher concentrations in limited quantities of materials. The measured signal in X-ray analysis is a distribution of counting rate R as a function of either 2θ angle (wavelength dispersive spectrometers), or as counts per channel as a function of energy (energy dispersive spectrometers). A measurement of a line at peak position gives a counting rate which, in those cases where the background is insignificant, can be used as a measure of the analyte concentration. However, where the background is significant the measured value of the analyte line at the peak position now includes a count rate contribution from the background. The analyte concentration in this case is related to the net counting rate. Since both peak and background count rates are subject to statistical counting errors, the question now arises as to the point at which the net peak signal is statistically significant. The generally accepted definition for the lower limit of detection is that concentration equivalent to two standard deviations of the background counting rate. A formula for the lower limit of detection LLD can now be derived (Jenkins and de Vries, 1970):

$$LLD = \frac{3}{m} \sqrt{\frac{R_b}{t_b}} \qquad (3\text{-}19)$$

Note that in Eq. (3-19) t_b represents one half of the total counting time. The detection limit expression for the energy dispersive spectrometer is similar to that for the wavelength dispersive system except that t_b now becomes the live-time of the energy dispersive spectrometer.

The sensitivity m of the X-ray fluorescence method is expressed in terms of the intensity of the measured wavelength per unit concentration, expressed in c/s per percent. Figure 3-7 shows the sensitivity (excitation factor) of a wavelength dispersive spectrometer and indicates that the sensitivity varies by about four orders of magnitude over the measurable element range, when expressed in terms of rate of change in response per rate of change in concentration. For a fixed analysis time the detection limit is proportional to $m/\sqrt{R_b}$ and this is taken as a figure of merit for trace analysis. The value of m is determined mainly by the power loading of the source, the efficiency of the spectrometer for the appropriate wavelength and the fluorescent yield of the excited wavelength. The value of R_b is determined mainly by the scattering characteris-

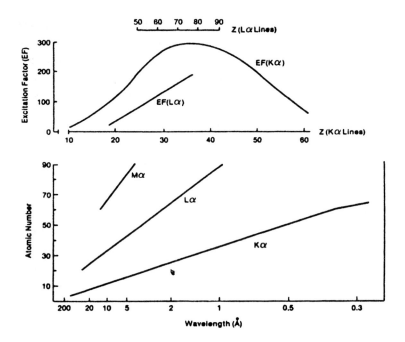

Figure 3-7. Sensitivity of the spectrometer as a function of atomic number.

tics of the sample matrix and the intensity/wavelength distribution of the excitation source.

It is important to note that not only does the sensitivity of the spectrometer vary significantly over the wavelength range of the spectrometer, but so too does the background counting rate. In general, the background varies by about two orders of magnitude over the range of the spectrometer. By inspection of Eq. (3-19) it will be seen that the detection limit will be best when the sensitivity is high and the background is low. Both the spectrometer sensitivity and the measured background vary with the average atomic number of the sample. While detection limits over most of the atomic number range lie in the low part per million range, the sensitivity of the X-ray spectrometer falls off quite dramatically towards the long wavelength limit of the spectrometer due mainly to low fluorescence yields and the increased influence of absorption. As a result, poorer detection limits are found at the long wavelength extreme of the spectrometer, that corresponds to the lower atomic numbers. Thus the detection limits for elements such as fluorine and sodium are at the levels of hundredths of one percent rather than parts per million. The detection limits for the very low atomic number elements such as carbon ($Z=6$) and oxygen ($Z=8$) are, however, very poor and are typically of the order of 3 to 5%.

The major factors effect the detection limit for a given element: first, the sensitivity of the spectrometer for that element in terms of the counting rate per unit concentration of the analyte element; second, the background (blank) counting rate; and third, the available time for counting peak and background photons. In comparing the energy and wavelength dispersive type systems, the absolute sensitivity of the wavelength dispersive system is almost always higher than the equivalent value for an energy dispersive system perhaps by one to three orders of magnitude. This is because modern wavelength dispersive systems are able to handle count rates up to 1 000 000 c/s compared to about 40 000 c/s (for the total output of the selected excitation range) in the case of most energy dispersive systems. This difference is due to the fact that in the case of the energy dispersive spectrometer the electronic pulses created by each X-ray must be sufficiently long for precise energy measurement, thus the maximum count rate occurs because of pulse pile up effects. In the case of the wavelength dispersive system the energy selection is done by the analyzing crystal and much faster detector electrons can be used. The ability of the wavelength dispersive system to work at high counting rates allows use of a high loading at the primary source.

Since it is not possible to measure the background directly, it is common practice to make a background measurement at a selected position close to the peak, the assumption being that the background at this position is the same as the background under the peak. This assumption will, of course, break down where the peak is superimposed on top of a variable background. In this case it is usual to measure the background either side of the peak. In energy dispersive X-ray fluorescence it is common practice to select a number of channels representing a net number of counts on the peak, superimposed on counts from the background. Two ranges of channels are chosen on either side of the peak giving the total background. A further complication occurs where the analyte line is partially overlapped by another line. In this instance the measured value of the peak includes a contribution from the background as before, but in addition, a contribution from the interfering peak and a line overlap correction must be included. The background that occurs at a selected characteristic wavelength or energy arises mainly from scattered source radiation. Since scatter increases with decrease in the average atomic number of the scatterer, it is found that backgrounds are much higher from low average atomic number specimens than from specimens of high average atomic number. To a first approximation, the background in X-ray fluorescence varies as $1/Z^2$. Since the spectral intensity from the X-ray source increases quite sharply as one approaches a wavelength equal to one half the minimum wavelength of the continuum, backgrounds from samples excited with bremsstrahlung sources are generally very high at short wavelengths (high energies), again, especially in the case of low average atomic number samples. The influence of the background is rather complicated because so many variables come into play. One of the major advantages of the secondary target energy dispersive system over the conventional system is that backgrounds are dramatically lower in the secondary excitation mode, because much of the background in a fluoresced X-ray spectrum comes from scattered primary continuum. The measured backgrounds for the wavelength and energy dispersive systems are very similar in the low atomic number regions of the X-ray spectrum, and for the secondary target energy dispersive system are lower by up to an order of magnitude in the higher energy (midrange atomic number) region. All of these factors taken together lead to detection limits for the secondary target energy dispersive system that are typically a factor of 3 to 8 times worse than the wavelength dispersive spectrometer. The actual detection limit measurable also depends upon the characteristics of the specimen itself, including such things as absorption, scattering power etc.

3.9.2 Analysis of Small Amounts of Sample

Conventional X-ray fluorescence spectrometers are generally designed to handle rather large specimens with surface areas of the order of several square centimeters, and problems occur where the sample to be analyzed is limited in size. The modern wavelength dispersive system is especially inflexible in the area of sample handling, mainly because of geometric constraints arising from the need for close coupling of the sample to X-ray tube distance, and the need to use an airlock of some kind to bring the sample into the working vacuum. The sample to be analyzed is typically placed inside a cup of fixed external dimensions that is, in turn, placed in the carousel. This presentation system places constraints not only on the maximum dimensions of the sample cup, but also on the size and shape of samples that can be placed into the cup itself. Primary source energy dispersive systems do not require the same high degree of focussing, and to this extent are more easily applicable to any sample shape or size, provided that the specimen will fit into the radiation protected chamber. In some instances the spectrometer can even be brought to the object to be analyzed. Because of this flexibility, the analysis of odd shaped specimens have been almost exclusively the purview of the energy dispersive systems.

In the case of secondary target energy dispersive systems, while the geometric constraints are still less severe than the wavelength system, they are much more critical than in the case of primary systems. This is due not just to the additional mechanical movements in the secondary target system, but also to the limitations imposed by the extremely close coupling of X-ray tube to secondary target. This is probably the reason that most energy dispersive spectrometer manufacturers generally offer a primary system for bulk sample analysis, and retain the secondary target system, generally equipped with a multiple specimen loader, for the analysis of samples that have been constrained to the internal dimensions of a standard sample cup during the specimen preparation procedure.

Where practicable, the best solution for the handling of limited amounts of material is invariably found in one of the specialized spectrometer systems. This is also generally true for the analysis of low concentrations. The TRXRF system is ideally suited for the analysis of small samples in those cases where the specimen can be dispersed as a thin film onto the surface of the reflector substrate. The high specific intensity synchrotron source offers sensitivities many times greater than the sealed X-ray tube source and because of the small beam divergence it is possible to obtain good signals from very small specimens.

3.10 References

Aiginger, H., Wobrauschek, P. (1985), *Adv. X-ray Anal. 28*, 1
Auger, P. (1925), *Journal de Physique 6*, 205.
Barbee, T. W. (1985), *Supperlattices Microstruct. 1*, 311
Bearden, J. A. (1964), *X-ray Wavelengths*, U. S. Atomic Energy Commission Report NYO-10586, p. 533.
Beattie, M. J., Brissey, R. M. (1954), *Anal. Chem. 26*, 980.
Bertin, E. P. (1975), *Principles and Practice of X-ray Spectrometric Analysis*, 2nd. ed. New York: Plenum Appendix 10.
Bird, J. R., Duerden, P., Clayton, E., Wilson, D. J., Fink, D. (1986), *Nucl. Instrum. and Phys. Res. Sect. B15*, 86.
Birks, L. S. (1963), *Electron Probe Microanalysis, Chemical Analysis Series*, Volume XVII. New York: Interscience.
Birks, L. S. (1976), *History of X-ray Spectrochemical Analysis*. Washington DC: American Chemical Society Centennial Volume, ACS.
Cesareo, M. (1982), *X-ray Fluorescence in Medicine*. Rome: Field Educational Italia. Claisse Inc., Sainte-Foy, Quebec. Claisse, F., Quintin, M. (1967), *Can. Spectrosc.* 159.
Claisse, F. and Quintin, M. (1967) *Can. Spectroscopy 12*, 159
Coster, D., Nishina, J. (1925), *Chem. News 130*, 149.
Criss, J. W., Birks, L. S. (1968), *Anal. Chem. 40*, 1080.

de Jongh, W. K. (1973), *X-ray Spectrom.* 2, 151.
Friedrich, W., Knipping, P., Von Laue, M. (1912), *Ann. Physik. 41*, 971.
Garbauskas, M. F., Goehner, R. P. (1983), *Adv. X-ray Anal. 26*, 345.
Garten, R. P. H. (1984), *Trends in Anal. Chem. 3*, no. 6, 152.
Giauque, R. D., Jaklevic, I M., Thompson, A. C. (1986), *AnaL Chem.* 58, 940–944.
Gilfrich, J. V. (1983), *Anal. Chem. 55*, 187.
Gilfrich, J. V., Birks, L. S. (1968), *AnaL Chem. 40*, 1077.
Glocker, R., Scheiber, H. (1928), *Ann. Physik. 85*, 1085.
Hadding, A., (1922), *Z. Anorg. Allgem. Chem. 122*, 195.
Iwanczyk, I-S., Dabrowski, A. I, Huth, G. C., Bradley, I-G., Conley, I M., Albee, A. L. (1986), *IEEE Trans. Nucl. Sci., NS33*, 355.
Jenkins, R. (1974), *An Introduction to X-ray Spectrometry*. London: Wiley/Heyden, Chap. 4.
Jenkins, R. (1988), *X-ray Fluorescence Spectrometry*. New York: Wiley-lnterscience.
Jenkins, R., de Vries, J. L. (1970), *Practical X-ray Spectrometry*, 2nd. ed. London: MacMillan, Chap. 4.
Jenkins, R., Gould, R. W., Gedcke, D. A. (1981), *Quantitative X-ray Spectrometry*. New York: Dekker, Sec. 4.3.
Jenkins, R., Hammell, B., Cruz, A., Nicolosi, J. A. (1985), *Norelco Reporter 32*, 1.
Knoth, I, Schwenke, H. (1978), *Fresenius Z. Anal. Chem. 301*, 200.
Lachance, G. R. (1985), *Introduction to Alpha Coefficients*, Paris: Corporation Scientifique.
Lachance, G. R., Traill, R. J. (1966), *Can Spectrosc. 11*, 43.
Lucas-Tooth, H.J, Pyne, C. (1964), *Adv. X-ray AnaL 7*, 523.
Malmqvist, K. G. (1986), *Nucl. Instrum. Meth. Phys. Res. Sect. B14*, 86.
Michaelis, W., Knoth, J., Prange, A., Schwenke, H. (1984), *Adv. X-ray Anal. 28*, 75.
Moseley, H. G. J. (1912), *Phil. Mag. 26*, 1024.
Moseley, H. G. J. (1913), *Phil. Mag. 27*, 703.
Nicolosi, J. A., Jenkins, R., Groven, J. P., Merlo, D. (1985), *Proc. SPIE 563*, p. 378.
Pella, P. A., Feng, L. Y., Small, I. A. (1985), *X-ray Spectrom. 14*, 125.
Pella, P. A., Newbury, D. E., Steel, E. B., Blackburn, D. H. (1986), *Anal. Chem. 56*, 1133.
Petersen, W., Ketelsen, P., Knoechel, A., Pausch, R. (1986), *Nucl. Instrum. Methods Phys. Sect. A, A246* (1–3) 731.
Rasberry, S. D., Heinrich, K. F. J. (1974), *Anal. Chem. 46*, 81.
Röntgen, W. C. (1898), *Ann. Phys. Chem. 64*, 1.
Sherman, J. (1955), *Spectrochim. Acta 7*, 283.

Shiraiwa, T., Fujino, N. (1967), *Bun. Chem. Soc. Japan. 40*, 2289.
Stephenson, D. A. (1971), *Anal. Chem. 43*, 310.
Sparks, C. J. Jr. (1980), *Synchrotron Radiation Research*: Winnick, H., Doniach, S. (Eds.). New York: Plenum Press, p. 459.
Tertian, R. (1986), *X-ray Spectrom. 15*, 177.
Von Hevesey, G. (1932), *Chemical Analysis by X-rays and its Application*. New York: McGraw-Hill.
Walter, F. J. (1971), *Energy dispersion X-ray analysis*, ASTM Special Technical Publication STP 485. Philadelphia: ASTM, p. 125.
West, T. S., Nurrenberg, H. W. (Eds.) (1988), *The Determination of Trace Metals in Natural Waters*. Oxford: Blackwell.
White, E. W., Johnson, G. G.Jr. (1970), *X-ray Emission and Absorption Wavelengths and Two-Theta Tables*, ASTM Data Series DS-37A. Philadelphia: ASTM.

General Reading – A Selection of Books Dealing with X-ray Spectrometry

Anderson, C. A. (Ed.), (1972), Microprobe Analysis. New York: Wiley-Interscience.
Azaroff, L. V. (Ed.), (1974), X-ray Spectroscopy. New York: McGraw Hill.
Bertin, E. P. (1975), Principles and Practice of X-ray Spectrometry Analysis, 2nd. ed. New York: Plenum.
Bertin, E. P. (1985), Introduction to X-ray Spectrometry Analysis. New York: Plenum.
Birks, L. S. (1969), X-ray Spectrochemical Analysis, 2nd ed New York: Wiley-Interscience.
Cauchois, Y., Bonnelle, C., Mande, C. (1982), Advances in X-ray Spectroscopy. Oxford: Pergamon.
Jenkins, R. (1974). An Introduction to X-ray Spectrometry. London: Heyden.
Jenkins, R. (1988), X-ray Fluorescence Spectrometry. New York: Wiley-Interscience.
Jenkins, R., de Vries, J. L. (1970), Practical X-ray Spectrometry, 2nd. ed. London: MacMillan.
Jenkins, R., Gould, R. W., Gedcke, D. (1981), Quantitative X-ray Spectrometry. New York: Dekker.
Liebhafsky, H. A., Pfeiffer, H. G., Winslow, E. H., Zemany, P. D. (1972), X-rays, Electrons, and Analytical Chemistry – Spectrochemical Analysis with X-rays. New York: Wiley-Interscience.
Pattee, H. H., Cosslett, V. E., Engstrom, (1963), X-ray Optics and X-ray Microanalysis. New York: Academic Press.
Russ, J. C., Shen, R. B., Jenkins, R. (1978), EXAM: Principles and Practice. Prairie View: Edax International.
Woldseth, R. (1973), X-ray Energy Spectrometry. Burlingame: Kevex Corporation.

4 Small-Angle Scattering of X-Rays and Neutrons

Claudine E. Williams

Laboratoire de Physique de la Matière Condensée, CNRS URA 792, Collège de France, Paris, France

Roland P. May

Institut Max von Laue–Paul Langevin (ILL), Grenoble, France

André Guinier

Laboratoire de Physique des Solides, Université Paris-Sud, Orsay, France

4.1	**Introduction**	213
4.2	**General Principles**	215
4.2.1	Small-Angle Intensity and the Notion of Contrast	216
4.2.2	The Invariant	218
4.2.3	The Porod Limit	219
4.3	**Structural Parameters**	220
4.3.1	Isolated Particles	220
4.3.2	Randomly Distributed Phases	221
4.3.3	Fractals	222
4.3.4	Polymer Chain as a Diffuse Object	223
4.3.5	Interacting Particles	224
4.4	**Contrast Variation**	225
4.4.1	Anomalous Scattering of X-Rays	225
4.4.1.1	The Underlying Physics	225
4.4.1.2	Contrast Variation by Anomalous Scattering	226
4.4.1.3	Design of an ASAXS Experiment	227
4.4.2	Isotopic Substitution in SANS	228
4.4.2.1	Solvent (Matrix) Scattering Density Variation	229
4.4.2.2	Specific Isotopic Labeling	230
4.4.2.3	Triple Isotopic Replacement	230
4.4.3	Magnetic Scattering of Neutrons	231
4.4.3.1	Spin-Contrast Variation	231
4.5	**Experimental Techniques**	232
4.5.1	Small-Angle Scattering of X-Rays	232
4.5.1.1	X-Ray Sources	232
4.5.1.2	SAXS Spectrometers	232
4.5.2	Small-Angle Scattering at Neutron Sources	235
4.5.2.1	Neutron Sources	235
4.5.2.2	SANS Diffractometers	235

4.5.2.3	Specific Properties of Neutrons Compared to X-Rays	236
4.5.2.4	Incoherent Scattering of Neutrons	237
4.5.3	Practical Aspects of Small-Angle Scattering	237
4.5.3.1	The Sample and its Environment	237
4.5.3.2	Data Reduction and Absolute Scaling	239
4.5.3.3	Detector-Response Correction	240
4.5.3.4	Estimation of the Background Level	241
4.5.3.5	Correction of Instrumental Effects	242
4.5.4	Data Analysis and Interpretation	243
4.5.4.1	Particle Mass Determination	243
4.5.4.2	Real-Space Considerations	243
4.5.4.3	Model Fitting	244
4.6	**Some Applications of Small-Angle Scattering**	244
4.6.1	Chain Conformation of Polymers, a Success of Selective Deuteration	244
4.6.2	Phase Separation in Binary Alloys: Towards a Shorter Time Resolution	246
4.6.3	Partial Structure Factors by Contrast Variation	248
4.6.4	Samples with External Constraints: Liquids Under Shear	250
4.6.5	Internal Structure of a Complex Particle by Label Triangulation	250
4.7	**References**	251

4.1 Introduction

This chapter deals with the scattering by matter of waves, especially electromagnetic ones in the case of X-rays. Many aspects of X-ray production, their detection and diffraction are also dealt with in chapter 1. This chapter is restricted to scattering in the small-angle range and describes the specific characteristics and the various possibilities of the technique.

Because X-rays interact with the electrons of an atom and neutrons with its nucleus, small-angle X-ray scattering (SAXS) is sensitive to inhomogeneities of the electron density whereas small-angle neutron scattering (SANS) detects the variation of the so-called *scattering length density*. However, the underlying principles are identical for both techniques.

Generally speaking, when a monochromatic plane wave in z direction (described by $\Psi = e^{ikz}$) illuminates a small sample, scattered waves of variable intensity are emitted in all directions of space. Every point in the sample creates a spherically symmetrical scattered wave of the form $\Psi = -(b/r)e^{ikr}$, where r is the distance from the respective point in the sample to the point of measurement. The (complex) scattering amplitude b has the dimension of a length and is therefore also called *scattering length* in neutron scattering, where the nuclei can be considered as point sources (Bacon, 1975). Both the incident wave and the wave scattered at an angle 2θ can be described by their wave vectors k_0 and k, respectively, of modulus $2\pi/\lambda$ and oriented perpendicular to the wave front (Fig. 4-1). We can define a vector q such that $k = k_0 + q$. q is the scattering vector which defines the geometry of the experiment. For small angles 2θ, its modulus is $2\pi\theta/\lambda$.

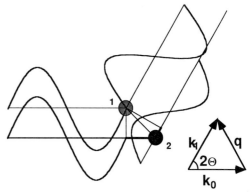

Figure 4-1. Scattering geometry: An incoming plane wave, depicted as two parallel sine functions (wave vector k_0), hits an object. Any two scattering centers 1 and 2 of equal volume in the object emit spherical waves of amplitudes proportional to their scattering-length densities ϱ_1 and ϱ_2. Due to the different distances of the two volumes from the wave fronts, the outgoing wave vectors k undergo different phase shifts. The superposition of waves of different amplitudes and phases results in a scattering intensity depending on the scattering angle 2θ and the momentum transfer q. It is assumed that the outgoing waves have the same energy as the incident ones, i.e. $|k| = |k_0|$ (elastic scattering).

A fundamental law in all scattering experiments tells us that the amplitude $A(q)$ of the scattered wave is the Fourier transform of the electron (or scattering length) density distribution $\varrho(r)$ in the sample. It means that the amplitude in a direction defined by the vector q is related to *one* term of the Fourier expansion of $\varrho(r)$, i.e. the density modulation perpendicular to q of wavelength $\Lambda = 2\pi/q$.

Let us discuss some orders of magnitude: for X-rays with a typical wavelength of 0.15 nm and a scattering angle 2θ of $1°$, $\Lambda = \lambda/\theta$ equals 17.2 nm. Since it is difficult to observe the scattered intensity at angles smaller than a few minutes with standard techniques, a practical upper limit of the size of the modulations that can be mea-

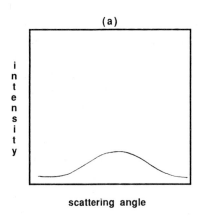
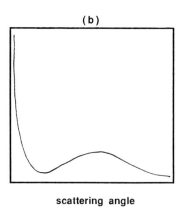

Figure 4-2. Small-angle X-ray scattering profiles of two different forms of amorphous silica, (a) glass and (b) silica gel.

sured with X-rays is of the order of 150 nm. On the other hand, the cold neutrons commonly used for SANS have a longer wavelength, up to 2 nm; consequently, sizes of 1000 nm are easily measured. The important point here is that SAS allows us to obtain structural information about inhomogeneities in materials, with a characteristic length of the order of tens to hundreds of nanometers whereas the usual crystallographic techniques (wide-angle scattering) describe the structure at an atomic scale, say, from 0.1 to 1 nm.

There is evidently a continuity between the small-angle and wide-angle techniques and one could reasonably assume that the complete diffraction pattern in the whole angular range should be used and no part should be isolated. Indeed this is true for some systems, such as soaps or proteins crystals, where the features appearing at small angles are interpreted as diffraction patterns but with a much larger periodicity than that in crystals of small molecules. But there are many cases where the small-angle scattering can be isolated from the rest of the scattering pattern, and this is the object of this chapter. Before developing it in details, let us illustrate this point with the example of pure amorphous silica. Figure 4-2 shows schematically the complete X-ray scattering patterns of two forms of this material, the amorphous glass and a fine powder (silica gel); no differences can be detected in the wide angle scattering of these samples because their structures are identical at atomic scale. However in the second case, the material is divided into small particles of mesoscopic size, and a strong scattering appears in the small-angle range; its intensity is large near the incident beam, then it decreases rapidly and is negligibly small above about 3°. Isolating the central pattern is equivalent to neglecting the electron density variations corresponding to the atomic scale; then the actual density is replaced by a *smooth* profile and the scattering at small angles is due to the modulations of the average electron density. As will be shown in the course of this chapter, the intensity depends on the size and shape of the individual grains, but is independent of the local structure of the grain itself, whether crystalline or amorphous. It is also independent of the local structure of the medium in which the grains might be imbedded. SAS appears only if the *average density* of that medium is different from that of the grains. Identical arguments hold for neutron scattering.

The scattered intensity is usually larger at small angles than at wide angles. In a fine powder of crystallized grains, for example, an individual grain will contribute to the diffraction only if its orientation corresponds to the Bragg condition, whereas all grains, whatever their orientation, scatter at small angles. Hence the SAS halo is usually more intense than a powder diagram.

Imaging methods such as transmission electron microscopy (TEM) also have the capability to characterize inhomogeneities of mesoscopic size in various materials and it is interesting to compare these two techniques. The main difference is that TEM provides an image in real space which SAS cannot do. For instance, with the powder sample referred to earlier, TEM gives an exact picture of individual grains, showing their sizes and shapes, whereas SAS gives stuctural information averaged over all the grains. Although precise, this information is usually insufficient to reconstruct the details of a complex object. But often an average description of the sample is precisely what the experimentalist needs: for instance, the granulometry of a catalyst is directly described in SAS by an average value and there is no need to visualise each individual particle. Also important is the fact that the constraints on the sample are much less stringent for SAS, especially with neutrons, than for TEM. Hence solutions of large molecules or molecular assemblies can only be investigated without difficulty by SAS; this explains the prominent use of the technique for biological studies.

Since small-angle scattering investigates distances that are large compared to interatomic distances, the sample is usually treated as a continuum and characterized by an average density and the fluctuations around it. Consequently, the scattering is described as arising from "scattering objects" of some scattering density embedded in a medium of another density. The scattering profiles contain information on the size of the particles as well as on the interactions between them. SAS is thus an appropriate technique to characterize a very large variety of samples, ranging from phase-separated alloys and glasses to polymers, colloids or proteins.

This chapter will first give the general principles of the theory of SAS; the emphasis will be more on the appropriate use of the essential formulae rather then on an exact derivation, which can be found in the monographs cited in the bibliography. Then a brief survey of the spectrometers and of the experimental techniques of SAXS and SANS will be given. Finally some examples will illustrate the use of the technique in various fields.

4.2 General Principles

Small-angle scattering measures spatial correlations in the scattering density, averaged over the time scale of the measurement. Hence, all the phenomena will be considered as elastic, i.e. it will be assumed that the incident and scattered photon/neutron have the same energy. For X-rays, because the energy transfer is small compared to the incident energy of the photon (≈ 10 keV), inelastic effects are small and difficult to measure anyway. On the other hand, neutrons have energies (≈ 10 meV) of the same order as that of the excitations in solids, and inelastic effects can be large. However, in the small angle range, their contribution is less meaningful and won't be dealt with in this monograph.

Scattering can still be coherent or incoherent. Coherent scattering, where scattered photons/neutrons from different atoms interfere, provide all information on the spatial distribution of atoms. Incoherent scattering produces a background signal which, we assume, has been corrected for. Fortunately, the main source of incoherent scattering of X-rays, Compton scattering, is negligible in the angular range of interest. With neutrons, the situation is more complex, since the incoherent signal can be much larger than the coherent one. It arises because there is a distribution of isotopes of an element in the sample (isotopic incoherent scattering), but mostly because the interaction of the neutron with the nucleus depends on spin orientation (spin incoherent scattering). Ways to correct for and also to make use of neutron incoherent scattering will be discussed later.

4.2.1 Small-Angle Intensity and the Notion of Contrast

In a scattering experiment, one counts the number of events $\mathcal{J}(q)$ received by a detector element (pixel)

$$\mathcal{J}(q) = \Phi_0 \, \Delta\omega \, \Delta t \, T \varepsilon \, \frac{d\Sigma(q)}{d\Omega} \qquad (4\text{-}1)$$

where Φ_0 is the incident beam flux at the sample (in neutrons per unit area and unit time), $\Delta\omega$ is the solid angle subtended by the detector element, i.e. its area divided by the square of the sample-to-detector distance, Δt is the measuring time, ε is the detector efficiency, and T is the sample transmission, i.e., the ratio of the direct beam intensities before and after the sample. $d\Sigma/d\Omega$ is the macroscopic (coherent) differential scattering cross section (with Ω a solid angle).

For samples containing N particles, an alternative expression for the experimental intensity is

$$\mathcal{J}(q) = \Phi_0 \, \Delta\omega \, \Delta t \, T \varepsilon \, N \, \frac{d\sigma(q)}{d\Omega} \qquad (4\text{-}2)$$

where $d\sigma/d\Omega$ is the cross section per unit solid angle and per particle. The prefactor, which depends on the experimental conditions, is often assumed to be equal to 1 when dealing with the physical properties of the material, as in most of this chapter. However, the complete formula has to be used for absolute calibrations (see Secs. 4.5.3.2 and 4.5.4.1). In what follows $I(q)$ stands for $d\Sigma/d\Omega(q)$ or $d\sigma/d\Omega(q)$.

As briefly stated in the introduction, general scattering theories tell us that the amplitude measured at a scattering vector q of a wave scattered from an atom situated at a point r is e^{iqr}, where $q = k - k_0$ (Fig. 4-1). Since we are dealing with elastic scattering, if the scattering angle is 2θ, and the wavelength λ, $k = k_0 = |k| = 2\pi/\lambda$, then

$$q = |q| = \frac{4\pi}{\lambda} \sin\theta \qquad (4\text{-}3)$$

It is important to note that different symbols can be found in the literature corresponding to the same definition of the scattering vector; q, h, Q, $2\pi s$ are identical.

The total amplitude at q is the sum of the waves scattered by all atoms in the sample. Because SAS deals with distances that are large compared to interatomic distances, it is not possible to separate the contributions of individual atoms to the scattering and the sum over discrete atoms can be replaced by an integral if one introduces a scattering density $\varrho(r) = \Sigma b_i/V_{res}$, where the scattering amplitudes b_i are summed over a volume V_{res} of the size of a resolution element. Then

$$A(q) = \iiint \varrho(r) \, e^{-iqr} \, dr \qquad (4\text{-}4)$$

which is the 3-d Fourier transform of $\varrho(r)$.

The scattered intensity is the product of $A(q)$ by its complex conjugate:

$$I(q) = |A(q)|^2 = A(q) A^*(q)$$
$$= \iiint dr_1 \iiint \varrho(r_1) \varrho(r_2) \, e^{-iq \cdot (r_1 - r_2)} \, dr_2$$
$$= V \iiint P(r) \, e^{-iq \cdot r} \, dr \qquad (4\text{-}5)$$

with

$$P(r) = \frac{1}{V} \iiint \varrho(r_0) \varrho(r_0 + r) \, dr_0$$
$$= \langle \varrho(r_0) \varrho(r_0 + r) \rangle \quad (4\text{-}6)$$

$P(r)$ is the autocorrelation function (Patterson function of crystallography). It expresses the correlation between the densities measured at two points separated by a vector r, averaged over the total irradiated volume V. It can be related to the structure of the scattering object.

To simplify the formalism, we assume here that the sample is statistically isotropic when averaged over the irradiated volume V and that there is no long-range order. In this case, $P(r)$ depends only on the distance between points and its value for large distances is the square of the average scattering density i.e. $P_\infty(r) = \bar{\varrho}^2$. It must be stressed that these assumptions are evidently not valid for all samples of interest in small angle scattering; for instance oriented lamellar phases are not isotropic or liquid and colloidal crystals often show long-range order. But the scattered intensity can still be calculated in these important cases and the derivation can be found in specialized textbooks.

To have a better picture of the type of sample we are dealing with in SAS and to get an insight in the meaning of the correlation function, let us consider first an isolated particle of constant scattering density ϱ_1 embedded in a matrix (solvent) of density ϱ_2 (Fig. 4-3a). Because the volume fraction of the matrix is so much bigger than that of the particle, the average density of the sample $\bar{\varrho} \simeq \varrho_2$ and the important quantity is $\varrho_1 - \varrho_2 = \Delta\varrho$. Now, formally replacing ϱ by $\varrho_1 - \bar{\varrho}$ will only add a δ-function at $q=0$ in the expression for $I(q)$, i.e. in the transmitted beam, which is of no importance for a measurement of the scattered intensity, since it would fall in the

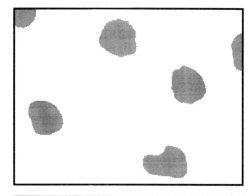

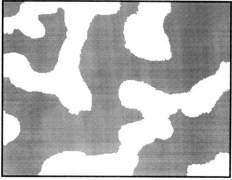

Figure 4-3. Two-dimensional representation of typical samples investigated by small-angle scattering: (top) isolated particles of scattering density ϱ_1 embedded in a matrix of scattering density ϱ_2; (bottom) two randomly distributed media of scattering densities ϱ_1 and ϱ_2. Characteristic sizes are of the order of 10 to a few 100 Å in both pictures.

beam-stop. This means that the scattering density of the particle is measured with respect to a reference medium, the solvent; $\Delta\varrho$ is referred to as the "contrast" of the particle. It can take negative or positive values.

The sample need not be particulate and more generally, it can be visualized as made of two media (noted 1 and 2) separated by an interface; each one is characterized by a constant scattering density ϱ and a volume fraction ϕ so that the average density of the sample is $\bar{\varrho} = \varrho_1 \phi_1 + \varrho_2 \phi_2$ with $\phi_1 + \phi_2 = 1$ (Fig. 4-3b). For each point in the sample, one can define the fluctuation of the scattering density with respect to its average value, η, that will take either of the 2 values:

$$\eta_1 = \varrho_1 - \bar{\varrho} = \phi_2(\varrho_1 - \varrho_2) \qquad (4\text{-}7)$$

or

$$\eta_2 = \varrho_2 - \bar{\varrho} = -\phi_1(\varrho_1 - \varrho_2) \qquad (4\text{-}8)$$

One can also look at the correlation between the fluctuations for all points distant of r (Debye, 1915). This is the correlation function $\gamma(r)$ first introduced by Debye and Bueche (1949) as:

$$\gamma(r) = \langle \eta(r_0) \eta(r_0 + r) \rangle = P(r) - \bar{\varrho}^2 \qquad (4\text{-}9)$$

The scattered intensity (Eq. (4-5)) takes the classic form

$$\frac{I(q)}{V} = \iiint \gamma(r) e^{-iqr} dr \qquad (4\text{-}10)$$

And, if we further assume spherical symmetry,

$$\frac{I(q)}{V} = 4\pi \int_0^\infty \gamma(r) \frac{\sin(qr)}{(qr)} r^2 \, dr \qquad (4\text{-}11)$$

As already noted, replacing $P(r)$ by $\gamma(r)$ implies that one adds the scattering of an infinitely large and perfectly uniform medium which would have a contribution exclusively in the transmitted beam, beyond the resolution of any experiment.

The correlation function is obtained by a Fourier transformation of the intensity, as

$$\gamma(r) = \frac{1}{2\pi^2 V} \int_0^\infty \frac{\sin(qr)}{(qr)} I(q) q^2 \, dq \qquad (4\text{-}12)$$

Direct application of this formula, although mathematically feasible, is not always straightforward because of the limited q-range experimentally accessible. Extending that range, mostly on the low-angle side for X-rays, but also on the high-angle side for neutrons, requires strict conditions on beam collimation and methods of detection (see Sec. 4.5.1).

4.2.2 The Invariant

To get a more physical insight into the meaning of the correlation function, let us imagine that we drop, at random, a stick of length r on the two-dimensional image of our sample (Fig. 4-3) and we count how many times both ends are in medium 1, or in medium 2, or one end in each medium. If the two media are distributed in a completely random way, points separated by r are not correlated, and the probabilities are proportional to the volume fraction of each phase, i.e.

$$P_{11} = \phi_1^2, \quad P_{22} = \phi_2^2, \quad P_{12} = P_{21} = \phi_1 \phi_2 \qquad (4\text{-}13)$$

If there is a correlation in the distribution of the two media, the deviation to these probabilities can be described by a function $\gamma_0(r)$ such that:

$$P_{11}(r) = \phi_1^2 + \gamma_0(r) \phi_1 \phi_2$$
$$P_{22}(r) = \phi_2^2 + \gamma_0(r) \phi_1 \phi_2$$
$$P_{12}(r) = P_{21}(r) = \phi_1 \phi_2 (1 - \gamma_0(r)) \qquad (4\text{-}14)$$

If $\gamma_0(r) > 0$, points at a distance r are more likely to be in the same region than if the medium is random; if it is < 0, they are more likely to be in different regions; if $\gamma_0(r) = 1$, the 2 points are in the same region. The correlation function becomes

$$\gamma(r) = \phi_1 \phi_2 (\varrho_1 - \varrho_2)^2 \gamma_0(r) \qquad (4\text{-}15)$$

For $r = 0$ the two ends are evidently in the same region and $\gamma_0(0) = 1$; setting $r = 0$ in Eq. (4-12) provides a relation between the parameters defining the sample and the scattered intensity as:

$$2\pi^2 \phi_1 \phi_2 (\varrho_1 - \varrho_2)^2 = \frac{1}{V} \int_0^\infty q^2 I(q) \, dq = Q \qquad (4\text{-}16)$$

This quantity is called the invariant Q because it is independent of the topology or the geometry of the sample. Indeed, a sample containing monodisperse spheres of density ϱ_1 in a matrix of density ϱ_2 will show a very different scattering profile from that of a bicontinuous arrangement of ϱ_1 and ϱ_2, but both samples have the same invariant, provided that ϱ_1, ϱ_2, and ϕ_1, ϕ_2 are the same.

Equation (4-16) is important because it *directly* relates quantities pertaining to the sample composition to the experimental scattered intensity. For instance, if the contrast between the phases can be calculated the invariant gives an evaluation of the total scattering volume in the sample, which can in turn be compared to structural models.

As such the invariant is a useful safeguard in the interpretation of scattering data. Discrepancies between the two evaluations can not only be due to the wrong model but also to the non-validity of the two-media approximation or to inadequate experimental conditions. Indeed a precise determination of Q requires data in an adequate q-range i.e. where all the scattering that characterizes the structure of the sample takes place. If the scattering particles are very large so that most of the scattering is confined to very small angles which cannot be reached by the experimental set-up, especially with X-rays, the invariant cannot be evaluated. With neutrons, on the other hand, one can not always reach large enough q with sufficient counting statistics. This emphasizes once more the importance of choosing adequate experimental conditions.

Measuring the time evolution of the scattering invariant has been most useful for the characterization of phase transitions such as the crystallisation of polymers from the melt when the temperature is lowered below the melting point: Theoretical models can be tested if the amount of material that has crystallized as time evolves is known (Elsner et al., 1985). A precise knowledge of the invariant is very valuable in materials where the exact composition is known, such as colloids and surfactants. For instance, the breaking up of micelles with the addition of short-chain alcohols can be clearly evidenced by the determination of the invariant (Hayter, 1985).

4.2.3 The Porod Limit

Let us look at the correlations over distances r much smaller than some typical length D in the sample (size, distance), i.e. $r \ll D$. If we drop a stick as we did previously, most of the drops will have both ends in the same phase and only those throws corresponding to the stick crossing an interface will be weighted by a non-zero $\Delta\varrho$ and will participate to the scattered intensity at large q's ($q \gg D^{-1}$). The large-angle range should thus reflect the structure of the interface.

An exact calculation for an infinitely sharp interface (Porod, 1951; Porod, 1982) shows that the surface area per unit volume $\mathscr{A} = S/V$ is related to the scattered intensity by

$$\mathscr{A} = \frac{1}{2\pi(\Delta\varrho)^2 V} \lim_{q\to\infty} q^4 I(q) = \frac{1}{2\pi(\Delta\varrho)^2 V} K \quad (4\text{-}17)$$

where K is usually referred to as the Porod limit. Using the invariant to eliminate scale constants,

$$\mathscr{A} = \pi \phi (1-\phi) \frac{K}{Q} \quad (4\text{-}18)$$

Fitting a horizontal line to $I(q) \cdot q^4$ versus q ("Porod plot") in *the correct region* yields a constant which is proportional to the internal surface. An alternative representation of $I(q) \cdot q^4$ versus q^4 is advantageous when the data include a constant background which appears as the slope of a straight line whose zero intercept again yields the internal surface. The residual constant background (including the self-term of the constant nuclear "form factor") may be used for slightly correcting the estimated background and consequently improving the quality of the data. For monodisperse particles, a particle surface can be deduced from the overall surface. The value of the surface area so determined depends on the maximal q to which the scattering curve can be obtained with good statistics. This depends also on the magnitude of the background. With SANS, the incoherent background level often becomes so large with respect to the coherent signal that the tail of the scattering curve cannot be interpreted in terms of the surface area.

Porod's law is valid at all distances for which the interface is seen as sharp. This assumption evidently breaks down for distances smaller than r_0, of the order of the interfacial thickness. Consequently Porod's law will deviate from the q^{-4} dependence at some $q \approx r_0^{-1}$. For a diffuse boundary, there is a gradual change of scattering density from one phase to the next. The intensity scattered at large angles has been calculated by convoluting the ideally sharp profile with various smoothing functions (Ruland, 1971; Koberstein et al., 1980). For a sigmoidal gradient model, the correction factor to the Porod law takes the form $\exp(-\sigma^2 q^2)$ where σ is the standard deviation. Thus, in principle, information about the interfacial thickness can be obtained in the Porod region. However, experimentally, the intensity scattered in that region is weak and the observed q-dependence relies very much on an exact background subtraction, so that it is sometimes difficult to even deduce an interfacial surface. Deviations to Porod's law can be best observed in a Porod plot.

As for the invariant, the validity of Porod's law is *independent of the geometry and topology of the sample*.

An elegant derivation of Porod's law in reciprocal space can be found in Auvray and Auroy (1991), where the influence of surface curvature and roughness are also discussed (see also Sec. 4.3.3).

4.3 Structural Parameters

Valuable informations can be gained from measurements of the invariant and the Porod limit but, as pointed out earlier, these are independent of the geometry of the two media in the sample and hence, are of no use to describe the structure. Fortunately, some structural parameters can be extracted directly from the intensity profile. We will first illustrate this point by considering isolated particles, then add interactions between them. Finally we will mention how the scattered intensity reveals the fractal behavior of a sample.

4.3.1 Isolated Particles

For particles with a simple well-defined shape, diluted in a matrix so that there is no interaction between them, the angular dependence of the scattered intensity can be calculated. For instance, for a homogeneous sphere of radius R,

$$I(q) = I_0 \left[3 \frac{\sin(qR) - (qR)\cos(qR)}{(qR)^3} \right]^2 \quad (4\text{-}19)$$

where I_0 contains the geometry independent terms. $I(q)$ has a series of well-defined zeros corresponding to $qR = 4.493, 7.725$, etc.

For an infinitely thin cylinder of length L,

$$I(q) = I_0 \left[\frac{2 \int_0^{qL} dt \frac{\sin t}{t}}{qL} - \frac{4 \sin\left(\frac{qL}{2}\right)}{q^2 L^2} \right] \quad (4\text{-}20)$$

It is clear from these two examples that the scattering profile is strongly dependent on the shape and size of a particle. However, Guinier (1938) pointed out that at very small angles, corresponding to inverse scattering vectors much larger than the size of the particle, the angular dependence of the scattering is a universal function of that size.

Indeed, if a particle of any shape, randomly oriented, is described by its radius of gyration, which in analogy to mechanics is defined as

$$R_G^2 = \frac{\sum_i b_i d_i^2}{\sum_i b_i} \qquad (4\text{-}21)$$

with b_i, the elementary scattering "masses" at distances d_i from the "centre of gravity", the intensity scattered in the angular range such that $qR_G \ll 1$, is given by (Guinier, 1938; Guinier and Fournet, 1955)

$$I(q) = I_0 \exp\left(-\frac{q^2 R_G^2}{3}\right) \qquad (4\text{-}22)$$

Therefore, a plot of $\ln I(q)$ vs q^2 should be linear if Guinier's law applies and R_G^2 is obtained from the slope of the interpolation line. For a known shape, R_G can be related to size parameters: for a sphere, $R = (5/3)^{1/2} R_G$. If the intensity is known on an absolute scale, I_0 is related to the relative molecular mass M_r of the particle.

Guinier's law has been valuable for determining the size of particles in metallurgy, biology and other fields for 50 years. However, its applicability relies on some approximations that should be kept in mind (it is unfortunate that results sometimes appear in the literature when it is clearly not applicable):

First of all, Guinier's law is strictly valid only in the range $qR_G \ll 1$ which requires to collect good-quality data at very small angles. Also, in the case of polydispersity, the (curved) Guinier plot starts off with a linear part for $qR_G^0 \ll 1$, where R_G^0 is the radius of gyration *of the largest particle*. Its slope gives an average radius of gyration which is strongly dominated by that of the larger particles. For $q > 1/R_G^0$, a meaningful R_G cannot be obtained, in general.

Finally, there must be no interaction between the particles. This last requirement is sometime difficult to meet if the sample is such that it cannot be diluted. If it can be, the effect of a weak interaction can be taken care of by extrapolation of experimental R_Gs to zero concentration.

At this point, it is worth stressing that materials science seldom deals with particles of uniform size and shape and that the scattered intensity results from a statistical average over all scattering elements. A model for the dispersion in size and shape has to be assumed but, even so, it is often impossible to discriminate between two models; for instance polydisperse spheres and monodisperse ellipsoids give the same scattering (Glatter, 1991). A further smearing may come from the experimental resolution. Numerous methods are available for correcting for instrumental smearing and polydispersity effects. The latter will be discussed in more detail later.

4.3.2 Randomly Distributed Phases

For a system such as that pictured in Fig. 4.3b, the interpretation of $\gamma(r)$ is not always clear. Without relaxing the hypothesis of a sharp interphase, Debye and Bueche (1949) have assumed an exponential damping of the spatial correlations, characterized by a correlation length a such that

$$\gamma(r) = e^{-r/a} \qquad (4\text{-}23)$$

The scattered intensity as a function of q decreases as

$$I(q) \propto \frac{1}{(1 + q^2 a^2)^2} \qquad (4\text{-}24)$$

The slope-to-intercept ratio in a plot of $I(q)^{-1/2}$ vs q^2 gives a measure of the correlation length a.

This model was actually first introduced to explain how the scattering of *visible light* in glass is due to large inhomogeneities at the micrometer scale. It was then transposed later to explain how finely divided oxide powders such as Al_2O_3 scatter X-rays. It illustrates well how the basic concepts are common to all scattering techniques, whether using light, X-rays, or neutrons (Debye and Bueche, 1949).

To give a more physical meaning to the correlation length a, Kratky (1966) has introduced the chord length l which measures, for any line drawn through the system, the length in each phase (see also Porod, 1948). If the average lengths are l_1 and l_2, then

$$l_1 = \frac{a}{\phi_1} \quad \text{and} \quad l_2 = \frac{a}{\phi_2} \qquad (4\text{-}25)$$

The analysis of scattering data of multiphase samples by the chord distribution function has been discussed in detail by Tchoubar (1991).

Note that the large-angle limit of the Debye-Bueche expression is different from the Porod relation since an exponential decay cannot describe a well-defined interphase.

4.3.3 Fractals

Fractals are characterized by self-similarity, i.e. their structure is independent of the length scale of observation, within some spatial range. If one measures the mass M of material in a volume R^3, it is related to that length scale by $M(R) = \text{const} \cdot R^{\mathscr{D}}$, where $\mathscr{D}$ is the fractal dimension, smaller than the dimension d of space. (For a homogeneous medium, $d = \mathscr{D}$.) In a fractal, the mass density varies as $R^{\mathscr{D}-d}$ and, by Fourier transform, one obtains a very simple scattering law, valid in the whole angular range for an ideal fractal:

$$I(q) = \text{const} \cdot q^{-\mathscr{D}} \qquad (4\text{-}26)$$

Equation (4-26) provides a convenient way to measure the fractal dimension; a log-log plot of $I(q)$ vs q yields $\mathscr{D}$ as the slope of the straight line fitting the experimental points. SAS is the most adequate technique to characterize fractal objects in the range below 1000 Å.

Although, mathematically, self-similarity is obeyed at all length scales, in actual facts, it breaks down at small scale when one sees individual scatterers of dimension r_0 (e.g. single particles in an aggregation process) and for some large size ξ when the process stops. In a SAS experiment, fractal behavior will be observed for $\xi^{-1} < q < r_0^{-1}$. The exponential scattering law should be obeyed for data extending over at least an order of magnitude in q before a fractal dimension can be deduced. Exact expressions for the scattering by fractal media can be found in numerous review papers, for instance Schmidt (1991); Teixeira (1988); Martin and Hurd (1987).

An object can also be characterized by the fractal character of its surface where self-similarity is observed. Examples are coals, lignites, and porous rocks. For fractal surfaces, as more generally for rough surfaces, the area depends on the scale of observation i.e. on q^{-1}. Remarkably, an analog of the Porod law exists; if $\mathscr{D}_S$ is the fractal surface dimension ($2 \leq \mathscr{D}_S \leq 3$) the scattering law becomes (Bale and Schmidt, 1984)

$$I(q) \propto q^{-(6-\mathscr{D}_S)} \qquad (4\text{-}27)$$

Thus the exponent of the power law is between 3 and 4; experimentally a precise determination is sometimes difficult. A complete description of the scattering by a rough surface can be found in Sinha et al. (1988).

In summary let us note that at intermediate q-values the decay of the scattered intensity, proportional to q^x, is related to the dimensionality of the structure: linear structures yield $x = -1$, platelets $x = -2$, dense structures with smooth interface $x = -4$; an intermediate exponent between 1 and 3 is obtained with mass fractals whereas surface fractals exponents are between 3 and 4.

4.3.4 Polymer Chain as a Diffuse Object

SAS is also a powerful technique to analyse the spatial correlations in diffuse objects such as a polymer chain or a macromolecule in solution. A polymer consists of a large number ($N \approx 10^5$ to 10^6) of elementary molecules or monomers covalently linked and immersed in a solvent (Fig. 4-4). Each monomer can be considered as an elementary scattering unit. The configuration of an ideal chain is that of a random walk, but the chain is continuously mobile so that a scattering measurement "sees" the average of all possible configurations. The average radius of gyration is $R_G^2 = N a^2/6$ if a is the length of a monomer unit (Flory, 1969). It must be stressed that the volume that comprises the macromolecule, of the order of $N^{3/2} a^3$, contains more solvent molecules than monomers (total volume $N a^3$). Polymers are diffuse, flimsy objects with no sharp interface, clearly different from the particles we have been dealing with up to now. Nevertheless, valuable structural features can be obtained by X-ray and neutron scattering and also, on a larger length scale, by light scattering.

The scattered intensity by a polymer chain has been calculated by Debye (1915) as

$$I_s(q) = I_0 \frac{1}{N^2} \sum_i \sum_j \left\langle \frac{\sin(q r_{ij})}{q r_{ij}} \right\rangle = I_0 \frac{2(e^{-x} + x - 1)}{x^2} \quad (4\text{-}28)$$

with $x = q^2 R_G^2$, and r_{ij}, the distance between monomers i and j. I_0 is a function of the molecular weight of the polymer (see Sec. 4.5.4.1).

For $q R_G \ll 1$, one finds Guinier formula. An alternative limited development of $I(q)$, known as the Zimm approximation (Zimm, 1948), yields

$$\frac{c}{I(q)} \approx \frac{1}{I_0}\left(1 - \frac{q^2 R_G^2}{3}\right) \quad (4\text{-}29)$$

The data are then plotted as $c I(q)^{-1}$ vs q^2, where c is the solute (polymer) concentration. Here again, SAS has been useful in evaluating the monomer-monomer correlations of macromolecules (des Cloizeaux and Jannink, 1990) and in understanding the statistical physics of polymers (de Gennes, 1979).

4.3.5 Interacting Particles

If the sample is made of interacting scattering entities, the observed intensity reflects both their geometry and the interactions between the various partners.

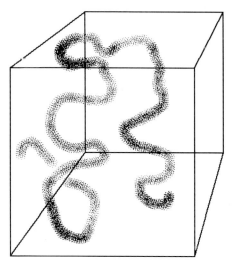

Figure 4-4. Schematic representation of a single polymer molecule in a solvent; the smooth curve is made of a large number of chained monomers, each acting as an elementary scattering unit.

For N identical, randomly oriented and *centrosymmetric* particles, the scattered intensity can be written as:

$$I(q) = N P(q) S(q) \qquad (4\text{-}30)$$

where $P(q)$ is the form factor of the particles which contains information on size, shape and internal structure and $S(q)$ the structure factor of the assembly. For uncorrelated particles, $S(q) \equiv 1$, and $I(q)$ simply describes isolated particles. For correlated particles, $S(q)$ characterizes the type of order and contains all informations about the interactions in the system.

The separation of these two contributions is a formidable problem of SAS, seldom solved completely. To illustrate the complexity of the problem, let us examine separately the resulting scattering curve and the form and structure factors related to a system of interacting micelles (Fig. 4-5): it is not possible to derive directly characteristic distances related to the particles (e.g. Guinier range no longer visible) or to their organisation [e.g. a peak is visible in $I(q)$ but is displaced with respect to its position in $S(q)$].

At this point, it is difficult to prescribe an analysis in all generality as the solution to this problem has to be adapted to the type of sample. Let us describe some methods.

1. Dilution method: If dilution is at all possible and if the particles remain identical upon dilution, extrapolation to zero concentration yields $P(q)$. At finite concentrations, $I(q)$ can be divided by $P(q)$ to yield $S(q)$.

2. Labeling method: If one can label only a point in each particle, the contrast of the rest being negligible, $S(q)$ will dominate the scattering.

3. Calculation of the scattering profile in a model where one assumes the shape of the particles and their interaction potential. For instance hard spheres with a Percus–Yevick pair potential could account for the scattering due to precipitates in some metallic alloys. The method of Hayter and Penfold (1981), originally devised for surfactant micelles is based on an analytical calculation of $S(q)$ in the mean spherical approximation and a choice of $P(q)$, including some parameters. $I(q)$ is calculated and compared to the experimental curves. The parameters are corrected until the solution converges. The form and structure factors shown in Fig. 4-5 were obtained by this method.

4.4 Contrast Variation

Most disciplines such as materials science or biology usually deal with complex multicomponent systems where a simple two-media description of the structure is seldom adequate. In all these systems, the scattered intensity at small angles contains the sum of the contributions of all components and thus yields information on the global structure only. If one wants to push further the structural description and for instance, investigate the spatial distribution of one definite species, it is necessary to "colour" it in a special way so that its contribution is distinguishable from that of the rest of the sample. The techniques for varying at will the colour of one compo-

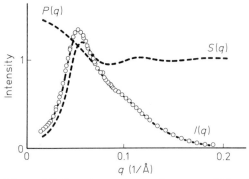

Figure 4-5. Experimental scattering curve $I(q)$ resulting from the product of the form factor $P(q)$ and the structure factor $S(q)$ for a system of interacting charged micelles.

nent without changing otherwise any of the characteristics of the sample are known as *contrast variation techniques*. They rely essentially on the use of either isotopic substitution in neutron scattering or anomalous dispersion effects in X-ray scattering.

4.4.1 Anomalous Scattering of X-Rays

X-rays interact with the electrons of an atom and substitution by another atom which would provide a change in the number of electrons is not possible in most cases without changing drastically the properties of the system. However, one can take advantage of the fact that for X-rays whose energy is close to the absorption edge of an element, the atomic scattering factor of that element is reduced by a few electrons from its value far from the edge, so that a variation of contrast is obtained without intervening on the sample by simply varying the energy of the incident X-ray photon. Although the method was known for quite a while and had been first tested in the laboratory, its development has been triggered recently by the availability of synchrotron radiation sources which provide monochromatic photons of high intensity over a broad energy range.

4.4.1.1 The Underlying Physics

Anomalous scattering occurs whenever the wavelength of the incident radiation is close to an allowed transition of the scatterer whether atomic or nuclear. For X-ray photons of energy E near the inner shell absorption edges of an atom, the coherent atomic scattering factor is the sum of three terms (James, 1965)

$$f(q, E) = f_0(E) + f'(q, E) + i f''(q, E) \quad (4\text{-}31)$$

The first term is the Fourier transform of the electron probability density $\varrho(r)$, proportional to the number of electrons in the atom and equal to Z in the small angle range. It is independent of energy and is the only term taken into account in standard X-ray scattering experiments.

The second and third terms are energy dependent and are known as the *anomalous scattering factors*; they have an appreciable amplitude only near each atom absorption edge and can be neglected in standard experiments.

In the small angle scattering range ($q a_B \ll 1$ with a_B, Bohr's radius), the q dependence of f can be neglected and Eq. (4-31) reduces to

$$f(E) = Z + f'(E) + i f''(E) \quad (4\text{-}32)$$

The imaginary part f'' is proportional to the absorption cross-section σ_a, and the real part f' is related to it through the Kramers–Kronig dispersion relation

$$f'(\omega) = (2/\pi) \int_0^\infty \frac{\omega' f''(\omega')}{(\omega'^2 - \omega^2)} \quad (4\text{-}33)$$

where ω is the X-ray frequency.

Thus f' is negative for photon energies in the immediate vicinity of the absorption edge. The values of f' and f'' have been tabulated by Sasaki (1983) and typical variations are shown in Fig. 4-6. The largest changes occur within 10 eV from the edge and amount to a few electrons for a K edge and more for the L and M edges (Wendin, 1980; Cromer and Libermann, 1970; Cromer and Libermann, 1981).

Since each element has its own characteristic edges, varying the energy of the photon near the K, L or M absorption edge provides a selective way of changing the contribution of that atom to the total intensity scattered by the sample. This effect is the basis of the contrast variation technique of X-ray anomalous scattering.

Although a detailed description of the atomic processes occurring when an X-ray photon interacts with an atom is outside

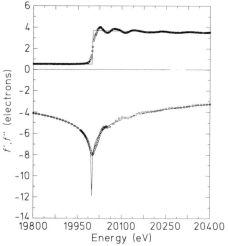

Figure 4-6. Typical energy dependence of the anomalous scattering factors f' and f'' near the K absorption edge of an element. Shown here is the molybdenum K edge at 20 keV. The full line corresponds to atomic Mo and the dots to Mo in $NiMoS_2$ (courtesy of J.-M. Tonnerre).

the scope of this paper, we give here an outline of some important aspects that will be necessary to define the experimental conditions and understand the origin of possible errors.

First let us note that an insight into the origin of anomalous scattering can be gained if one considers a single classical electron as a damped harmonic oscillator in the field of the incident photon. Resonance occurs at the frequency of the oscillator and the amplitude of the radiation emitted by the oscillator contains corrective terms which can be written as Eq. (4-32). Hence the term *resonant scattering* is more descriptive than *anomalous scattering*.

f' and f'' can be associated with scattering via absorption and reemission of photons, resulting in a phase shift. Amongst the emission processes taking place at the edge, fluorescence and Raman inelastic scattering are of importance since they produce parasitic background which has to be corrected for or avoided by a proper choice of experimental conditions.

Since the atom is usually surrounded by others, the absorption edge may be displaced from its free atom value and its detailed features may be altered by different local environments. Hence it may be necessary to determine experimentally the exact absorption profile and deduce the precise values of f' for the investigated sample by a Kramers–Kronig transformation.

4.4.1.2 Contrast Variation by Anomalous Scattering

The anomalous effects briefly described so far apply generally to all scattering techniques. We will now limit ourselves to anomalous small-angle X-ray scattering (ASAXS).

To gain insight into the possibilities of this contrast variation technique, let us consider a hypothetical two-media system described by an electron density profile as pictured in Fig. 4-7. The scattered intensity is proportional to $(\Delta\varrho)^2$, the square of the contrast between, say, matrix and aggregates. Suppose now that an element M is distributed unevenly between the two phases. When the energy of the incident

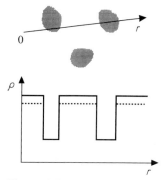

Figure 4-7. Electron density profile along a line drawn across a two-media sample, at photon energies far from (——) and close to (- - -) the edge of an element present exclusively in the matrix. In this example, the electron density remains unchanged in the particles, whereas it decreases in the matrix, so that the anomalous effects reduce the contrast.

photon approaches that of an absorption edge of M, the effective electron density of each phase is going to decrease by an amount proportional to the concentration of that element in the phase and to the square of the atomic scattering factor at that energy. Depending on the distribution of M between the two phases, the total scattered intensity can increase or decrease. Comparison of the intensity variation to models for the suspected structure provides information on the distribution of M.

Let us illustrate this with a problem in metallurgy; a CuTi alloy can be electrolytically hydrogenated up to 40 at.% with the appearance of a broad peak in the SAXS pattern (Goudeau et al., 1987). In order to ascertain the origin of the peak and deduce the location of hydrogen, experiments were conducted near both the Cu and Ti K edges. It was found that the peak intensity *decreased* by 53% at the Cu edge (8980 eV) and *increased* by 28% at the Ti edge (4964 eV) with no change in the shape of the profile. These findings immediately rule out the possibility that the scattering entities are H_2 bubbles. Indeed, if so, both types of anomalous atoms would be in the matrix and its effective average electron density would decrease with no change in that of the bubbles. Thus the peak intensity would decrease at both edges. If, on the other hand, it is assumed that both atoms may be distributed everywhere and that the electron density of the scattering entities is lower than that of the matrix, the increased contrast with decreasing contribution of Ti indicates that this element is predominantly in these entities. Quantitative comparison with possible models showed that hydrogen is contained exclusively in TiH_2 clusters.

Such straightforward applications are useful in multiphase materials when little is known of the entities giving rise to a SAXS signal; more complex applications pertaining to the extraction of partial structure factors will be found in Sec. 4.5 and in specialized articles in metallurgy (Epperson and Thiyagarajan, 1988; Simon and Lyon, 1989) and biology (Stuhrmann, 1989; Phillips and Hodgson, 1980).

4.4.1.3 Design of an ASAXS Experiment

Let us list now a few important points that one must keep in mind when designing an ASAXS experiment:

(i) Each element is characterized by different threshold energies corresponding to the K, L, M edges. On the high energy side, the energy range available for X-ray scattering is limited by the characteristics of the synchrotron radiation source (e.g. high energy machines, wigglers, etc.) and by the performance of position sensitive detectors; 15 keV is probably a practical limit for the moment. On the low energy side, the high attenuation of the beam by absorption in the sample and in all elements in the flight path gives a practical limit of about 4000 eV for all standard experiments. The easily accessible elements have Z values between 20 and 36 for the K-edges and between 50 and 82 for the L-edges. Consequently, the most frequent atoms encountered in polymer science or biology (C, O, H, and N) cannot be used as anomalous scatterers. However, by careful design of a SAXS spectrometer, Stuhrmann et al. (1989) have been able to reach the sulfur edge at 2470 eV. This involves the use of a single crystal monochromator, an ultra-high vacuum flight path and very thin samples which renders the experiments very delicate.

(ii) Because the anomalous effects are largest in a narrow energy range near the absorption edge, a high-resolution monochromator ($\Delta E/E < 10^{-3}$) is a prerequisite to any ASAXS experiment. Recently most synchrotron radiation centres have developed specially designed spectrometers which should make the technique available to more non-specialists (Dubuisson et al., 1986; Haubold et al., 1989).

(iii) ASAXS experiments are usually performed at energies below the edge where the value of f'' is very low and hardly varies with energy. It avoids experimental complications above the edge, due to the fact that the sample absorption is large and that inelastic processes give rise to a strong isotropic background, which can even saturate the detector. (Note that, up to now, position-sensitive detectors with suitable resolution in position *and* energy do not exist.) Although the anomalous effects are more pronounced very close to the edge, the existence below the edge of resonant Raman scattering (Sparks, 1974) and of some fluorescence due to imperfect monochromation of the beam limits the practical energy range to 5 eV or more from the edge.

(iv) Quantitative analysis of the anomalous effects requires a precise knowledge of f' and f''. Their values have been listed for all elements and are usually sufficient for energies 10 eV or more from the edge. Very close to the edge, f' has to be derived from experimental values of the absorption of the same sample, using the Kramers–Kronig relation.

(v) The necessity of obtaining absolute intensities at different energies with a precision better than a few percent requires more than just the standard corrections. It involves evaluation of the detector efficiency at the different energies and precise subtraction of the inelastic background close to the edge, before absolute intensities can be obtained. It must also be emphasized that sample absorption is a crucial measurement for these experiments.

4.4.2 Isotopic Substitution in SANS

While the X-ray atomic scattering length increases linearly with the atomic number of an element, the neutron coherent-scattering length (which gives rise to the interference scattering necessary for structural investigations) varies irregularly along the periodic table (see Bacon, 1975). Furthermore, because neutrons interact with the nucleus of an atom, isotopes of the same element can have considerably different amplitudes. On the other hand, different isotopes have (almost) the same chemical and physical properties. Selectively substituting a nucleus by one of its isotopes does not introduce a significant perturbation in the material, but can alter considerably its scattering intensity. Mixing the isotopes in different proportion allows one to vary the average scattering length at will. The most prominent example is that of the two stable isotopes of hydrogen, ^{1}H and ^{2}H (deuterium); the coherent scattering-length of ^{2}H is positive and of a value similar to that of most other elements in organic matter, whereas that of ^{1}H is negative, i.e. for ^{1}H there is a 180° phase shift of the scattered neutrons with respect to other nuclei. This large difference has been taken advantage of in the fields of colloid and polymer science (e.g. Wignall, 1987) and structural molecular biology (e.g. Timmins and Zaccai, 1988) and is at the origin of important scientific breakthroughs. In these fields two complementary aspects of the technique have been used, contrast variation of the matrix by changing its average scattering density and specific isotopic labeling of some parts of a sample.

In metallurgy, other isotopes are being used for similar purposes, e.g. the nickel isotope ^{62}Ni which also has a negative scattering length, or the silver isotopes ^{107}Ag and ^{109}Ag (see the review of Kostorz, 1988). A set of samples with variable isotope content have to be prepared. This can be a problem when the final state of the material depends on the sample history.

In what follows the emphasis will be on the physical chemistry of complex multicomponent systems rather than on metallurgical problems, only because the former offers many more opportunities for using subtle contrast variation techniques.

4.4.2.1 Solvent (Matrix) Scattering Density Variation

The most straightforward method of contrast variation with neutrons is the variation of the bulk scattering density. This technique was introduced into SANS by Ibel and Stuhrmann (1975) on the basis of previous results of X-ray crystallography (Bragg and Perutz, 1952), SAXS (Stuhrmann and Kirste, 1965), and light scattering (Benoit and Wippler, 1960). For scattering "particles" in solution, bulk contrast variation is obtained with mixtures of "light" and "heavy" solvents available in protonated and deuterated forms (H_2O, ethanol, cyclohexane, etc.). The scattering-length density of H_2O, for instance, varies between -0.562×10^{10} cm^{-2} for natural-abundance water, which is nearly pure ^1H_2O, and 6.404×10^{10} cm^{-2} for pure heavy water.

The scattering-length densities of different kinds of material are different from each other and from pure protonated and deuterated solvents and can be matched by $^1H/(^1H + ^2H)$ mixture ratios characteristic for their chemical compositions. This mixture ratio (or the corresponding absolute scattering-length density) is called the scattering-length-density match-point, or, semantically incorrect, contrast match-point. However, if a molecule contains non-covalently bound hydrogens, these may be exchanged for solvent hydrogens. This exchange is proportional to the ratio of all labile ^1H and ^2H present; in dilute aqueous solutions, it is dominated by the solvent hydrogens. A plot of the scattering-length density of a particle *versus* the $^1H/(^1H + ^2H)$ ratio in the solvent shows a linear increase if there is exchange; hence the value of the match point depends on solvent exchange as well. The fact that many particles have a high contrast with respect to ^2H_2O or other deuterated solvents makes neutrons superior to X-rays for studying small particles at low concentrations.

The scattering-length density $\varrho(r)$ can be written as a sum

$$\varrho(r) = \varrho_0 + \varrho_F(r) \tag{4-34}$$

where ϱ_0 is the average scattering-length density of the particle at zero contrast, $\Delta\varrho = 0$, and $\varrho_F(r)$ describes the fluctuations about this mean. $I(q)$ can then be written

$$I(q) = (\varrho_0 - \varrho_s)^2 I_c(q) + \\ + (\varrho_0 - \varrho_s) I_{cs}(q) + I_s(q) \tag{4-35}$$

$I_s(q)$ is the scattering intensity due to the fluctuations at zero contrast. The cross-term $I_{cs}(q)$ also takes into account all solvent-exchange phenomena in the most general sense (including solvent water molecules bound to the particle surface, which can have a density different from that of bulk water). This expression is mathematically correct, since one can assume that solvent exchange is proportional to $\Delta\varrho$. The term $I_c(q)$ is due to that part of the volume where the scattering density is independent of the solvent (Luzzati et al., 1976; see also Witz, 1983).

The method is very valuable, since it allows calculation of the scattering at any given contrast on the basis of at least three measurements at well chosen $^1H/(^1H + ^2H)$ ratios (including data near, but preferentially not exactly at the minimal contrast).

It is sometimes limited by ^{2}H-dependent aggregation effects.

4.4.2.2 Specific Isotopic Labeling

Especially in the polymer field, specific isotope labeling is a method which has created unique applications of SANS. Again, it is mainly concerned with the exchange of ^{1}H by ^{2}H in the particles to be studied at hydrogen positions which are not affected by exchange with solvent atoms, e.g. carbon-bound hydrogen sites.

With this technique, single polymer chains can be studied amongst others which are identical except for their hydrogen isotope content. Even if there is some risk that the thermodynamics of the system might be slightly modified in some rare cases, there is no other method that could replace neutrons in this field (see Sec. 4.6.1).

Inverse contrast variation forms an intermediate between matrix scattering density variation and specific isotopic labeling. The contrast with respect to the solvent of a whole particle or of well-defined components of a particle, e.g. a macromolecular complex, is changed by varying its degree of deuteration keeping that of the solvent constant. Since solvent-exchange effects remain practically identical for all samples, the measurements can be more precise than in the "classical" contrast variation (Knoll et al., 1985).

4.4.2.3 Triple Isotopic Replacement

Biological molecules in particular often are complexes consisting of several unique components ("subunits") of identical or different chemical composition (proteins, proteins, and nucleic acids, lipids or sugars). Sometimes, these components can be separated and put back together to form the original complex by chemical or physical methods. Pavlov and Serdyuk (1987) have proposed an elegant way of determining the structure of a component inside a molecular complex. Their method is based on measuring the scattering curves from three preparations. Two or them contain the complex to be studied at two different levels of labeling, ϱ_1 and ϱ_2, and are mixed together to yield sample 1, the third one contains the complex at an intermediate level of labeling, ϱ_3 (sample 2). If the condition

$$\varrho_3(r) = (1-\delta) \cdot \varrho_1(r) + \delta \cdot \varrho_2(r) \qquad (4\text{-}36)$$

is satisfied by δ, the relative concentration of particle 2 in sample 1, then the difference between the scattering from the two samples only contains contributions from the single component. The concentrations of the three differently labeled complexes must be identical. Additionally, the contributions from contamination, aggregation and interparticle effects are suppressed provided that they are exactly the same in the three samples, i.e. also independent of the partial-deuteration states.

In the case of small complexes, δ can be obtained by measuring the scattering curves $I_1(q)$, $I_2(q)$ and $I_3(q)$ of the three particles as a function of contrast and by plotting the differences of the zero-angle scattering $I_1(0) - I_3(0)$ and $I_2(0) - I_3(0)$ versus δ. Both curves intercept at the correct ratio δ_0.

The triple-isotopic-replacement method is most easily applied using ^{1}H and ^{2}H as labels; it can be considered as a special case of systematic inverse contrast-variation of a selected component. When the scattering from the third sample is subtracted from that of the mixture of samples 1 and 2, the difference curve is exclusively due to the nuclei in the component, whereas the rest of the complex vanishes (at any contrast). This was shown for a complex of the ribosomal elongation factor EF-TU · GDP with tRNA by Pavlov et al. (1991). Mathematically, the difference curve is independent of the contrast of the rest of the complex with respect to the solvent. In practice, one should use low contrast, like in the case of triangulation (see Sec. 4.6.5).

In biological systems, an identical subunit composition of the differently labeled complexes can be assumed as guaranteed, but the method can also work with block copolymers, for example.

4.4.3 Magnetic Scattering of Neutrons

The neutron possesses a magnetic moment; it is sensitive to the orientation of spins in the sample (see e.g. Abragam et al., 1982). Especially in the absence of any other (isotopic) contrast, an inhomogeneous distribution of spins in the sample is detectable by neutron low-q scattering. The neutron spins need not be oriented themselves, although important contributions can be expected from measuring the difference between the scattering of neutron beams with opposite spin orientation. At present, several low-q instruments are being planned or even built including neutron polarization and polarization analysis.

Studies of magnetic SANS without (and rarely with) neutron polarization include dislocations in magnetic crystals and amorphous ferromagnets (see the review by Kostorz, 1988).

Janot and George (1985) have pointed out that it is important to apply contrast variation for suppressing surface-roughness scattering and/or volume scattering in order to isolate magnetic scattering contributions by matching the scattering-length density of the material with that of a mixture of heavy and light water or oil, etc.

4.4.3.1 Spin-Contrast Variation

As far as "nonmagnetic" matter is concerned, the magnetic properties of the neutron have been neglected for a long time. Spin-contrast variation, proposed by Stuhrmann (Knop et al., 1986; Stuhrmann et al., 1986), takes advantage of the different scattering lengths of the hydrogen atom in its spin-up and spin-down states. Normally, these two states are mixed, and the cross-section of unpolarized neutrons with the undirected spins gives rise to the usual value of the scattering amplitude of hydrogen. If, however, one is able to orient the spins of a given atom, and especially hydrogen, then the interaction of *polarized* neutrons with the two different oriented states offers an important contribution to the scattering amplitude:

$$A = b + 2B\boldsymbol{I} \cdot \boldsymbol{s} \qquad (4\text{-}37)$$

where b is the isotropic nuclear scattering amplitude, B is the spin-dependent scattering amplitude, s is the neutron spin, and I the nuclear spin. For hydrogen, $b = -0.374 \times 10^{-12}$ cm, and $B = 2.9 \times 10^{-12}$ cm.

The sample protons are polarized at very low temperatures (order of mK) and high magnetic fields (several Tesla) by dynamic nuclear polarization, i.e. by spin-spin coupling with the electron spins of a paramagnetic metallo-organic compound present in the sample which are polarized by a resonant microwave frequency. It is clear that the principles mentioned above also apply to other than biological and chemical material.

4.5 Experimental Techniques

4.5.1 Small-Angle Scattering of X-Rays

4.5.1.1 X-Ray Sources

SAXS experiments have been performed in laboratories for more than 50 years using sealed tubes or rotating anodes as sources of X-rays; the fixed wavelength is determined by the target material e.g. 1.54 Å for copper or 0.709 Å for molybdenum. As a rule, data are restricted to systems that scatter moderately well, where the inhomogeneities are not too large (<100 Å) and which can be described by time-averaged structural features. Recently, the availability of synchrotron-radiation (SR) sources with intense, brilliant beams of variable wavelength has opened new opportunities. Increase in the flux by several orders of magnitude has made possible the study of very dilute samples or weak scatterers; for average scatterers, the measuring time has been reduced in such a way that time-resolved measurements can be performed. Since one can afford to loose a reasonable amount of incident X-ray photons on the sample, better collimation is obtained, leading to the possibility of investigating larger structural features (>1000 Å). Another interesting aspect of SR sources is the broad energy spectrum of the photons. The precise wavelength of the incident beam, fixed by a monochromator, can be adapted to the characteristics of the sample (thickness, composition, desirable q-range). Furthermore the possibility of varying precisely that wavelength has provided a means of performing anomalous scattering experiments.

Although SR sources offer unique opportunities and pioneering SAXS experiments are still being developed, their availability is evidently limited; conventional experiments often provide the required answer to most problems. Because the characteristics of the source, whether it is a bending magnet, a wiggler or an undulator (Koch, 1983), impose the most stringent requirements on the optical elements, the following description of the experimental set-ups will be limited to those using SR X-ray sources.

4.5.1.2 SAXS Spectrometers

Any typical SAS set-up (Fig. 4-8a) includes elements that select the suitable wavelength, i.e. a monochromator/mirror assembly, elements that define the beam size and shape on the sample, i.e. slits and focusing elements, and a device to detect the scattered X-rays. There are many ways by which these functions can be performed, depending on the requirements of the actual SAXS experiment. A recent review (Russell, 1991) gives a detailed description of the different possibilities which will be only briefly summarized here.

The final set-up will depend mostly on whether the emphasis is on high flux or on high angular and energy resolution. For most applications where one tries to optimize the number of photons on the sample, a double focussing mirror/monochromator is used (Rosenbaum and Holmes, 1980). With this design, a cylindrically or elliptically bent mirror focuses the beam vertically; a typical mirror is made of quartz coated with gold or platinum, and also acts as a filter for the shorter wavelength X-rays contained in the white beam that are not reflected by the mirror. A bent perfect-crystal monochromator, such as a triangular shaped germanium single crystal, focuses the beam in the horizontal plane and selects photons of a fixed preselected wavelength. The monochromator can be placed before or after the mirror, depending on the cooling system chosen for the first elements. This geometry has been used both with rotating anodes in the laboratory, providing about 5×10^7 photons s^{-1} and for SR applications where a flux of 10^{11} photons s^{-1} can be obtained. The beam size is typically 1 mm^2, providing cameras with pin-hole collimation.

If more stringent requirements are put on resolution, double crystal monochro-

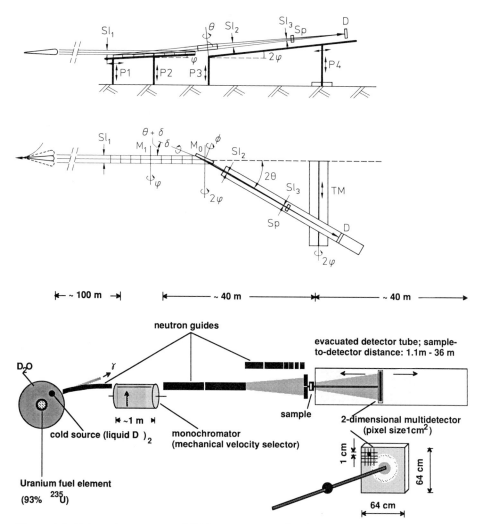

Figure 4-8. Schematic geometry of SAS spectrometers for the examples of (top) the mirror-monochromator camera X11 at DORIS, Hamburg (after Holmes, 1982) and (bottom) the instrument D11 of the Institut Laue–Langevin in Grenoble, France (after Lindner, 1992). Note the different length scales.

mators combined with doubly focussing mirrors are often used (Koch, 1988). With this arrangement, the wavelength can be varied with little change in the geometry of the following optical elements since the outgoing beam remains parallel to the incident one. Such set-ups are best suited to anomalous scattering experiments.

With these types of focussing optics, the spatial resolution is practically limited to sizes of the order of a few 10^3 Å. If very large spacings must be resolved, longer wavelengths could be used. This is not very practical for X-rays, since the transmission of the beam traversing the sample is a strong function of the wavelength. An alternative solution pioneered by Bonse and Hart (1966), is to use non-focussing optics. The monochromator is a multiple-bounce monolithic single crystal which provides a clean parallel beam on the sample. A second similar crystal is placed after the sample to analyse the scattered beam. The scattered intensity is measured by a single point detector (e.g. a scintillator) in a serial mode. Spatial resolution of the order of microns can be obtained. However, multiple reflections and lack of focussing elements reduce dramatically the photon throughput which needs to be compensated by the high brilliance of the source.

Recently, efforts have been devoted to the development of multilayers, i.e. synthetic compounds where the electron density varies periodically with depth. If the periodicity is appropriate, the multilayers reflect X-rays of a given wavelength. They can also be bent to provide a focused monochromatic beam. Higher flux is obtained at the expense of wavelength resolution. A $\Delta E/E \approx 0.01$ is still adequate for some SAXS experiments as will be illustrated in the section on applications.

The choice of a monochromator is dictated by the amount of the incident beam of a given wavelength with an acceptable dispersion (10^{-3} for most applications) that it delivers. It depends on the angular acceptance of the monochromator, the type of crystal used and its mosaic spread.

A critical element in a SAXS spectrometer is the detector. Several types are currently available, but none fulfils all the requirements of a SAXS experiment. In choosing a detector, several parameters have to be taken into account, namely the spatial and energy resolution, the linearity in response, the geometry (1D, 2D, annular, etc.), the background noise, the counting rate capability and the dynamic range. For experiments with SR, the last two points are the limiting factors for using very high flux, and efforts to develop new detectors are going on.

The most common detectors to-date are gas-filled ones, where the incident photon interacts with an ionizing gas and produces an electron-ion pair. A voltage applied results in a current that is proportional to the total number of photons and can be localized in space. The count rate limitation to a few hundred kHz is a major drawback of gas detectors. Vidicon cameras and photodiode arrays have a relatively high background noise. High energy resolution can be obtained with solid-state detectors, but their spatial resolution is inadequate for SAXS experiments. The recently developed imaging plates have a very high dynamic range and good spatial resolution, but the stored information has to be read out in a long process afterwards, similarly to the first detectors, photographic films.

4.5.2 Small-Angle Scattering at Neutron Scources

4.5.2.1 Neutron Sources

The neutrons produced by the fuel element of a reactor (or by a spallation source) are moderated by the heavy water surrounding the core. Normally, the neutrons leave the reactor with a thermal velocity distribution. So-called cold sources, small liquid-deuterium filled vessels in the reactor tank, decrease the average neutron velocity and thereby produce neutron wavelengths in the range of 4 to 20 Å ("cold" neutrons) which are more useful for SANS. They can be transported through neutron guides which also serve the purpose of changing the distance of the apparent source from the sample.

A narrow wavelength band is usually selected for SANS either by an artificial multilayer monochromator with a relatively broad bandpass ($\Delta\lambda/\lambda \approx 5$ to 10%) or – more frequently, due to the slow speed of cold neutrons – by a velocity selector. This is a rotating drum with a large number (about 100) of helical slots at its circumference. Only neutrons of the suitable velocity are able to pass through this drum. The wavelength resolution $\Delta\lambda/\lambda$ of velocity selectors is usually between 5% and 40% (full width at half maximum, FWHM); 10% and 20% are frequently used values.

On pulsed neutron sources, time-of-flight (TOF) SANS cameras have been developed (e.g. Hjelm, 1988). They use short bunches (about 100 μs long) of neutrons with a "white" wavelength spectrum produced by a pulsed high-energy proton beam impinging on a tungsten or uranium target with a repetition rate of the order of 10 ms, or by rendering an undercritical reactor critical for very short time fractions by a mechanically driven reflector device. The wavelength and, consequently, the q value, of a scattered neutron are determined by its flight time, assuming the scattering to be elastic. Especially in the high-q limit, the dynamic q-range of TOF SANS instruments is rather large, due to the large number of rapid neutrons in the pulse. The pulse-repetition rate of the source determines the low-q limit, because of frame overlap with the following pulse. The latter can be decreased, if necessary, with choppers turning in phase with the pulse production and selecting only every nth pulse. For reactor-based TOF SANS cameras, this disadvantage does not exist; the pulse-repetition rate can be optimally adapted to the chosen maximal and minimal wavelength. TOF SANS has a principal problem in the "upscattering" of cold neutrons, i.e. their gain in energy, by ^{1}H-rich samples: The background scattering may not reach the detector at the same time as the elastic signal, and may thus not be attributed to the correct q value (Hjelm, 1988; Ghosh and Rennie, 1990).

4.5.2.2 SANS Diffractometers

Low q need not necessarily mean small angles for typical neutron wavelengths: For an inhomogeneity of dimension D, the interesting q-range can be estimated as $1/D < q < 10/D$. The scattering angle 2θ corresponding to the upper q limit for $D = 100$ Å is 1.4° for Cu K$_\alpha$ radiation ($\lambda = 1.54$ Å), but amounts to 9.1° for neutrons of 10 Å wavelength. It is therefore preferable to speak of low-q rather than of small-angle neutron scattering.

As with SR, the cameras most frequently used for SANS are of the "pin-hole"-type; an example is the instrument D11 (Fig. 4-8b) at the Institut Max von Laue–Paul Langevin in Grenoble, France (Ibel, 1976; Lindner et al., 1992), from which some of the numbers below are quoted. For intensity reasons, the cross-section of the primary beam is usually chosen to be rather large (e.g. 3 cm × 5 cm). Consequently, pinhole instruments tend to be big: D11 with its overall length of 80 m is unique, but all classical SANS facilities are of an impressive size. At least one SANS camera exists at nearly every research reactor in the world.

The smallest q value that can be measured at a given distance is just outside the image of the direct beam on the detector (which either has to be attenuated or is hidden behind a beamstop, a neutron-absorbing plate of several 10 cm^2, e.g. a cadmium plate). Very small q's thus require long sample-to-detector distances. The area detector of D11 (ILL, Grenoble) has 64 × 64 resolution elements (pixels) of 1 cm^2. It moves within an evacuated tube of 1.6 m diameter and a length of 40 m. Thus, a q range of 5×10^{-4} to 0.5 Å^{-1} can be covered by different instrument settings.

The geometrical resolution is determined by the length of the free neutron flight-path in front of the sample, moving sections of neutron guide into or out of the beam ("collimation") and the cross-section of the guide or that of a beam-limiting aperture. In general, the collimation length is chosen roughly equal to the sample-to-detector distance. Thus the geometrical and wavelength contributions to the q resolution match at a certain distance of the scattered beam from the direct beam position in the detector plane. In order to resolve scattering patterns with very detailed features, e.g. of particles with high symmetry, a smaller beam divergence is sometimes required at the expense of intensity.

Double-crystal neutron diffractometers (corresponding to the X-ray instruments described above) are able to reach the very small q values of some 10^{-5} Å^{-1} typical of static light scattering. The high wavelength resolution leads to a very low neutron yield limiting the application of neutron Bonse-Hart cameras, in which slit geometry can be used, but not 2D-detectors.

Recently, an ellipsoidal-mirror SANS camera has been developed: The mirror, necessarily of very high surface quality (roughness of a few Å), focuses the divergent beam from a small (several mm) source onto a detector with a resolution of the order of 1×1 mm. Due to the more compact beam image, all other dimensions of the SANS camera can be reduced drastically (Alefeld et al., 1989) to values close to those of synchrotron X-ray cameras. Whether or not there is a gain in intensity as compared to pinhole geometry is strongly determined by the maximal sample dimensions. Long mirror cameras (e.g. 20 m) are always superior to double-crystal instruments in this respect (Alefeld et al., 1989), and can also reach the light-scattering q domain.

4.5.2.3 Specific Properties of Neutrons Compared to X-Rays

Neutrons interact with the point-like nuclei, and they are used for low-q scattering essentially for the same reasons as for other neutron experiments. These reasons are:

1) neutrons are sensitive to the isotopic composition of the sample;
2) neutrons possess a magnetic moment and, therefore, can be used as a magnetic probe of the sample; and
3) because of their weak interaction with and consequent deep penetration into matter, neutrons allow us to investigate bulk properties of thick samples.
4) For similar reasons, strong materials transparent for neutrons are available as sample environment equipment (such as pressure cells or shear-flow apparatus).

The fact that the kinetic energies of thermal and cold neutrons are comparable to those of excitations in solids, which is the main reason for the use of neutrons for inelastic scattering, is not of importance for SANS, with the exception of time-of-flight SANS (see Sec. 4.5.2.1).

SANS offers an optimal choice of the wavelength: With sufficiently large wavelengths, for example, first-order Bragg scattering (and therefore the contribution of multiple Bragg scattering to small-angle scattering) can be suppressed: The Bragg condition written as $\lambda/d_{max} = (2 \sin \Theta)/n < 2$ cannot hold for $\lambda \geq 2 d_{max}$, where d_{max} is the largest interreticular distance in a crystalline sample. For the usually small scattering angles in SANS, even quite small λ will not produce first-order peaks.

Long wavelengths (≥ 10 Å) are easily obtained with neutrons by using cold sources. For X-rays, they require synchrotrons, with costs similar to that of neutrons. Unlike neutrons, however, X-ray scattering suffers from the strong absorption of X-rays by matter and its strong wavelength dependence ($\propto \lambda^3$).

Very low q-values are more easily obtained with long wavelengths than with very small angles, as it is necessary with X-rays, since the same q value can be observed further away from the direct beam. Objects of linear dimensions of several 1000 Å, e.g. opals, where spherical particles of amorphous silica form a closed-packed lattice with cell dimensions of up to several thousands of Å, can still be investigated easily with neutrons. X-ray double-crystal diffractometers (Bonse and Hart, 1966) may also reach very low q.

4.5.2.4 Incoherent Scattering of Neutrons

Incoherent scattering is produced by the interaction of neutrons with nuclei that are not in a fixed phase relation with that of other nuclei. It arises, for example, when not all molecules contain the same isotope of an element (isotopic incoherent scattering). The most important source of incoherent scattering in SANS, however, is the spin-incoherent scattering from protons. It is due to the fact that only protons and neutrons with identical spin direction can form an intermediate (unstable) compound nucleus (Bacon, 1975). The statistical probabilities of the parallel and antiparallel spin orientations, the similarity in size of the scattering lengths for spin-up and spin-down and their opposite sign result in an extremely large incoherent scattering cross section for ^{1}H, together with a coherent cross-section of normal magnitude (but negative sign). Incoherent scattering contributes a background which can be orders of magnitude more important than the coherent signal, especially at larger q. On the other hand, it can be used for the calibration of the incoming intensity and of the detector efficiency (see below).

4.5.3 Practical Aspects of Small-Angle Scattering

4.5.3.1 The Sample and its Environment

X-ray spectrometers with pin-hole collimation use typically a beam of 1 mm by 1 mm at the sample position. The optimum sample thickness is of the order of $1/\mu$, where μ is the sample linear absorption coefficient, which increases with the atomic number of the elements in the sample and with the X-ray wavelength; it varies from a few μm for metallic samples to 1 mm for solutions of polymers or colloids at a λ of 1.5 Å. Hence, the sample volume probed in a SAXS experiment is of the order of 1 mm^3 or less. For weak scatterers and low-intensity sources, it is often necessary to increase the probed volume by using a semi-linear collimation with a beam of, say, 1 mm by 10 mm. However, this geometry, most often used with conventional equipment, requires a collimation-desmearing of the data.

In order to partially compensate for the scarcity of neutrons, even at high-flux reactors, compared to X-ray sources, relatively large sample sizes are required. Typical beam dimensions at the sample are of the order of 25 to 400 mm^2. With a sample thickness which can reach up to 10 mm, the irradiated volume is considerably larger than with X-rays. There is often a conflict between optimal scattering conditions and the difficulty of providing large samples for reasons of cost and/or availability of large quantities, e.g. of labeled compounds.

Measuring times depend, of course, on the sample size and on the properties of the system to be studied (particle dimensions, contrast, concentration), on the type of radiation and, not least, on the flux available at a given instrument. For neutron scattering, a typical measuring time for one sample at D11 (ILL, Grenoble) is of the order of 20 min for optimal conditions, but many hours may be needed for highly diluted (≤ 1 mg/ml) solutions; in the case of critical scattering, a few minutes can be plenty. The total time allocation one may expect to get for a well written, highly rated proposal is counted in days. Therefore, the success of an experiment depends crucially on a thorough preparation, and it is difficult to make up for mistakes or misfortune. Similar remarks apply for SR experiments, but the counting time rarely exceeds one hour. The emphasis is more on reaching short counting times, of the order of a millisecond, to perform time-resolved measurements on a system that is evolving with external perturbations.

In SAXS and SANS, the sample environment can be in air in the majority of cases, because the vacuum which is needed in the beam path to optimize intensity loss and to reduce parasitic scattering can be interrupted by windows made of weakly scattering material (quartz or sapphire, a few mm thick, for neutrons; mica, mylar or kapton sheets of 10–100 μm for X-rays). Similar materials are used for the windows of the solution cells.

Neutrons also permit the use of relatively thick (several mm) sample container walls, e.g. for high-pressure cells, cryostats, ovens, and Couette-type shearing apparatus (Lindner and Oberthür, 1985; Lindner and Oberthür, 1988 – see Sec. 4.6.3). New fields of low-q scattering, like dynamic studies of polymers in a shear gradient, time-resolved studies of samples under periodic stress or under high pressure have thereby become accessible.

Unlike with standard X-rays, samples can be relatively thick, and nevertheless be studied at low q values. This is particularly evident for metals, where X-rays are usually restricted to thin foils, but neutrons can easily accept 1–10 mm thick samples.

High energy X-rays (corresponding to shorter wavelengths) can overcome the limitations of thickness of the sample or of the container. Thus photons of 20–30 KeV are used currently at synchrotron facilities to study high pressure or shearing effects for instance. Another interesting feature of the newest synchrotron sources is the high brillance which allows the use of micrometer size beams to investigate very small monocrystals, or fibers and also to scan a larger heterogeneous sample.

4.5.3.2 Data Reduction and Absolute Scaling

Before any structural analysis may be performed, the raw data have to be corrected for all contributions to the scattered intensity that are not due to the sample itself. These include parasitic scattering, i.e. spurious scattering in the beam path and from the sample container, background noise due to the detector and its electronics, and the detector response as a function of the position and energy of the scattered photon or neutron.

Let us assume that for each detector element (pixel) i, measurements are made of the intensity with the sample in the beam, $I_S(i)$; of the intensity with the sample removed from the beam (or with an empty cell if the sample is a liquid), $I_P(i)$; of the intensity with the beam turned off, $I_E(i)$. I_S is the total scattering, I_P the parasitic scattering and I_E the electronic noise. All other experimental conditions are kept identical. If all measurements are normalised to the same counting time t, the corrected scattering intensity $I_c(i)$ is given by (4-38)

$$I_c(i) = \{I_S(i) - I_E(i)\} - T\{I_P(i) - I_E(i)\}$$

where T is the attenuation factor (transmission) of the sample. If all detector elements do not have the same sensitivity, further corrections have to be applied which are discussed in the next section. A final normalization to identical number of incident photons or neutrons is necessary before different intensity profiles may be compared. For that purpose, beam monitors are placed in the beam path before the sample. This normalization is especially important for synchrotron radiation, since the number of X-ray photons emitted by a storage ring decreases with time.

Conversion of detector pixel position to angle and q-value is easily performed if the beam wavelength, the sample-to-detector distance and the size of one pixel are known. Corrected scattering profiles are obtained as intensities as a function of q-values. Successive scattering profiles may be directly compared.

Determination of a particle molecular weight, measurement of scattering density differences, or comparison of scattering data to theoretical models require the knowledge of all scattered intensities on an absolute scale. In other words, the number of neutrons or photons incident on the sample has to be measured precisely. Numerous methods have been used over the years, better adapted to either X-rays or neutrons. Each one has its advantages and its problems.

Wignall and Bates (1987) compare different methods for the absolute calibration of SANS data. Methods implying incoherent scattering are most common. Vanadium is frequently used although its incoherent scattering cross-section is more than an order of magnitude lower than that of protons. Moreover, the Vanadium sample must be handled with precaution to avoid scratches that would contribute much to the scattering, and it has to be sealed air-tight to avoid hydrogen incorporation (Wignall and Bates, 1987). Water is also commonly used. Indeed, the coherent cross sections of the two protons and one oxygen of light water add up to a nearly vanishing coherent-scattering-length density, whereas the incoherent scattering cross-section of the water molecule is very high. The (quasi-) isotropic incoherent scattering from a thin, i.e. about 1 mm or less, sample of ^1H_2O, therefore, is an ideal means for determining the absolute intensity of the sample scattering (Jacrot, 1976; Stuhrmann et al., 1976), on the condition that the sample-to-detector distance L is not too large, i.e. up to about 10 m, depending on the neutron flux at the sample. Deviations from the isotropic behavior due to inelastic incoherent scattering contributions of ^1H_2O and the influence of the wavelength dependence of the detector response have to be corrected for (May et al., 1982).

A technique which is often used with X-rays is the calibrated-filter method (Luzzati, 1960): The incident beam is attenuated by a series of thin metal foils of known thickness and whose attenuation coefficient at the wavelength of the experiment has been determined independently. The transmitted intensity is measured after traversing an increasing number of such filters. The extrapolation to zero thickness of the natural logarithm of intensity provides a measurement of the intensity of the incident beam. The method is to be used only with monochromatic beams because attenuation would change any wavelength distribution.

Measurement of the angle-independent scattering by a gas or a liquid (water) has also been applied in SAXS. The intensity due to the thermal density fluctuations is directly related to the known isothermal compressibility at the temperature and pressure of the experiment. Unfortunately, this method requires long counting times.

It is also possible to calibrate the beam by measuring the scattering of a stable suspension of colloidal particles of known composition. Since the scattering density of both particles and solvent, and the concentration of the particles in solution are known, a determination of the scattering invariant Q (Eq. (4-14)) gives the intensity of the incident beam. The relatively short counting time required to get good quality data is a great asset of this method and other variants.

Previously calibrated secondary standards are also very useful. They are, in general, specimens showing a broad scattering of similar intensity to that of the sample being investigated in a particular experiment (so as to avoid detector corrections due to different count rates). The time stability of the secondary sample has to be checked. Slabs of semi-crystalline polymers or glassy ionomers are currently used (Kratky, 1963; Schaffer and Hendricks, 1974; Russell, 1983).

4.5.3.3 Detector-Response Correction

As mentioned previously, all elements of a position-sensitive detector may not have the same sensitivity. In order to correct for spatial inhomogeneities, the response of each pixel when illuminated by a constant number of photons or neutrons has to be determined. With the small detectors and shorter distances used in SAXS, a radioactive source may be placed at the sample position or far enough away to provide a constant illumination of the detector. A ^{55}Fe source, for instance, provides a radiation at 5.9 keV and can be conveniently used to calibrate the detector when the X-ray beam is not available. Alternatively, the sample can be replaced by a specimen containing elements that fluoresce strongly at the energy of the beam.

For specific experiments such as those using photons of different energies to enhance anomalous effects, the response of the detector as a function of energy has to be determined. For this purpose, the scattering of a sample which should be independent of energy is measured at all the relevant energies.

In SANS, samples that scatter neutrons with the same probability into all angles (at least in the forward direction), i.e. vanadium and thin water-filled cells, are most commonly used. If these samples have the same size and thickness as the sample being investigated, geometrical corrections are taken into account simultaneously. At sample-to-detector distances larger than about ten meters, the scattering from water (and even more that from vanadium) is no longer sufficiently strong for correcting sample scattering curves obtained with the same settings. Experience shows that it is possible in this case to use a calibration curve measured at a shorter sample-to-detector distance, providing the distance is not so short that the flat surface of the detector deviates too much from the spherical shape of the scattered waves.

It should also be recalled that all detectors have a limited dynamic range and are usually not capable of counting consecutive events if they are too close in time. This effect can be very important when comparing two samples with very different count rates and dead-time corrections may have to be applied.

4.5.3.4 Estimation of the Background Level

In SANS, it is necessary to subtract the level of incoherent scattering from the scattering curve which is initially a superposition of the (desired) coherent sample scattering, electronic and neutron background noise, and incoherent scattering. The latter contribution may even be dominant if the sample contains, for instance, a large number of protons.

A frequently used technique is the subtraction of a reference sample that has the same level of incoherent scattering, but lacks the coherent scattering from the inhomogeneities under study. This is particularly simple in the case of solutions when mixtures of solvent can be prepared.

For dilute aqueous solutions, there is a procedure using the sample and reference transmissions for estimating the incoherent background level (May et al., 1982): The incoherent scattering level from the sample, $I_{i,s}$ can be estimated as

$$I_{i,s} = \frac{I_{H_2O} f_\lambda (1 - T_s)}{1 - T_{H_2O}} \quad (4\text{-}39)$$

where I_{H_2O} is the scattering from a water sample, T_{H_2O} its transmission, and T_s that of the sample. f_λ is a factor depending on the wavelength, the detector sensitivity, the solvent composition, and the sample thickness; it can be determined experimentally by plotting $I_{i,s}/I_{H_2O}$ versus $(1 - T_s)/(1 - T_{H_2O})$ for a number of partially deuterated solvent mixtures.

This procedure is justified because of the overwhelming contribution of the incoherent scattering of ^1H to the macroscopic scattering cross section of the solution, and therefore to its transmission. The procedure should also be valid for organic solvents. The precision of the estimation is limited by the precision of the transmission measurement, the relative error of which can hardly be much better than about 0.005 for reasonable measuring times and currently available equipment, and by the (usually small) contribution of the coherent cross-section to the total cross-section of the solution. A modified version of Eq. (4-39) can be used if a solvent with a transmission close to that of the sample has been measured, but the factor f_λ should not be omitted.

An equation similar to Eq. (4-39) holds for systems with a larger volume occupation ϕ of particles in a (protonated) solvent with a scattering level I_{inc} in a cell with identical pathway (without the particles):

$$I_{i,s} = I_{inc} \frac{1 - T_{inc}^{1-\phi}}{1 - T_{inc}} \quad (4\text{-}40)$$

In this approximation, the particles' cross-section contribution is assumed to be zero, i.e. the particles are considered as bubbles.

In the case of dilute systems of monodisperse particles, the residual background (after initial corrections) can be quite well estimated from the zero distance value of the distance-distribution function calculated by indirect Fourier transformation (Glatter, 1979).

Similar problems arise in SAXS in the (rare) cases when there is some fluorescence in the sample that cannot be avoided by a change of the photon energy. The correction can be particularly crucial with ASAXS and depends very much on the type of sample being studied. For instance, if the scattering profiles have been analysed in the larger angle range in the absence of fluorescence, i.e. at a different energy, it is possible to use the known q-dependence of the elastic signal (e.g. $I \propto q^{-4}$) to determine the superimposed background.

4.5.3.5 Correction of Instrumental Effects

All theoretical developments of the first chapters assume that the source is point-like and that the detector elements are infinitely small. This is evidently not the case in actual experiments, and the finite sizes of source and detecting elements result in a broadening of all scattering features. Classical SAXS experiments use a source that can either be approximated by an infinitely long slit or by a narrow rectangular beam.

All data have to be "desmeared"; the various procedures are discussed in detail by Glatter (1982). Typical experiments with synchrotron sources use small quasi-circular beams; the source can be approximated as point-like, and in most cases no corrections are necessary, except at very small angles where the same desmearing methods can be applied. The finite size and shape of the detector elements act in a similar way and their effect can also be corrected for.

Resolution errors affect SANS data in the same way as SAXS data. Since SANS cameras usually work with pin-hole collimation, the influence of the effective source dimensions on the scattering pattern is small; even less important is, in general, the pixel size of 2D-detectors. A correction specific to neutrons is the effect of gravity, which may affect the data, but only in rare cases (Boothroyd, 1989). The preponderant contribution to the resolution of the neutron-scattering pattern, especially at larger angles, is the wavelength distribution. The situation is more complicated for TOF-SANS (Hjelm, 1988).

As has been shown in an analytical treatment of the resolution function by Skov Pedersen et al. (1990), wavelength resolution broadening has a negligible influence on the results of the data analysis for scattering patterns with a smooth intensity variation. However, the wavelength distribution has to be considered for scattering patterns with distinct features like those resulting from spherical latex particles (Wignall et al., 1988) or from viruses (Cusack, 1984). The same is true if the particle-size distribution in the sample is to be determined; size distribution and wavelength distribution in the beam have similar effects on the data.

Since measured scattering curves contain errors and have to be smoothed before they can be desmeared, iterative indirect methods are, in general, superior: A guessed solution of the scattering curve is convoluted with known smearing parameters and iteratively fitted to the data by a least-squares procedure. The guessed solution can be simply a parametrized scattering curve, without knowledge of the sample (Schelten and Hossfeld, 1971), but it is of more interest to fit the smeared Fourier transform of the distance-distribution function (Glatter, 1979) or the radial density distribution (e.g. Cusack et al., 1981) of a real-space model to the data.

4.5.4 Data Analysis and Interpretation
4.5.4.1 Particle Mass Determination

In the case of a dilute solution of identical particles, the relative molecular mass M_r of a particle can be determined from the intensity extrapolated to zero angle, $I(0)$ (Jacrot and Zaccai, 1981).

A thermodynamic approach to the particle size problem, in view of the complementarity of different methods, has been given by Zaccai et al. (1986), on the basis of the theory by Eisenberg (1981). It permits the determination of the molecular mass, of the hydration and of the amount of bound salts.

4.5.4.2 Real-Space Considerations

The scattering from a large number of randomly oriented particles at *infinite dilution* (and, as a first approximation, that of particles at sufficiently high dilution) is completely determined by a function $p(r)$ in real space, the distance distribution function. It describes the probability p of finding a given distance r between any two volume elements within the particle, weighted by the scattering densities of these volume elements.

Theoretically, $p(r)$ can be obtained by an infinite sine Fourier transform of the isolated-particle scattering curve

$$I(q) = \int_0^\infty p(r) \frac{\sin(qr)}{qr} dr \qquad (4\text{-}41)$$

$p(r)$ relates to $\gamma(r)$ of Eq. (4-12) as

$$p(r) = 4\pi r^2 V \gamma(r) \qquad (4\text{-}42)$$

In practice, the scattering curve can neither be measured to $q = 0$ (but an extrapolation is possible to this limit), nor to $q \to \infty$. "Indirect" iterative methods have been developed that fit the experimental scattering curve by the *finite* Fourier transform

$$I_{\text{IFT}}(q) = \int_0^{D_{\max}} p(r) \frac{\sin(qr)}{qr} dr \qquad (4\text{-}43)$$

of a $p(r)$ function described by a limited number of parameters between $r = 0$ and a maximal chord length $D_{\max}$ within the particle.

This procedure was termed "indirect Fourier transformation (IFT)" method by Glatter (1979) who uses equidistant B-splines in real space that are correlated by a Lagrange parameter, thus reducing the number of independent parameters to be fitted. Errors in determining a residual flat background only affect the innermost spline at $r = 0$; the intensity at $q = 0$ and the radius of gyration are not influenced by a (small) flat background.

Another IFT method by Moore (1980) uses an orthogonal set of sine functions in real space. This procedure is more sensitive to the correct choice of $D_{\max}$ and to a residual background that might be present in the data.

Rather than deconvoluting the scattering intensities with respect to the wavelength distribution and to geometrical smearing, IFT calculates the ideal scattering curve by fitting the experimental data with the smeared theoretical scattering curve obtained from the real-space model. In fact, it is possible to convolute the scattering curves obtained from the single splines which are calculated only once at the beginning of the fit procedure. The convoluted constituent curves are then iteratively fitted to the experimental scattering curves.

Once the function $p(r)$ is determined, the zero-angle intensity and the radius of gyration can be calculated from its integral and its second moment, respectively.

4.5.4.3 Model Fitting

The experimental scattering curves can be compared to those of more or less elaborate theoretical models. This can be rather straightforward in the case of highly symmetric particles like icosahedral viruses that can be regarded as spherical at low resolution. The scattering curves of such viruses are easily reproduced by an assembly of spherical shells where different scattering-length densities have been assigned to the different shells (e.g. Cusack, 1984). Neutron contrast variation helps decisively to distinguish between the shells.

Fitting complicated models to the scattering curves is more critical because of the averaging effect of small-angle scattering. While it is correct and easy to show that the scattering curve produced by a model coincides with the measured curve, in general a unique model cannot be deduced from the scattering curve alone. Stuhrmann (1970) has presented a procedure using Lagrange polynomials to calculate low-resolution real-space models directly from the scattering information. It has been applied successfully to the scattering curves from ribosomes (Stuhrmann et al., 1976). These methods are well adapted to biological systems where well-defined particles exist in dilute solution.

4.6 Some Applications of Small-Angle Scattering

Nowadays, a single issue of any specialized scientific journal in polymer science contains four or five papers where the conclusions rely on SAS experiments; the situation is similar in metallurgy or in the physical chemistry of colloids and, to a lesser extent, in biology. An exhaustive review of the various applications is evidently outside the scope of this chapter that will rather focus on a few selected examples of recent developments that make use of sophisticated contrast variation techniques or benefit from the availability of more intense sources.

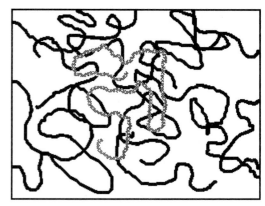

Figure 4-9. Entangled polymer chains in a melt or in a semi-dilute solution. One chain has been hatched differently to distinguish it from its neighbours. It interacts with a large number of other chains.

4.6.1 Chain Conformation of Polymers, a Success of Selective Deuteration

One of the most spectacular achievements of SANS allied to isotopic substitution is the description of a single polymer chain amongst similar ones in a melt (i.e. without solvent). The prediction by Flory (1949) that the long flexible polymer chains are entangled and that each chain interacts with many other chains (Fig. 4-9) was followed by almost 25 years of heated arguments due to the lack of a direct experimental method of verification. Indeed, an amorphous polymer with no solvent is characterized by uniform composition and uniform monomer density and hence it does not scatter radiation. However, if one were to differentiate some chains, for example by preparing a dilute solid solution of deuterated chains in hydrogenated ones, it would become possible to "see" an individual polymer coil. What appears now as an evident technique was first recognized in the seventies and led to three almost simultaneous publications based on experiments performed at the newly created Institut Laue–Langevin in Grenoble, France (Cotton et al., 1974a; Cotton et al., 1974b; Kirste et al., 1975; Wignall et al., 1974). The data (Fig. 4-10) show clearly that the distribution of monomers is described by gaussian statistics as predicted by Flory and that the overall dimensions of the coil is a universal function of its length, simply expressed as $R_G \propto N^{0.5}$, where R_G is the radius of gyration and N the number of monomers. These pioneering investigations have been followed by a rapid development of polymer physics with a rare cross-fertilisation between theory and experiments.

In the first experiments dealing with overlapping chains in the bulk or in concentrated solutions, the single-chain scattering signal was obtained by extrapolation to zero deuterium concentration of the experimental data of samples containing few deuterated chains (dilute solutions of labeled chains in an unlabeled matrix, the total monomer concentration being kept constant). This methodology had been developed earlier for the light scattering of dilute polymer solutions ("Zimm plots"). It

4.6 Some Applications of Small-Angle Scattering

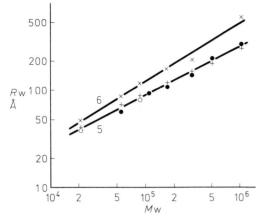

Figure 4-10. Radii of gyration of deuterated polystyrene chains amongst protonated ones as a function of molecular weight (●) in the bulk compared to those of single chains (×) in a good solvent and (+) in a θ solvent. The bulk data were obtained from an extrapolation to zero deuterium content of a concentration series. The slope of 0.5 in the double logarithmic plot obtained for bulk and θ solvent corresponds to a random walk, whereas the value of 0.6 for the good solvent data corresponds to a more open structure (after Cotton et al., 1974a).

was soon recognized that for a non-diluted polymer, the single chain scattering function is better obtained with a 50/50 H/D mixture (Boué et al., 1982; Daoud et al., 1975). Indeed, the scattered intensity per chain of N identical monomers can be written as

$$I(q) = \varrho^2 \{ N S_1(q) + N(N+1) S_2(q) \} \quad (4\text{-}44)$$

where the intramolecular scattering factor $S_1(q)$ reflects the correlations between monomers on the same chain, the intermolecular scattering factor $S_2(q)$ those between monomers on different chains. The fact that $I(q) \equiv 0$ at small angle for an incompressible polymer melt means that

$$S_2(q) = -\frac{S_1(q)}{N} \quad (4\text{-}45)$$

When a mole fraction x_1 of the chains is labeled, the scattered intensity simply reads

$$I(q) = x_1(1-x_1)(\varrho_1 - \varrho_2)^2 N S_1(q) \quad (4\text{-}46)$$

So the scattered intensity is maximum for $x_1 = x_2 = 0.5$, and it measures directly the single chain scattering function $S_1(q)$.

The method of selective deuteration relies on the assumption that there is no repulsive interaction between hydrogenated and deuterated monomers and that there is complete mixing. In actual fact, the interaction parameter χ is very small, but finite, and for certain polymers of high molecular weight it is possible to reach, even at room temperature, the critical point for phase separation of a symmetric binary mixture, given by $\chi N = 2$ (de Gennes, 1979). Indeed, the onset of phase separation has been observed and investigated in detail near the critical point for demixing of a binary mixture of polybutadiene (Bates et al., 1985), and it turned out that these are ideal samples to investigate the processes of phase separation in polymers. Fortunately, demixing occurs rarely, and high-concentration labeling can be used without artifacts for most moderate-molecular-weight polymer mixtures.

In semi-dilute solutions, when the chains overlap, high-concentration labeling has to be combined with an adequate choice of the solvent average contrast (also by mixing labeled and unlabeled species) to yield separately the conformation and the interactions of the chains (Williams et al., 1979; Akcasu et al., 1980). Monomer–monomer correlations within a chain can also be probed by selectively labeling some monomers of a chain and so synthesize a pseudo-copolymer (Duplessix et al., 1979).

These are only a few examples of the unique features of deuterium labeling for the understanding of the statistical physics of organic polymers. Many more applications can be found in Higgins and Benoit (1993); here the interested reader will appreciate the subtleties and intricacies of the technique.

4.6.2 Phase Separation in Binary Alloys: Towards a Shorter Time Resolution

Since the seminal investigations by Guinier (1939) and Preston (1938) in the late thirties, of the precipitation of coherent aggregates (GP zones) in an AlZn alloy, SAXS and SANS have been indispensable tools for analyzing the microstructure of metallic solid solutions after unmixing. Scattering techniques are most often used in conjunction with imaging techniques (Bowen and Hall, 1975) [transmission electron microscopy, field-ion microscopy (Cerezo and Smith, 1993; Müller and Tsong, 1969)] to characterize size, shape, dispersion, and composition of the aggregates of precipitated phases, which greatly control the macroscopic properties of materials.

To initiate phase separation, the alloy is homogenized at high temperature and subsequently brought into the two-phase region of the phase diagram, either by slow cooling (nucleation and growth) or by quenching (spinodal decomposition). Diffusion-controlled transformations take place as the system proceeds towards thermodynamic equilibrium, and the microstructure evolves as a function of both, time and temperature. Theoretical developments and numerical simulations have attempted to describe the unmixing kinetics of statistically isotropic binary mixtures at various stages of their evolution (Gunton and Droz, 1984). Scattering functions are often derived which can be directly compared to experimental data from metallic alloys, glasses, simple liquids, polymers, etc. Wagner and Kampmann (1991) give a detailed account of the progress made in the understanding of the kinetics of phase separation.

Experimentally, phase-separating alloys are characterized by a scattering halo whose position q_m shifts to smaller angles and whose amplitude increases with time, whatever the quenching conditions. Generally speaking, q_m is related to an average spacing between solute-rich clusters. Although the detailed description is quite complex, one can envisage two limiting cases: one is the nucleation, growth and coalescence of solute-rich clusters in the matrix, with some spatial periodicity; the other one is the appearance of periodic composition fluctuations that evolve in amplitude and spatial extension. The resulting structures are somewhat similar to the ones presented in Figs. 4-3a and 4-3b, respectively.

Ideally, the kinetics of the transformation should be analyzed from the onset of phase-separation (where properties specific of the particular system may be important) to the long aging times (where a "universal" behavior is more likely to exist). If the latter are easily obtained in the laboratory, the short-time evolution is difficult to catch, and continued experimental effort has been devoted to in-situ experiments with a time resolution better than, say, one second. These require an intense source (high-flux reactor, spallation source, or synchrotron radiation), a perfect control of the sample thermal treatment and a fast detecting system. First, let us note that the choice between X-rays or neutrons as a source depends very much on the type of alloy being investigated; factors such as scattering lengths of the atoms, existence of fluorescence or incoherent scattering, sample absorption, or of magnetic effects, etc. have to be considered.

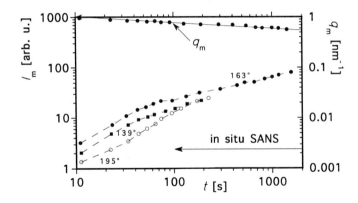

Figure 4-11. Time evolution of the position of the maximum of the small-angle X-ray scattering profile q_m and its intensity I_m for an Al12%Zn alloy aged in situ at temperatures given in the plot. The time range where neutron data could be obtained is also indicated (after Simon and Lyon, 1989).

Let us first illustrate recent experimental improvements in time resolution with the example of Al-rich AlZn alloys which can be investigated by both SAXS and SANS. When the temperature-concentration conditions are such that the time evolution is slow, the samples are quenched at different moments after phase-separation sets in and analyzed with standard SAXS laboratory equipment. Each sample then represents a different time slice. When the system evolves rapidly, heat treatment has to be performed *in situ* so that the concentration fluctuations can be detected as they appear. In an early experiment using synchrotron radiation (Naudon et al., 1979), a scattering spectrum was recorded every 100 seconds. With equipment easily available at most facilities, a time resolution of the order of 10 seconds is commonly reached. Such was the case for an Al12%Zn sample, isothermally aged at 139°C (Simon and Lyon, 1989), where the time evolution of the position and the intensity of the maximum in the scattering function shows two sequential regimes (Fig. 4-11). Most of the nucleation and growth takes place during the first minute, followed by a regime of slow coalescence. In a parallel SANS experiment, the first data could only be recorded after a few minutes, and the first regime could not be detected (Simon et al., 1984). However, these results, obtained with a 2-D detector on single crystals, give complementary information on the morphology changes and the anisotropy of the GP zones. Note that for long aging times, averaged SANS data (Komura et al., 1985) as well as SR SAXS data (Hoyt et al., 1989) on a series of AlZn samples of different compositions and aging temperatures have provided a good test for the theoretical models that predict spatial self-similarity and dynamic scaling.

The improvement in the brightness of SR sources and the parallel development of appropriate monochromators and fast 2-D detectors has permitted to reach even better time resolution. Recently, Osamura et al. (1988) have obtained new results, with a resolution of 200 ms, on the growth of Guinier-Preston zones for early stages where the effect of the elastic misfit energy for coherent precipitate should be felt.

Similarly, there have been many time-resolved SAXS and SANS studies of early- and late-stage phase separation behavior in silicate and borate glasses (see, for instance, Jantzen et al., 1981; Yokota and Nakajima, 1985; Craievich et al., 1986; Stephenson, 1984; Stephenson et al., 1991). Apart from their technological interest, these non-crystalline systems are believed to be good models to test the basic theory of phase-separation in liquids. Indeed, the relatively long-range interaction between molecules in the system corresponds better to the hypotheses of the model, and the slower diffusion coefficients make them more amenable to experimentation. Although late stage results agree well with the theory based on dynamic scaling, recent data with a short time resolution (Stephenson et al., 1991) have suggested that the kinetic theory had to be revised to include the effect of the stress due to the unequal mobilities of the diffusing species and of the relaxation of this stress through deformation.

4.6.3 Partial Structure Factors by Contrast Variation

Phase separation of ternary systems involves a supplementary degree of freedom which can lead to phenomena different from those observed in binary or pseudo-binary mixtures. A complete description of the material requires a knowledge of the pair-correlations between atoms of each species or of their Fourier transforms, the partial structure factors S_{ij}, where i j describe the type of atom. Quite generally, the scattered intensity can be written as

$$I(q) = \sum (c_i c_j)^{1/2} \Delta\varrho_i \Delta\varrho_j S_{ij}(q) \qquad (4\text{-}47)$$

where c_i is the concentration and ϱ_i the scattering density of each component. The S_{ij} can only be obtained by varying the contrast $\Delta\varrho_i \Delta\varrho_j$ (see Sec. 4.4). The choice between neutron scattering with isotopic substitution or anomalous X-ray scattering will depend very much on the system being investigated.

In the ternary alloy AlZnAg, for example, the Zn–Zn, Zn–Ag and Ag–Ag partial structure factors have been extracted from three independent neutron scattering experiments on three samples with different silver isotopes, and (hopefully) identical sample treatment (de Salva-Ghilarducci et al., 1983). Multi-wavelength X-ray experiments have also been performed on the same system, close to the zinc K-edge (Lyon and Simon, 1986) and have allowed a comparison of the two techniques. Briefly, one can say that SANS relies on the sample invariance upon isotopic substitution and thermal treatment; using the same sample is a definite advantage of ASAXS. However, with this technique, the accessible contrast variation, smaller than in SANS, can lead to large errors in the S_{ij}; as a rule, experiments at the edge of two elements are necessary to determine the partial structure factors (Fig. 4-12). Investigations of FeCrCo and CuNiFe have shown that a pseudo-binary description of ternary alloys may not always be adequate (see for instance Kostorz, 1991).

The concept of partial structure factors can be extended to more complex ternary systems such as charged micelles in solution, described by the micelle–micelle, micelle–counterion and counterion–counterion correlations. SANS is well suited to describe the colloidal particles and their interactions, but it is insensitive to the contribution of most atomic counterions due to their low scattering amplitude and their small size; their effect appears indirectly in the interaction potential used to describe the scattering profiles. If one wants to look at the counterion distribution, it is necessary to substitute a labeled bulky counterion such as the tetramethylammonium cation (with no guarantee of the invariance of the system) (Derian et al., 1988) or to use ASAXS to vary the contrast of suitable counterions such as bromine (Derian and Williams, 1988; Derian and Williams, 1990). Both techniques have given a de-

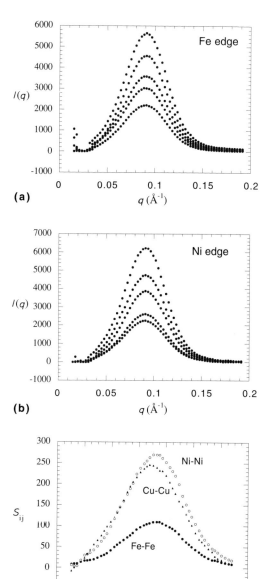

Figure 4-12. Experimental scattering profiles for a CuNiFe alloy, recorded at variable energies: (a) below the iron K-edge, between 6875 and 7104 eV, and (b) below the nickel K-edge, between 8199 and 8322 eV. For both sets of data, the intensity increases as the edge is approached. Figure (c) shows the partial structure factors corresponding to the Cu–Cu, Ni–Ni and Fe–Fe pair correlations that have been extracted from these data (O. Lyon, private communication; to be published). Intensities are in units of (electrons)^2/at. and structure factors, in (at.)$^{-1}$.

scription of the condensation of counterions near the surface of spherical micelles which could be compared to the available theoretical models. Similar reasoning applies to polyelectrolyte solutions (Jannink, 1986).

4.6.4 Samples with External Constraints: Liquids Under Shear

The availability of synchrotron X-ray and neutron sources for small-angle scattering boosted scattering experiments involving time resolution and/or external forces. Time-resolved studies are preferably performed with high-intensity synchrotron sources, while studies necessitating sophisticated (especially bulky) sample environment are more adapted to neutrons since these are much less absorbed by most materials. It is therefore possible to build sample containers that can withstand external forces such as pressure or shear. The recent possibility to use high-energy X-rays (Siddons et al., 1990) will undoubtedly lead to new experimental developments, since it alleviates the absorption problems. On the other hand, the use of very short wavelengths makes it more difficult to reach the very-low q region.

Typical examples for the use of neutrons can be found in investigations of polymer chains under shear, either in a Couette type shear cell or in a turbulent pipe flow apparatus (Lindner, 1991).

Couette cells for neutrons are made of a fixed and a rotating glass cylinder, leaving a gap between them which is filled by the liquid sample. The apparatus of Lindner (1991) has a fixed inner, laterally liquid-cooled piston (stator) with an outer diameter of either 46 or 47 mm and an outer rotating beaker (rotor) of 48 mm inner diameter, leaving a gap of 0.5 or 1 mm, respectively. The wall thickness of both cylinders is 2.5 mm. The neutron beam passes the sample twice. The corresponding smearing of the scattering pattern is negligible due to the large sample-to-detector distances (≈ 10 m) involved for studying long polymer chains with neutrons of about 10 Å. Sample volumes of about 4.5 ml and 7 ml are needed for gaps of 0.5 and 1 mm, respectively. The apparatus serves for studying liquid systems in a *laminar* shear flow. Shear gradients $\gamma = \partial v_z / \partial x$ up to $\gamma = 12\,000$ s^{-1} can be reached for solution viscosities of 10 mPas.

Laminar Couette flow leads to a combined rotation into the direction of the flow and to a deformation of the polymer molecules due to viscous drag (friction forces) in the streaming solution (Kuhn and Kuhn, 1943). These modifications in real space are reflected in a modified scattering behavior: the isotropic scattering pattern of the polymer solution at rest changes into an anisotropic one under shear.

4.6.5 Internal Structure of a Complex Particle by Label Triangulation

Biological macromolecules are often composed of several well-defined constituents. Ribosomes, the protein factories in all living cells, for example, consist of two subunits, each of which in turn is made of a multitude of proteins and one or two nucleic acid chains, each.

The spatial arrangement of the components of such a complex can be reconstructed if their coordinates are known; these can be obtained by measuring the distances between pairs of the components. SANS is an elegant way of determining these distances by a technique known as Label Triangulation (May, 1991). It consists in specifically labeling single components of a multi-component complex, measuring the small-angle scattering curves from (a) particles with two labeled components, (b) and (c) particles with either of the two components labeled, and (d) a (reference) particle which is not labeled at all. Subtracting the scattering from (b) + (c) from that of (a) + (d) yields information on the scattering originating exclusively from vectors combining volume elements in one component with volume elements in the other component.

The distances between the centres of mass of the components can be calculated from the scattering difference curve. A table of such distances yields the spatial arrangement of the components. In total, $n \cdot (n-1)/2$ distances exist between n components; in order to reconstruct their spatial arrangement, at least $4n - 10$ distance values must be known for $n \geq 4$:

Three distances define a basic triangle; three more yield a basic tetrahedron. Its handedness is arbitrary and has to be determined by independent means. At least four more distances are required to fix any further component in space; they define a new tetrahedron which must not be degenerated to a plane. Mathematically, the information contained in the fourth distance value is higher than the "one bit" information needed to decide the handedness of the additional tetrahedron, because every new component location requires only three coordinates.

Heavy-metal distances in organo-metallic compounds were first determined by X-ray scattering (Kratky and Worthmann, 1947). Hoppe (1972) proposed the application of this method to large biological complexes; Engelman and Moore (1972) first saw the advantage of neutrons. The need to mix preparations (a) plus (d) and (b) plus (c) for obtaining the desired scattering-difference curve in the case of high concentrations and/or inhomogeneous complexes (consisting of different classes of matter) has been shown by Hoppe (1973). It is preferable to measure the scattering curves at the scattering-density matching point of the reference particle for reducing undesired contributions.

Capel et al. (1987) presented the complete map of all protein positions within the small subunit from *E. coli* ribosomes based on this method. Fitting the Fourier transform of "moving splines" to the scattering curves provides an alternative approach for obtaining the distance information contained in the scattering curves from pairs of proteins (May and Nowotny, 1989). The naturally inhomogeneous ribosomal subunits were rendered homogeneous for neutrons by specific partial deuteration, and the reference particles were made "invisible" in a buffer solution containing nearly 100% ^2H_2O (Nierhaus et al., 1983). This procedure is being used for studying the protein positions in the large (50S) ribosomal subunit from E. coli (May et al., 1991).

4.7 References

Abragam, A., Bacchella, C. L., Coustham, J., Glättli, H., Fourmond, M., Malinowski, A., Meriel, P., Pinot, M., Roubeau, A. (1982), *J. Phys. 43 C7*, 373–381.
Akcasu, A. Z., Sommerfield, G. C., Jahshan, S. N., Han, C. C., Kim, C. Y., Yu, H. (1980), *J. Polym. Sci., Polym. Phys. Ed. 18*, 865.
Alefeld, B., Schwahn, D., Springer, T. (1989), *Nucl. Instr. and Meth. A274*, 210–216.
Auvray, L., Auroy, P. (1991), in: *Neutron, X-ray and Light Scattering: Introduction to an Investigative Tool for Colloidal and Polymeric Systems:* Lindner, P., Zemb, T. (Eds.). Amsterdam: North-Holland, pp. 199–221.
Bacon, G. E. (1975), *Neutron Diffraction.* Oxford: Clarendon Press.
Bale, H. D., Schmidt, P. (1984), *Phys. Rev. Lett. 53*, 596.

Bates, F. S., Wignall, G. D., Koehler, W. C. (1985), *Phys. Rev. Lett. 55*, 2425.
Benoit, H., Wippler, C. (1960), *J. Chim. Phys. 57*, 524–527.
Bonse, M., Hart, M. (1966), *Z. Phys. 189*, 151–162.
Boothroyd, A. T. (1989), *J. Appl. Crystallogr. 22*, 252–255.
Boué, F., Nierlich, M., Leibler, L. (1982), *Polymer 23*, 29–35.
Bowen, D. K., Hall, C. R. (1975), *Microscopy of Materials.* London: MacMillan Press.
Bragg, W. L., Perutz, M. F. (1952), *Acta Crystallogr. 5*, 277–283.
Capel, M. S., Engelman, D. M., Freeborn, B. R., Kjeldgaard, M., Langer, J. A., Ramakrishnan, V., Schindler, D. G., Schneider, D. K., Schoenborn, B. P., Sillers, I. Y., Yabuki, S., Moore, P. B. (1987), *Science 238*, 1403–1406.
Cerezo, A., Smith, G. D. W. (1993), in: *Characterization of Materials:* Lifshin, E. (Ed.). Weinheim, VCH, Vol. 2 B of *Materials Science and Technology – A Comprehensive Treatment:* Cahn, R. W., Haasen, P., Kramer, E. (Eds.), Chapter 18.
Cotton, J. P., Decker, D., Benoit, H., Farnoux, B., Higgins, J., Jannink, G., Ober, R., Picot, C., des Cloizeaux, J. (1974a), *Macromolecules 7*, 863–872.
Cotton, J. P., Decker, D., Farnoux, B., Jannink, G., Ober, R., Picot, C. (1974b), *Phys. Rev. Lett. 32*, 1170–1172.
Craievich, A. F., Sanchez, J. M., Williams, C. E. (1986), *Phys. Rev. B34*, 2762.
Cromer, D. T., Libermann, D. A. (1970), *J. Chem. Phys. 53*, 1891.
Cromer, D. T., Libermann, D. A. (1981), *Acta Crystallogr. A37*, 267.
Cusack, S. (1984), in: *Neutrons in Biology:* Schoenborn, B. P. (Ed.). New York: Plenum Press, pp. 173–188.
Cusack, S., Mellema, J. E., Krijgsman, P. C. J., Miller, A. (1981), *J. Mol. Biol. 145*, 525–543.
Daoud, M., Cotton, J. P., Farnoux, B., Jannink, G., Sarma, G., Benoit, H., Duplessix, R., Picot, C., de Gennes, P. G. (1975), *Macromolecules 8*, 804.
de Gennes, P. G. (1979), *Scaling Concepts in Polymer Physics.* Ithaca: Cornell University Press.
de Salva-Ghilarducci, A., Simon, J. P., Guyot, P., Ansara, I. (1983), *Acta Metall. 31*, 1705–1713.
Debye, P. (1915), *Ann. Phys. Leipz. 46*, 809–823.
Debye, P., Bueche, A. M. J. (1949), *J. Appl. Phys. 20*, 718–725.
Derian, P. J., Belloni, L., Drifford, M. (1988), *Europhys. Lett. 7*, 243.
Derian, P. J., Williams, C. E. (1988), in: *Ordering and Organization in Ionic Solutions:* Ise, N., Sogami, I. (Eds.). Singapore: World Scientific, pp. 233.
Derian, P. J., Williams, C. E. (1990), in: *Progress in X-Ray Synchrotron Radiation Research:* Balerna, A., Bernieri, E., Mobilio, S. (Eds.). Bologna: SIF, pp. 945.

des Cloizeaux, J., Jannink, G. (1990), Polymers in Solution: Their Modelling and Structure. Oxford: Clarendon Press; New York: Oxford University Press.
Dubuisson, J. M., Dauvergne, J. M., Depautex, C., Vachette, P., Williams, C. E. (1986), *Nucl. Instr. Meth. A246*, 636.
Duplessix, R., Cotton, J. P., Benoit, H., Picot, C. (1979), *Polymer 20*, 1181.
Eisenberg, H. (1981), *Quart. Rev. Biophys. 14*, 141–172.
Elsner, G., Riekel, C., Zachmann, H. G. (1985), *Adv. Polym. Sci. 67*, 1.
Engelman, D. M., Moore, P. B. (1972), *Proc. Natl. Acad. Sci. USA 69*, 1997–1999.
Epperson, J. E., Thiyagarajan, P. (1988), *J. Appl. Crystallogr. 21*, 652.
Flory, P. J. (1949), *J. Chem. Phys. 17*, 303.
Flory, P. J. (1969), *Statistical Mechanics of Chain Molecules*. New York: Interscience Publishers.
Ghosh, R. E., Rennie, A. R. (1990), in: *Neutron Scattering Data Analysis:* Johnson, M. W. (Ed.). Bristol: IOP Publishing, pp. 233–244.
Glatter, O. (1979), *J. Appl. Crystallogr. 12*, 166–175.
Glatter, O. (1982), in: *Small-angle X-ray scattering:* Glatter, O., Kratky, O. (Eds.). London: Academic Press, pp. 167–196.
Glatter, O. (1991), in: *Neutron, X-ray and Light Scattering: Introduction to an Investigative Tool for Colloidal and Polymeric Systems:* Lindner, P., Zemb, T. (Eds) Amsterdam: North-Holland, pp. 33–82.
Goudeau, P., Naudon, A., Chamberod, A., Rodmacq, B., Williams, C. E. (1987), *Europhys. Lett. 3*, 267.
Guinier, A. (1938), *Nature 142*, 569.
Guinier, A. (1939), *Ann. Phys. 12*, 161.
Guinier, A., Fournet, G. (1955), *Small-Angle Scattering of X-rays*. New York: Wiley.
Gunton, J. D., Droz, M. (1984), *Introduction to the Theory of Metastable and Unstable States*. Berlin: Springer Verlag.
Haubold, H. G., Grünhagen, K., Wagener, M., Jungbluth, H., Heer, H., Pfeil, A., Rongen, H., Brandenberg, G., Möller, R., Matzerath, J., Hiller, P., Halling, H. (1989), *Rev. Sci. Instrum. 60*, 1943–1946.
Hayter, J. B. (1985), in: *Physics of Amphiphiles: Micelles, Vesicles and Microemulsions:* Degiorgio, V., Corti, M. (Eds.). Amsterdam: North Holland, pp. 59–93.
Hayter, J. B., Penfold, J. (1981), *Mol. Phys. 42*, 109–118.
Higgins, J. S., Benoit, H. (1993), *Neutron Scattering of Polymers*. Oxford: Oxford University Press.
Hjelm, R. P. (1988), *J. Appl. Crystallogr. 21*, 618–628.
Holmes, K. C. (1982), in: *Small-Angle X-Ray Scattering:* Glatter, O., Kratky, O. (Eds.). London: Academic Press, pp. 85–103.
Hoppe, W. (1972), *Isr. J. Chem. 10*, 321–333.
Hoppe, W. (1973), *J. Mol. Biol. 78*, 581–585.
Hoyt, J. J., Clark, B., de Fontaine, D., Simon, J. P., Lyon, O. (1989), *Acta Metall. 37*, 1597–1609.
Ibel, K. (1976), *J. Appl. Crystallogr. 9*, 296–309.
Ibel, K., Stuhrmann, H. B. (1975), *J. Mol. Biol. 93*, 255–265.
Jacrot, B. (1976), *Rep. Prog. Phys. 10*, 911–953.
Jacrot, B., Zaccai, G. (1981), *Biopolymers 20*, 2413–2426.
James, R. W. (1965), *The Optical Principles of the Diffraction of X-Ray*. Ithaca: Cornell University Press.
Jannink, G. (1986), *Makromol. Chem., Makromol. Symp. 1*, 67.
Janot, C., George, B. (1985), *J. Physique Lett. 46*, L85–L88.
Jantzen, C. M., Schwahn, D., Schelten, J., Herman, H. (1981), *Phys. Chem. Glasses 22*, 122.
Kirste, R. G., Kruse, W. A., Ibel, K. (1975), *Polymer 16*, 120–124.
Knoll, W., Schmidt, K., Ibel, K. (1985), *J. Appl. Crystallogr. 18*, 65–70.
Knop, W., Nierhaus, K. H., Nowotny, V., Niinikoski, T. O., Krumpolc, M., Rieubland, J. M., Rijlart, A., Schärpf, O., Schink, H.-J., Stuhrmann, H. B., Wagner, R. (1986), *Helvetica Phys. Acta 59*, 741–746.
Koberstein, J. T., Morra, B., Stein, R. S. (1980), *J. Appl. Crystallogr. 13*, 34.
Koch, E. E. (Ed.) (1983), *Handbook on Synchrotron Radiation, 1*. Amsterdam: North Holland.
Koch, M. H. J. (1988), *Makromol. Chem., Makromol. Symp. 15*, 79.
Komura, S., Osamura, K., Fuji, H. T. T. (1985), *Phys. Rev. B31*, 1278.
Kostorz, G. (1988), in: *Materials Science Forum:* Elcombe, M. M., Hicks, T. J. (Eds.). Aedermannsdorf (Switzerland): Trans. Tech. Publications, pp. 325–344.
Kostorz, G. (1991), *J. Appl. Cryst. 24*, 444–456.
Kratky, O. (1963), *Progr. Biophys. 13*, 105.
Kratky, O. (1966), *J. Pure and Appl. Chem. 12*, 483.
Kratky, O., Worthmann, W. (1947). *Mh. Chem. 76*, 263–281.
Kuhn, W., Kuhn, H. (1943), *Helv. Chim. Acta 26*, 1394.
Lindner, P. (1991), *Physica A174*, 74–93.
Lindner, P. (1992), *New Scientist 133*, 38–41.
Lindner, P., May, R. P., Timmins, P. A. (1992), *Physica B 180 & 181*, 967–972.
Lindner, P., Oberthür, R. C. (1985), *Colloid and Polymer Sci. 263*, 443–453.
Lindner, P., Oberthür, R. C. (1988), *Colloid and Polymer Sci. 266*, 886–897.
Luzzati, V. (1960), *Acta Crystallogr. 13*, 939–945.
Luzzati, V., Tardieu, A., Mateu, L., Stuhrmann, H. B. (1976), *J. Mol. Biol. 101*, 115–127.
Lyon, O., Simon, J. P. (1986), *Acta Metall. 34*, 1197.
Martin, J. E., Hurd, A. J. (1987), *J. Appl. Crystallogr. 20*, 61–78.

May, R. P. (1991), in: *Neutron, X-ray and Light Scattering: Introduction to an Investigative Tool for Colloidal and Polymeric Systems:* Lindner, P., Zemb, T. (Eds.). Amsterdam: North Holland, pp. 119–133.
May, R. P., Ibel, K., Haas, J. (1982), *J. Appl. Crystallogr. 15,* 15–19.
May, R. P., Nowotny, V. (1989), *J. Appl. Crystallogr. 22,* 231–237.
May, R. P., Nowotny, V., Nowotny, P., Voß, H., Nierhaus, K. H. (1991), *EMBO J. 11,* 373–378.
Moore, P. B. (1980), *J. Appl. Crystallogr. 13,* 168–175.
Müller, E. W., Tsong, T. T. (1969), *Field Ion Microscopy.* New York: Elsevier.
Naudon, A., Lemonnier, M., Rousseaux, F. (1979), *C. R. Acad. Sci. (Paris) B21,* 288.
Nierhaus, K. H., Lietzke, R., May, R. P., Nowotny, V., Schulze, H., Simpson, K., Wurmbach, P., Stuhrmann, H. B. (1983), *Proc. Natl. Acad. Sci. USA 80,* 2889–2893.
Osamura, K., Okuda, H., Hashizume, H., Amemiya, Y. (1988), *Met. Trans. 19A,* 1973–1980.
Pavlov, M. Y., Rublevskaja, I. N., Serdyuk, I. N., Zaccai, G., Leberman, R., Ostanevitch, Y. M. (1991), *J. Appl. Crystallogr. 24,* 243–254.
Pavlov, M. Y., Serdyuk, I. N. (1987), *J. Appl. Crystallogr. 20,* 105–110.
Phillips, J. C., Hodgson, K. O. (1980), in: *Synchrotron Radiation Research:* Winick, H., Doniach, S. (Eds.). New York: Plenum Press, pp. 565–604.
Porod, G. (1948), *Acta Phys. Austriaca 2,* 133.
Porod, G. (1951), *Kolloid-Z. 124,* 83–114.
Porod, G. (1982), in: *Small-Angle X-ray Scattering:* Glatter, O., Kratky, O. (Eds.). London: Academic Press, pp. 17–51.
Preston, J. (1938), *Proc. Roy. Soc. (London) A167,* 526.
Rosenbaum, G., Holmes, K. C. (1980), in: *Synchrotron Radiation Research:* Winick, H., Doniach, S. (Eds.). New York: Plenum Press, pp. 533–564.
Ruland, W. (1971), *J. Appl. Crystallogr. 4,* 70.
Russell, T. P. (1983), *J. Appl. Crystallogr. 16,* 473.
Russell, T. P. (1991), in: *Handbook on Synchrotron Radiation:* Brown, G., Moncton, D. E. (Eds.). Amsterdam: North Holland, pp. 381–469.
Sasaki, S. (1983), *Anomalous Scattering Factors for Synchrotron Radiation Users, Calculated Using Cromer and Liberman's Method.* National Laboratory for High Energy Physics, Tsukuba, Japan (KEK report 83-22).
Schaffer, L. B., Hendricks, R. W. (1974), *J. Appl. Crystallogr. 7,* 157.
Schelten, J., Hossfeld, F. (1971), *J. Appl. Crystallogr. 4,* 210–223.
Schmidt, P. W. (1991), *J. Appl. Crystallogr. 24,* 414–435.
Siddons, P., Riekel, C., Hastings, J. (1990), *J. Appl. Crystallogr. 23,* 401.
Simon, J. P., Guyot, P., de Salva, A. (1984), *Phil. Mag. A49,* 151.
Simon, J. P., Lyon, O. (1989), *J. de Chim. Phys. 86,* 1523–1530.
Sinha, S. K., Sirota, S. B., Garoff, S. (1988), *Phys. Rev. B38,* 2297–2311.
Skov Pedersen, J. S., Posselt, D., Mortensen, K. (1990), *J. Appl. Crystallogr. 23,* 321–333.
Sparks, C. J. (1974), *Phys. Rev. Lett. 33,* 262.
Stephenson, G. B. (1984), *J. Non-Cryst. Solids 66,* 393.
Stephenson, G. B., Warburton, W. K., Haller, W., Bienenstock, A. (1991), *Phys. Rev. B43,* 13417–13437.
Stuhrmann, H. B. (1970), *Acta Crystallogr. A26,* 297–306.
Stuhrmann, H. B. (1989), in: *Synchrotron Radiation and Biophysics:* Hasnain, S. S. (Ed.). New York: Wiley, pp. 223.
Stuhrmann, H. B., Bartels, K. S., Boulin, C., Dauvergne, F., Gabriel, A., Görigk, G., Munk, B. (1989), in: *Biophysics and Synchrotron Radiation:* Bianconi, A., Congiu-Castello, A. (Eds.). Berlin: Springer Verlag, pp. 257.
Stuhrmann, H. B., Haas, J., Ibel, K., De Wolf, B., Koch, M. H. J., Parfait, R., Chrichton, R. R. (1976), *Proc. Natl. Acad. Sci. USA 73,* 2379–2383.
Stuhrmann, H. B., Kirste, R. G. (1965), *Z. Phys. Chem. Neue Folge 46,* 247–250.
Stuhrmann, H. B., Schärpf, O., Krumpolc, M., Niinikoski, T. O., Rieubland, M., Rijllart, A. (1986), *Eur. Biophys. J. 14,* 1–6.
Tchoubar, D. (1991), in: *Neutron, X-ray and Light Scattering: Introduction to an Investigative Tool for Colloidal and Polymeric Systems:* Lindner P., Zemb, T. (Eds.). Amsterdam: North-Holland, pp. 157–174.
Teixeira, J. (1988), *J. Appl. Crystallogr. 21,* 781–785.
Timmins, P. A., Zaccai, G. (1988), *Eur. Biophysics, J. 15,* 257–268.
Wagner, R., Kampmann, R. (1991), in: *Phase Transformation in Materials:* Haasen, P. (Ed.). Weinheim: VCH Verlagsgesellschaft, pp. 213–304.
Wendin, G. (1980), *Physica Scripta 21,* 535.
Wignall, G. D.(1987), in: *Encyclopedia of Polymer Science and Engineering:* Kroschwitz, J. I. (Ed.). New York: Wiley, pp. 112–184.
Wignall, G. D., Ballard, D. G. H., Schelten, J. (1974), *Eur. Polymer J. 10,* 861–865.
Wignall, G. D., Bates, F. S. (1987), *J. Appl. Crystallogr. 20,* 28–40.
Wignall, G. D., Christen, D. K., Ramakrishnan, V. (1988), *J. Appl. Crystallogr. 21,* 438–451.
Williams, C. E., Nierlich, M., Cotton, J. P., Jannink, G., Boué, F., Daoud, M., Farnoux, B., Picot, C., de Gennes, P. G., Rinaudo, M., Moan, M., Wolf, C. (1979), *J. Polym. Sci. Polym. Lett. Ed. 17,* 379.
Witz, J. (1983), *Acta Crystallogr. A39,* 706–711.
Yokota, R., Nakajima, H. (1985), *J. Non-Cryst. Solids 70,* 343.
Zaccai, G., Wachtel, E., Eisenberg, H. (1986), *J. Mol. Biol. 190,* 97–106.
Zimm, B. H. (1948), *J. Chem. Phys. 16,* 1093.

General Reading

Scattering methods:

Baruchel, J., Hodeau, J.-L., Lehmann, M. S., Schlenker, C., Regnard, J.-R. (Eds.) (1993) *HERCULES: Neutron and Synchrotron Radiation for Condensed Matter Studies,* Heidelberg: Springer Verlag.

Feigin, L. A., Svergun, D. I. (1987), *Structure Analysis by Small Angle X-ray and Neutron Scattering.* New York: Plenum Press.

Glatter, O., Kratky, O. (Eds.) (1982), *Small Angle X-ray Scattering,* London: Academic Press.

Guinier, A., Fournet, A. (1955), *Small Angle Scattering of X-rays.* New York: Wiley.

Snyder, R. L. (1992), in: *Characterization of Materials:* Lifshin, E. (Ed.). Vol. 2 A of *Materials Science and Technology – A Comprehensive Treatment:* Cahn, R. W., Haasen, P., Kramer, E. (Eds.). Chapter 4. Weinheim: VCH.

Wilson, A. J. C. (Ed.) (1992), *International Tables for Crystallography, Vol. C,* Dordrecht: Kluwer, Academic Publishers.

Metallurgical applications:

Kostorz, G. (1988), in: *Materials Science Forum:* Elcombe, M. M., Hicks, T. J. (Eds.). Aedermannsdorf (Switzerland): Trans. Tech. Publications, pp. 325–344.

Kostorz, G. (1991), *J. Appl. Cryst. 24,* 444–456.

Polymer and colloid applications:

des Cloizeaux, J., Jannink, G. (1990), Polymers in Solution: Their Modelling and Structure. Oxford: Clarendon Press; New York: Oxford University Press.

Hayter, J. B. (1988), *Materials Science Forum 27/28,* 345–358.

Higgins, J. S., Benoit, H. (1993), *Neutron Scattering of Polymers.* Oxford: Oxford University.

Lindner, P., Zemb, T. (Eds.) (1991), *Neutron, X-ray and Light Scattering: Introduction to an Investigative Tool for Colloidal and Polymeric Systems, Delta-Series,* Amsterdam: North Holland.

Wignall, G. D. (1987), in: *Encyclopedia of Polymer Science and Engineering:* Kroschwitz, J. I. (Ed.). New York: Wiley, pp. 112–184.

Biological applications:

Jacrot, B. (1976), *Rep. Progr. Phys. 39,* 911–953.

Timmins, P. A., Zaccai, G. (1988), *Eur. Biophyics J. 15,* 257–268.

The proceedings of the International Conferences on Small Angle Scattering, held every three years, have been published in the Journal of Applied Crystallography, with the exception of the first one edited as a separate book: Brumberger, H. (Ed.). (1967), *Small Angle Scattering.* New York: Gordon & Breach and the 1994 conference published as a special issue of *Journal de Physique.*

A chapter on SAS is usually included in the proceedings of most schools and workshops in those specialized fields where SAXS and SANS are basic techniques of analysis.

All neutrons and synchrotron radiation facilities have websites where up-to-date informations and reports may be found.

Index

absolute sensitivity, X-ray fluorescence 187
absorptiometry 172
absorption
– electromagnetic waves 12 f
– X-rays 179
– X-ray fluorescence analysis 172, 196, 200
absorption-diffraction method, quantitative phase analysis 73, 81
absorption edges
– oxidation dependence 118
– X-ray spectrum 8
accuracy
– X-ray diffractometers 63 ff
– X-ray fluorescence 194 ff
acoustic modes, lattice vibrations 9
aegerine glass 118
Ag anodes 182
Al-Zn alloys 247
Al-Zn-Ag ternary alloys 248
alloy compositions, third element effect 197
aluminum, fluorescent yield 176
aluminum oxide 221
ammonium dihydrogen phosphate 148
ammonium sulfate 153 f
amorphization 115
amorphography 47
amorphous materials, X-ray absorption spectroscopy 49
amorphous phase analysis 82
amorphous silica 214
analog output, X-ray spectrometers 193
analytical profile fitting, X-ray powder diffraction pattern 88 ff
angular dispersion 188
angular quantum number 176
angular speed 194
anodes 182
anomalous dispersion, X-ray scattering 40 f
anomalous X-ray scattering 224
anorthic systems, crystallography 20
arsenic, structural parameters 116
atomic motion effects, X-ray diffraction patterns 41
atomic scattering 39 f
atomic structures 177
attrition fragments, growth rate influences 139
Au anodes 182

Auger process 174 f
autocorrelation function 216
automated X-ray diffractometers 53, 60 ff
– single crystals 33

background
– small-angle scattering experiments 240
– X-ray diffraction profiles 92 f
– X-ray fluorescence 201
background counting, X-ray fluorescence 195, 206
background determination, X-ray powder diffractometers 60 f
barium, fluorescent yields 176
bending magnets, synchrotron radiation 109
beryllium, K lines 187
beta-filters, elctromagnetic waves 13
binary alloys, phase separation 246 f
binary mixtures, X-ray diffraction analysis 82
birefringent-effect liquid crystal displays 149
blank counting, X-ray fluorescence 206
body-centered unit cells 22
borosilicate glasses 117 f
bound electrons, X-ray scattering 39
Bragg angle 181, 194
Bragg diffraction 180
Bragg reflection 79
Bragg–Brentano optics 53, 56, 89
Bravais lattices 21 f
bremsstrahlung 6
– synchrotron source X-ray fluorescence 192
– X-ray fluorescence analysis 175, 185
brightness, synchrotron radiation 110
bromine 152 f

calcite, thermal decomposition 160
calibration
– automated X-ray diffractometers 64
– small-angle scattering 239
– X-ray diffraction profile fitting 94
– X-ray fluorescence 197 f
cement, structural changes 129
chain conformation, polymers, small-angle scattering 244 f
chalcogenide glasses 149 f

characteristic radiation, X-ray fluorescence analysis   175
charge coupled devices (CCD)   34
chord length   221
chromium/iron/nickel system   197
cimetidine   125
Claisse–Quintin influence correction models, X-ray fluorescence   203
coherent scattering   215, 228
– electromagnetic waves   11 f
– X-radiation   179
cold neutrons   234
collimators
– wavelength dispersive systems   188
– X-ray fluorescence   184
colloidal particles   248
compressibility measurements   130
Compton scattering   9, 11, 216
– X-radiation   180
– X-ray fluorescence   196
concentration correction, X-ray fluorescence   201
constrained phase analysis   76
continuous radiation synchrotron topographs   157
continuous radiation, X-ray fluorescence analysis   175
continuous synchrotron radiation spectrum   109
contrast, small-angle scattering   224 ff
copper, K-edge absorption spectrum   48
copper ions, habit modifiers   153 f
copper-rich precipitates, structure investigation   113 f
correction methods, X-ray fluorescence   203
correlation function   218
correlation length   221
corrosion, borosilicate glasses   117 f
Couette cells   249
counting gas, P10   15
counting statistical errors, X-ray fluorescence   194
counting time, X-ray fluorescence   195
Cr anodes, X-ray fluorescence   182
crack tip, dislocation motions   163
Criss–Birks model, X-ray fluorescence   204
crossed effects, influence correction models   203
Crystal Data database   72
crystal diffraction techniques   32 ff
crystal lattice, X-ray fluorescence analysis   172
crystal/solution interface structure   148
crystal structure   42
crystal structure constraints, quantitative phase analysis   78 ff
crystal structure determination
– Laue diffraction patterns   141
– X-ray diffraction   96 ff
crystal systems   21
crystallite size effects   37 f
crystallization, in liquid environment   133
crystallizers, habit modifiers   153
crystallographic symmetry elements   18 ff
Cu–Ti alloys, small-angle X-ray scattering   227
cubic systems   21

Darwin profile, X-ray powder diffraction   90
Darwin width   38
dead time, X-ray fluorescence   200
Debye–Scherrer camera   52
Debye–Waller temperature factor   41
defects, X-ray topography   155 ff
deflection field, magnetic   193
density, average   214
derived correction methods, X-ray fluorescence   203
detector-reponse correction, small-angle scattering   240
detectors   13 ff
– SAXS spectrometers   234
– X-ray fluorescence   182 f
deuterated monomers   245
deuterium, coherent scattering   228
diamond anvil cells   131
diesel fuel, wax crystallization   125
diffraction   24 ff, 180
– wavelength dispersive spectrometers   186
– accuracy   66
diffractometer optics   55
diffractometers
– Bragg–Brentano parafocusing   53
– Parrish geometry   55
– powder   124
– small-angle neutron scattering   235
– very high resolution   57 ff
digital filters, automated X-ray diffractometers   62
digital output, X-ray spectrometers   193
dislocation motion, synchrotron topography   159
dislocation velocity studies, silicon   163
dispersion, anomalous   40 f
divergence, synchrotron radiation   109
double crystal neutron diffractometers   236
double crystal topography   162 ff
dynodes   17

electrode interface structure examinations   146
electron synchrotron storage rings   107
elemental and interplanar spacings (EISI) index   72
ellipsoidal-mirror small-angle neutron scattering cameras   236
emission lines, X-ray spectrum   7
empirical correction technique, X-ray fluorescence   201
energy bands, X-ray fluorescence analysis   172
energy dispersive instruments   185, 189
energy dispersive X-ray diffraction (EDXRD)   128 ff
energy dispersive spectrometry (EDS)   18, 173
energy resolution, sealed gas proportional counters   15
environmental chamber, X-ray topography   159
epitaxial layers, double crystal topography   164
equatorial diffraction scan   142
error statistics, X-ray fluorescence   194 ff
errors, lattice parameter determination   88
Ettringite, cement hydration   130
Ewald sphere   30 f

exhaustive parameter space indexing 86
extended X-ray absorption fine structure (EXAFS) 48, 111
external standard method, X-ray diffractometer calibration 64
extinction, X-ray diffraction 44 ff

face-centered unit cells 22
fast ions, proton excited X-ray fluorescence 193
figures of merit, lattice parameter determination 83 f
filter loadings, X-ray diffraction analysis 81
β-filters, elctromagnetic waves 13
fine focus X-ray tubes 7
fine structure, X-ray absorption spectrum 48
Fink method, powder diffraction data 68
flash X-ray devices 8
flowing gas proportional counters 15
fluorescence 9 f
fluorescence analysis 171–209
fluorescence measurements 112
fluorescent screens 14
fluorescent yield 11, 176
fluorine, K lines 174, 187
focussing optics, SAXS spectrometers 233
Fourier transform, small-angle scattering 216
fractals 222
Fresnel reflection, X-ray absorption spectroscopy 50
full-pattern fitting, X-ray powder diffractometers 77 f
fundamental correction methods, X-ray fluorescence 203 f

GaAs(100), surface oxide 115
gain, X-ray fluorescence 184
gallium, structural parameters 116
gamma-rays, X-ray fluorescence 181
Gandolfi camera 52
gas-ionization detectors 14 ff
Gaussian distribution, X-ray fluorescence 194
Ge/Si(001) strained layer semiconductor interfaces 144
Geiger counter 173
Geiger-Müller counters 13, 16
geometrical resolution, synchrotron topography 158
germanium crystals, X-ray detectors 16
glancing angle X-ray absorption spectroscopy 113
glasses, chalcogenide 149 f
glide planes 22
gold, surface examinations 146 f
gold anodes 182
goniometers 188, 194
Goss texture 37
grazing incidence, X-ray reflectometry 50
group III-V quantum-well structures 144
group III-V semiconductors, surface oxides 115
Guinier camera 56
Guinier law 220

habit modifiers, industrial crystallizers 153 f
Hanawalt method, powder diffraction data 68
Hart–Parrish high resolution diffractometer 58
Hermann–Maugin symbols 19
heterogeneity, sample preparation 198
hexagonal systems 21
hexamethylbenzene 86
high-intensity X-ray devices 7 f
high-pressure structural phase transformations 130 ff
high-resolution diffractometers 57 ff
high-resolution powder diffraction analysis 124 ff
high-temperature superconductivity 131
homogenization, sample preparation 198
hydraulic cement, structural changes 129
hydrocarbon crystallization 125 f
hydrogen, coherent scattering 228
hydrogenated monomers 245

improper axis, rotational symmetry 19
incoherent scattering 11, 179, 215, 237, 241
indexing, lattice parameter determination 83 ff
indirect Fourier transformation 243
indium/tin oxide (ITO) 149
influence correction methods, X-ray fluorescence 201 f
InGaAs quantum well, surface diffraction investigation 145
inner orbital electron, X-ray fluorescence analysis 172
insertion devices, synchrotron radiation 109
intensities
– X-ray diffraction 39, 43 f
– X-ray fluorescence 201
interacting particles, small-angle scattering 223
interdiffusion, photostimulated 151
inter-element effects 198
interface structure, crystal/solution 148
interfaces, Porod's law 219
interference, X-rays 152, 180
internal standard method
– generalized 76
– quantitative phase analysis 73 f
– X-ray diffractometer calibration 64
– X-ray fluorescence 202
intrinsic X-ray diffraction profiles 90
ionization, X-ray fluorescence 183
ionization chamber 15
iron foil, K-edge spectrum 111
iron/nickel/chromium system 197
irradiation area, X-ray fluorescence analysis 172
irreducible representations 19
isotropic incoherent scattering 216
isotropic substitutions, small-angle neutron scattering 228 f
isotropic temperature factor 42

K edge spectrum, iron foil 111
K lines 174, 187

Index

K series, X-ray fluorescence analysis 174 f
Kiessig fringes 142
Kossel single crystal technique 35 f
K_2ZnCl_4, phase transformation 120 f

L series, X-ray fluorescence analysis 174
La-Ba-Cu-O system, superconductivity 131
Lachance–Claisse influence correction models 203
Lachance–Trail model 201 ff
laminar shear flow 250
lattice defects 155 ff
lattice parameter determinations 83 ff
lattice parameter refinement 86 f
lattice vibrations 9
Laue diffraction
– synchrotron radiation 134 ff
– X-ray fluorescence analysis 172
Laue single crystal technique 35 f
leaching mechanisms, metallic ions from glass surfaces 117
line broadening, X-ray diffraction profile modeling 95
line intensity, X-ray fluorescence analysis 173
line overlap, X-ray fluorescence analysis 201
linear absorption coefficient 12
linear influence correction models 203
liquids, shear 249
lithium doped silicon crystals, X-ray detectors 16
long fine focus X-ray tubes 7
long-mirror cameras 236
Lorentz factor 43
low-concentration trace analysis, 205
lower limit of detection (LLD) 205
lower symmetry X-ray diffraction patterns 85
lutetium, K series 174

magnesium, K lines 187
magnetic deflection 193
magnetic quantum numbers 176
magnets, synchrotron radiation 109
manganese, L series 174
martensitic strain distortion 58
mass attenuation, X-radiation 179
matrix effects 179, 196
matrix scattering density variation, small-angle scattering 229
mechanical alloying 115 ff
microabsorption, X-ray diffraction 45 f
microfocus X-ray tubes 7
microstress effects, diffraction patterns 38
Miller indices 24
molecular structure determinations 135
molybdenum, K edge 115, 225
molybdenum anodes 182
molybdenum target X-ray tube 6
monochromators
– small-angle X-ray scattering spectrometers 232
– X-ray powder diffraction 55

monoclinic space groups 23
monoclinic systems 21
mosaic spread, crystal growth 148
Moseley law 175, 178, 181
Moseley spectrometer 173
multiple element methods, X-ray fluorescence 199

NaCl, X-ray diffraction pattern 54
NaI crystals, X-ray detectors 17
neutron scattering, small-angle 211–254
Ni-Ti-V alloys, electron diffraction pattern 20
nickel, K-edge EXAFS spectrum 115
nickel/chromium/iron system, third element effect 197
nickel formate reduction 121 f
Niggli cell 24
$Ni_{50}Mo_{50}$, mechanical alloying 115
noise reduction, automated X-ray diffractometers 62
nomenclature, X-ray wavelengths 178
nondestructive testing 193
nonelectronic X-ray detectors 14
NRLXRF program, X-ray fluorescence analysis 204
nuclear reactor pressure vessel 113
nuclear waste management 117

octylcyanobiphenyl films 151
optic modes, lattice vibrations 9
orbitals, X-ray fluorescence analysis 172, 177
organic crystals 137
organic materials, white radiation synchrotron topography 158
organo-metallic compounds 250
orientation determination, single crystal X-ray diffraction 35 f
orthorhombic systems 21
oxygen, K lines 187

P10 counting gas 15
parafocusing geometry, Bragg–Brentano 53
parallel beam diffractometer 58
Parrish geometry 53, 55
partial structure factors, small-angle scattering 248
particle accelerators 5
particle mass determination 242
particle size effects, X-ray fluorescence 198
Patterson function 216
Patterson map 47
peak intensities, X-ray fluorescence 185
peak removal techniques, X-ray diffractometers 63
Pearson function 92
Peierls barrier 159
pelletizing, sample preparation 198
Peltier cooler, energy dispersive systems 190
Penrose tile 20
phase detectability limit 66 f
phase identification 67 ff
phase separation, polymers 245

phase transition studies, Laue topographic technique   140
phonons   9
phosphor scintillator   183
photodissolution, silver   149 f
photoelectric absorption, X-radiation   179
photoelectrons   112, 175
photographic films   14
photomultiplier   183
photons
– short wavelength limit   7
– X-ray fluorescence analysis   174, 206
photostimulated interdiffusion   151
pin-hole cameras, small-angle neutron scattering   235
plane multiplicity factor   43
plastic deformation studies,   159
point groups, crystallography   20 f
polarization, synchrotron radiation   110
polarization factor, X-ray electron scattering   39
pole figure measurements   37
polished stainless steels, oxidation   118 f
polybutadiene   245
polychromatic radiation   172, 185
polymer chains
– shear   249
– small-angle scattering   222, 244 f
polymorphs mixtures, X-ray diffraction analysis   81
polyphenylene films, structure investigations   149
Porod limit   219
position sensitive detectors (PSD)   14, 56 f
potassium tetrachlorozincate (KZC)   120 f
powder diffraction analysis   51 ff, 66, 124
powder diffraction data base   68
precession camera   33
precipitation
– irradiation-induced   113
– wax   125
preconcentration techniques, sample preparation   198
preferred orientation, crystallites   37
pressure vessel   113
primary extinction   45 f
primitive unit cell   22
profile analysis, transition aluminas   98
profile fitting, X-ray powder diffraction   88 ff
proper axis, rotational symmetry   19
proton excitation   193
proton induced X-ray emission spectrometers (PIXE)   186, 193
protons, X-ray fluorescence   181
pseudo-Voigt function   92
Pt/Re catalysts, sulfidation processes   122 f
pulse height analyzers (PHA)   17
pure intensity measurements, X-ray fluorescence   201

qualitative analysis, X-ray spectrometers   193
quantitative analysis
– X-ray fluorescence   198 f
– X-ray powder diffraction   73 ff

quantum numbers   176 f
quartz, surface acoustic wave observations   161
quick scanning technique   121

Rachinger diffraction peak removal   63
radial distribution function (RDF)   47
radiation damage, synchrotron radiation methods   135, 160
radio isotopes, X-ray fluorescence   181
radius of gyration   220
random counting error   194
randomly distributed phases   221
Rasberry–Heinrich influence correction models   203
rate meter   193
reciprocal lattice   26 ff
reduced cell   23 f
reference intensity ratio (RIR)   75, 80
reference standards, X-ray fluorescence   202
reflection, Ewald sphere   30 f
reflectivity coefficient   113, 142
reflectometry   50
regression correction methods   203
residual stress
– crystallites   36
– diffraction patterns   38
resolution
– small-angle scattering data   242
– synchrotron topography   158
– X-ray fluorescence   183
resonance absorption   8
resonant scattering   226
Rh anodes   182
rhombohedral systems   21
ribosomes, small-angle scattering techniques   250
Rietveld analysis
– quantitative phase analysis   78
– transition aluminas   99 f
– X-ray diffraction profiles   96
Rietveld refinement   125
Rochelle salt crystals   139
Röntgen radiation   172
rotational symmetry, crystallographic   19

sample environment, small-angle scattering   237 f
sample preparation, X-ray fluorescence   197
satellite lines, X-ray fluorescence analysis   177
scaling, small-angle scattering   238 f
scattering   216
– small-angle range   213
– X-ray fluorescence analysis   172, 179, 205
Scherrer equation   38
Schottky barrier   183
scintillation detectors   17
scintillator, X-ray fluorescence   183
screw axes, crystallography   22
sealed gas proportional counters   15
second phase precipitation, irradiation-induced   113
secondary extinction   45 f

secondary nucleated particles, growth rate   139
secondary photons   9
Seemann–Bohlin transmission geometry   56
selection rules, X-ray fluorescence analysis   176 f
selective deuteration   244 f
semiconductor interfaces, strained layers   144
sensivity, X-ray fluorescence   187, 205
shear, liquids   249
Sherman modell   201, 204
shielding constant, X-ray fluorescence analysis   176
short wavelength limit   7
Si(Li) detector   183
Siegbahn nomenclature   178
signal processing   17 f
silica, amorphous   214
silicon, dislocation velocity studies   163
silicon crystals   16
silver anodes   182
silver   149 f
single crystal diffraction techniques   32 ff
single element methods   199
size analysis, X-ray diffraction profiles   96
size effects, crystallites   37
skin, X-ray effects   14
small amount analysis, X-ray fluorescence   207
small-angle scattering   211–254
solid state chemical transformations   129 f
solid state detectors   16
solid state polymerization   137 f
solutions, small-angle scattering   241
solvent scattering density variation, small-angle
     scattering   229
sources, X-ray fluorescence   181
space groups, crystallography   22
spatial resolution   233
specific isotropic labeling   229
specimen preparation, X-ray fluorescence   197
spectral distribution, X-ray diffraction profiles   90
spectral range, white radiation topographs   156 f
spectrometer ranges   174, 178
spectrometers
– small-angle X-ray scattering   232
– wavelength dispersive (WDS)   11
– X-ray fluorescence   185
spiking method, quantitative phase analysis   73, 82
spin incoherent scattering   216
spin-contrast variation, small-angle neutron
     scattering   231
split-Pearson function, X-ray diffraction profile
     fitting   94
spray drying, X-ray powder diffraction   66
stainless steel, oxidation   118 f
standard additions method   73, 82
Standard Reference Materials (SRM) 1832/1833   203
standardless quantitative phase analysis   76
standards, X-ray fluorescence   202
standing wave field   152
statistical errors, X-ray fluorescence   194
step scanning, X-ray spectrometers   194
storage rings, synchrotron   107 f

strain analysis, X-ray diffraction profiles   96
strain effects, attrition fragment growth rates   139
strain mapping, synchrotron double crystal
     topography   163
stress effects   36, 38
stroboscopic synchrotron topography   161
structural investigations   123 ff
structural properties, small-angle scattering   220 ff
structure analysis, crystals   96 ff
structure factor   42, 44
structures, atomic   177
subgrain misorientation   157
sulfidation processes   122 f
sulfur, K lines   187
superconducting magnets   109
superconductivity, high temperature   131
surface acoustic wave observations   161
surface coordination, adsorbed bromine   152 f
surface diffraction techniques   142 ff
surface oxides, group III-V semiconductors   115
surface scattering, bulk materials   142 ff
surfacing, sample preparation   198
symmetry elements, crystallographic   18 ff
synchroton source spectrometers (SSXRF)   186, 191
synchrotron radiation   5
– X-ray fluorescence   181
synchrotron storage rings   107 f
synchrotron X-radiation, applications   105–169

ternary alloys, partial structure factors   248
tetragonal systems, crystallography   21
thallium acid phthalate (TAP), wavelength dispersive
     systems   189
thermal atomic motion effects, X-ray diffraction
     patterns   41
thermal decomposition, synchrotron topography   160
thin films, X-ray absorption spectroscopy   50
third element effect, X-ray fluorescence   196, 200
Thomson equation   39
TiB_2, compressibility measurements   130
time-of-flight, small-angle neutron scattering
     cameras   235
time-resolved small-angle scattering   247
tomography   51
topography   34, 155 ff
total reflection X-ray fluorescence (TRXRF)
     spectrometers   186, 190
total reflection   50
Townsend avalanche   16
trace analysis
– proton induced X-ray emission spectrometers   193
– total reflection spectrometers   190
– X-ray fluorescence   205 ff
transition aluminas   98
transition groups   177
translational symmetry, crystal lattice   21
transmission experiments, X-ray absorption
     spectra   112
triclinic systems, crystallography   20

triple isotropic replacement   230
tungsten anodes   182
tungsten cathodes   5
tungsten, lattice parameter determination   124
tungsten filaments   175
tungsten wires   181
two-media approximation, scattering theories   218
type standardizations, X-ray fluorescence   202

ultraviolet spectra   193
unconstrained X-ray diffraction profile fitting   93 f
undulators   5, 109
unit cells, crystal structure   21 f
uranium, L series   174

vibration effects, X-ray diffraction patterns   41
vitrification processes, nuclear waste management   117
Voigt function   92

wavelength accuracy, automated X-ray diffractometers   64
wavelength dispersive spectrometry (WDS)   11, 18, 173, 185, 188
wavelengths   172, 175 ff
wavenumber   4
wax crystallization   125
weight fraction term, X-ray fluorescence   203
Weissenberg camera   33
white line, extended X-rax absorption fine structure   117

white radiation, X-ray fluorescence analysis   175
white radiation synchrotron topography   155 ff
wide-angle scattering, amorphous silica   214
wigglers   5, 109
wollastonite   66

X-radiation, synchrotron   105–169
X-ray absorption near edge structure (XANES)   111
X-ray absorption spectroscopy (XAS)   48 ff, 110 ff
X-ray diffraction   1–103
X-ray fluorescence analysis   171–209
X-ray powder diffraction   51 ff
X-ray scattering, small-angle   211–254
X-ray sources   4
X-ray spectrometers, small-angle scattering   231 f
X-ray standig wave spectroscopy (XSW)   152 ff
X-ray tomography   51
X-ray topography   34
X-ray tubes   5 f

Y-Ba-Cu-O system, X-ray structure analysis   97 f

zero background holders (ZBH), X-ray diffractometers   65
Zimm approximation   223
Zimm plots   244
zirconium hydroxide, precipitation   129
zirconium, absorption edges   13
ZnS, fluorescent screens   14